INTRODUCTION TO
DIGITAL AUDIO CODING
AND STANDARDS

THE KLUWER INTERNATIONAL SERIES
IN ENGINEERING AND COMPUTER SCIENCE

INTRODUCTION TO DIGITAL AUDIO CODING AND STANDARDS

Marina Bosi
Stanford University

Richard E. Goldberg
The Brattle Group

KLUWER ACADEMIC PUBLISHERS
Boston / Dordrecht / London

This Edition Authorized by:

Kluwer Academic Publishers, Dordrecht,

The Netherlands

Sold and Distributed in:

People's Republic of China, Hong Kong, Macao, Taiwan

By: Sci-Tech Publishing Company LTD.

TEL: 02-27017353   FAX: 02-27011631

http://sci-tech.com.tw

 Electronic Services <http://www.wkap.nl>

Library of Congress Cataloging-in-Publication Data

Bosi, Marina
Introduction to Digital Audio Coding and Standards / by Marina Bosi and Richard E. Goldberg
p. cm. –(The Kluwer international series in engineering and computer science; SECS 721
Includes bibliographical references and index.
ISBN 1-4020-7357-7 (alk. paper)

Contents

FOREWORD...xiii

PREFACE... xvii

PART I: AUDIO CODING METHODS

Chapter 1. INTRODUCTION
    1. Representation of Audio Signals..3
    2. What is a Digital Audio Coder?.. 4
    3. Audio Coding Goals.. 5
    4. The Simplest Coder – PCM 7
    5. The Compact Disk 8
    6. Potential Coding Errors..9
    7. A More Complex Coder.. 10
    8. References.. 12
    9. Exercises 12

Chapter 2. QUANTIZATION
    1. Introduction.. 13
    2. Binary Numbers 14
    3. Quantization 20
    4. Quantization Errors 34
    5. Entropy Coding 38
    6. Summary 43
    7. References... 44
    8. Exercises 44

Chapter 3. REPRESENTATION OF AUDIO SIGNALS
    1. Introduction 47
    2. Notation 48
    3. Dirac Delta 49
    4. The Fourier Transform 51
    5. Summary Properties of Audio Signals 53
    6. The Fourier Series 59
    7. The Sampling Theorem 61
    8. Prediction 63
    9. Summary 68
   10. Appendix – Exact Reconstruction of a Band-Limited,
        Periodic Signal from Samples within One Period 68
   11. References 69
   12. Exercises 70

Chapter 4. TIME TO FREQUENCY MAPPING PART I: THE PQMF
    1. Introduction 75
    2. The Z Transform 77
    3. Two-Channel Perfect Reconstruction Filter Banks 84
    4. The Pseudo-QMF Filter Bank, PQMF 90
    5. Summary 99
    6. References 100
    7. Exercises 101

Chapter 5. TIME TO FREQUENCY MAPPING PART II: THE MDCT
    1. Introduction 103
    2. The Discrete Fourier Transform 104
    3. The Overlap-and-Add Technique 113
    4. The Modified Discrete Cosine Transform, MDCT 124
    5. Summary 143
    6. References 144
    7. Exercises 146

Chapter 6. INTRODUCTION TO PSYCHOACOUSTICS
    1. Introduction 149
    2. Sound Pressure Levels 150
    3. Loudness 150
    4. Hearing Range 151
    5. Hearing Threshold 153
    6. The Masking Phenomenon 156
    7. Measuring Masking Curves 160
    8. Critical Bandwidths 164
    9. How Hearing Works 168

　　10. Summary 174
　　11. References 175
　　12. Exercises 177

Chapter 7. PSYCHOACOUSTIC MODELS FOR AUDIO CODING
　　1. Introduction 179
　　2. Excitation Patterns and Masking Models 180
　　3. The Bark Scale 182
　　4. Models for the Spreading of Masking 183
　　5. Masking Curves 190
　　6. "Addition" of Masking 192
　　7. Modeling the Effects of Non-Simultaneous (Temporal)
　　　　Masking 195
　　8. Perceptual Entropy 196
　　9. Masked Thresholds and Allocation of the Bit Pool 197
　　10. Summary 198
　　11. References 198
　　12. Exercises 200

Chapter 8. BIT ALLOCATION STRATEGIES
　　1. Introduction 201
　　2. Coding Data Rates 202
　　3. A Simple Allocation of the Bit Pool 204
　　4. Optimal Bit Allocation 205
　　5. Time-Domain Distortion 214
　　6. Optimal Bit Allocation and Perceptual Models 216
　　7. Summary 218
　　8. References 219
　　9. Exercises 219

Chapter 9. BUILDING A PERCEPTUAL AUDIO CODER
　　1. Introduction 221
　　2. Overview of the Coder Building Blocks 221
　　3. Computing Masking Curves 223
　　4. Bitstream Format 230
　　5. Business Models and Coding Secrets 233
　　6. References 235
　　7. Exercises 235

Chapter 10. QUALITY MEASUREMENT OF PERCEPTUAL AUDIO CODECS
　　1. Introduction 237
　　2. Audio Quality 239
　　3. Systems with Small Impairments 240
　　4. Objective Perceptual Measurements of Audio Quality 251

   5. What Are We Listening For? 255
   6. Summary 257
   7. References 257
   8. Exercises 261

PART II: AUDIO CODING STANDARDS

Chapter 11. MPEG-1 AUDIO
   1. Introduction 265
   2. Brief History of MPEG Standards 266
   3. MPEG-1 Audio 268
   4. Time to Frequency Mapping 273
   5. MPEG Audio Psychoacoustic Models 278
   6. MPEG-1 Audio Syntax 296
   7. Stereo Coding 307
   8. Summary 310
   9. References 310

Chapter 12. MPEG-2 AUDIO
   1. Introduction 315
   2. MPEG-2 LSF, "MPEG-2.5" and MP3 315
   3. Introduction to Multichannel Audio 318
   4. MPEG-2 Multichannel BC 321
   5. Summary 330
   6. References 330

Chapter 13. MPEG-2 AAC
   1. Introduction 333
   2. Overview 333
   3. Gain Control 338
   4. Filter Bank 340
   5. Prediction 343
   6. Quantization and Coding 346
   7. Noiseless Coding 350
   8. Bitstream Multiplexing 353
   9. Temporal Noise Shaping 355
   10. Joint Stereo Coding 358
   11. Test Results 359
   12. Decoder Complexity Evaluation 363
   13. Summary 367
   14. References 367

Chapter 14. DOLBY AC-3
    1. Introduction 371
    2. Main Features 372
    3. Overview of the Encoding process 374
    4. Filter Bank 377
    5. Spectral Envelope Coding 382
    6. Multichannel Coding 385
    7. Bit Allocation 390
    8. Quantization 394
    9. Bitstream Syntax 395
   10. Performance 396
   11. Summary 397
   12. References 398

Chapter 15. MPEG-4 AUDIO
    1. Introduction 401
    2. MPEG-4: What is it? 402
    3. MPEG-4 Audio Goals and Functionalities 405
    4. MPEG-4 Audio Tools and Profiles 408
    5. MPEG-1 and 2 Versus MPEG-4 Audio 422
    6. The Performance of the MPEG-4 Audio Coding Tools 424
    7. Intellectual Property and MPEG-4 425
    8. Summary 426
    9. References 426

INDEX 431

About the Authors

Marina Bosi is a Consulting Professor at Stanford University's Computer Center for Research in Music and Acoustics (CCRMA) and Chief Technology Officer of MPEG LA®, a firm specializing in the licensing of multimedia technology. Past president of the Audio Engineering Society, Dr. Bosi is the author of numerous articles and the holder of several patents in audio technology. Dr. Bosi has been involved in the development of MPEG, Dolby, and DTS audio coders.

Richard E. Goldberg is a Partner at The Brattle Group, a management consulting firm specializing in economics and finance issues. Dr. Goldberg's practice focuses on business valuation and risk management. Dr. Goldberg has a Ph.D. in Physics from Stanford University and an A.B. in Astrophysics from Princeton University. Audio coding technology and related business applications have long been areas of interest for him.

Foreword

THE RISE OF DIGITAL AUDIO

Leonardo Chiariglione – Telecom Italia Lab, Italy

Analogue speech in electrical form has a history going back more than a century and a quarter to the early days of the telephone. However, interest in digital speech only gathered momentum some 40 years ago when the telecommunications industry started a global project to digitize the telephone network. The technology trade-off of the time in this infrastructure-driven project led to a preference for adding transmission capacity over finding methods to reduce the bitrate of the speech signal so the use of compression technology for speech remained largely dormant. When in the late 1980s the ITU-T standard for visual telephony became available enabling compression of video by a factor of 3,000, the only audio format in use to accompany this highly compressed video was standard telephone quality 64 kb/s PCM. It was only where transmission capacity was a scarce asset, like in the access portion of radiotelephony, that speech compression became a useful tool.

Analogue sound in electrical form has a history going back only slightly more than a century ago when a recording industry began to spring up around the gramophone and other early phonographs. The older among us fondly remember collections of long playing records (LPs) which later gave way to cassette tapes as the primary media for analogue consumer audio. Interest in digital audio received a boost some 20 years ago when the

Consumer Electronics (CE) industry developed a new digital audio recording medium: a 12 cm platter – the compact disc (CD) – carrying the equivalent of 70 minutes of uncompressed stereo digital audio. This equivalent of one long playing (LP) record was all that the CE industry needed at the time and compression was disregarded as the audio industry digitized.

Setting aside some company and consortium initiatives, it was only with the MPEG-1 project in the late 1980s that compressed digital audio came to the stage. MPEG-1 had the ambitious target of developing a single standard addressing multiple application domains: the digital version of the old compact cassette, digital audio broadcasting, audio accompanying digital video in interactive applications, the audio component of digital television and professional applications were listed as the most important.

The complexity of the task was augmented by the fact that each of these applications was targeted to specific industries and sectors of those industries, each with their own concerns when it comes to converting a technology into a product. The digital version of the old compact cassette was the most demanding: quality of compressed audio had to be good, but the device had to be cheap; in digital audio broadcasting quality was at premium, but the device had to have an affordable price; audio in interactive audio-visual applications could rely on an anticipated mass market where a high level of silicon integration of all decompression functionalities could be achieved; a similar target existed for audio in digital television; lastly, many professional applications required the best quality possible at the lowest possible bitrates.

It could be anticipated that these conflicting requirements would make the task arduous, and indeed the task turned out to be so. But the Audio group of MPEG, in addition to being highly competitive, was also inventive. Without calling them so, the Audio group was the first to define what are now known as "profiles" under the name of "layers". And quite good profiles they turned out to be because a Layer I bitstream could be decoded by a Layer II and a Layer III decoder in addition to its own decoder, and a Layer II bitstream could be decoded by a Layer III decoder in addition to its own decoder.

The MPEG-2 Audio project later targeted multichannel audio, but the story was a complicated one. With MPEG-1 Audio providing transparent quality at 256 kb/s for a stereo signal with Layer II coding and the same quality at 192 kb/s with Layer III coding, it looked like a natural choice that MPEG-2 Audio should be backwards compatible, in the sense that an MPEG-1 Audio decoder of a given layer should be able to decode the stereo component of an MPEG-2 Audio bitstream. But it is a well-known fact that backwards compatible coding provides substantially lower quality compared to unconstrained coding. This was the origin of the bifurcation of the multichannel audio coding work: Part 3 of MPEG-2 specifies a backward compatible multichannel audio coding and Part 7 of MPEG-2 (called Advanced Audio Coding – AAC) a non backward compatible or unconstrained multichannel audio coding standard.

AAC has been a major achievement. In less than 5 years after approving MPEG-1 Audio layer III, the MPEG Audio group produced an audio compression standard that offered transparency of stereo audio down to 128 kb/s.

This book has been written by the very person who led the MPEG-2 AAC development. It covers a gap that existed so far by offering both precious information on digital audio in general and in-depth information on the principles and practice of the 3 audio coding standards MPEG-1, MPEG-2 and MPEG-4. Its reading is a must for all those who want to know more, for curiosity or professional needs, about audio compression, a technology that has led mankind to a new relationship with the media.

Preface

The idea of this book came from creating and teaching a class for graduate students on Audio Coding at Stanford University's Computer Center for Research in Music and Acoustics (CCRMA). The subject of audio coding is a "hot topic" with students wanting to better understand the technology behind the MP3 files they are downloading over the internet, their audio choices on their DVDs, the digital radio proposals in the news, and the digital television offered by cable and satellite providers. Now in its sixth year, the class attracts a wide range of participants including music students, engineering students, and industrial professionals working in telecommunications, hardware design, and software product development.

In designing a course for such a diverse group, it is important to develop a shared vocabulary and understanding of the basic building blocks of a digital audio coder so that the choices made in any particular coder can be discussed using a commonly understood language. In the course, we first address the theory and implementation of each of the basic coder building blocks. We then show how the building blocks fit together into a full coder and how to judge the performance of such a coder. Finally, we discuss the features, choices, and performance of the main state-of-the-art coders in commercial use today.

The ultimate goal of the class, and now of this book, is to present the student and the reader with a solid enough understanding of the major issues in the theory and implementation of perceptual audio coders that they are able to build their own simple audio codec. MB is always very pleasantly surprised to hear the results of her student's work. As a final project for the class, they are able to design and implement perceptual audio coding schemes equivalent to audio coding schemes that were state-of-the-art only a few years ago. It is our hope that this book will allow advanced readers to achieve similar goals.

The book is organized in two parts: The first part consists of Chapters 1 through 10 which present the student with the theory of the major building blocks needed to understand the workings of a perceptual audio coder. The second part consists of Chapters 11 through 15 in which the most widely used perceptual audio coders are presented and their major features discussed. Typically, the students start their final project (building their own perceptual audio coder) at the transition from the first part to the second. In this manner, they are confronting their own trade-offs in coder design while hearing how these very same trade-offs are handled in state-of-the-art commercial coders. The particular chapter contents are as follows:

Chapter 1 serves as an introductory chapter in which the goals and high-level structure of audio coders are discussed.

Chapter 2 discusses how to quantize sampled data so that it can be represented with a finite number of bits for storage or transmission. Errors introduced in the quantization process are discussed and compared for uniform and floating point quantization schemes. The ideas of noiseless (entropy) coding and Huffman coding are introduced as means for further reducing the bit requirement for quantized data.

Chapter 3 addresses sampling in the time domain and how to later recover the original continuous time input signal. The basics of representing audio signals in the frequency domain via Fourier Transforms are also introduced.

Chapters 4 and 5 present the main filter banks used for implementing the time to frequency mapping of audio signals. Quadrature Mirror filters and their generalizations, Discrete Fourier Transforms, and transforms based on Time Domain Aliasing Cancellation are all analyzed. In addition, methods for designing time variant filter banks are illustrated.

Chapters 6 addresses the fundamentals of psychoacoustics and human hearing. Chapter 7 then discusses applications of frequency and temporal masking effects to develop masking curves for use in audio coding.

Chapter 8 presents methods for allocating bits to differing frequency components so as to maximize audio quality at a given bitrate. This chapter shows how the masking curves discussed in the previous chapter can be exploited to reduce audio coding bitrate.

Chapter 9 discusses how the pieces described in the previous chapters fit together to create a perceptual audio coding system. The standardization process for audio coders is also discussed.

Chapter 10 is devoted to the understanding of methods for evaluating the quality of audio coders.

Chapter 11 gives an overview MPEG-1 Audio. The different audio layers are discussed as well implementation and performance issues. MPEG Layer III is the coding scheme used to create the well-known MP3 files.

Chapters 12 and 13 present the second phase of MPEG Audio, MPEG-2, extending the MPEG-1 functionality to multichannel coding, to lower sampling frequencies, and to higher quality audio. MPEG-2 LSF, MPEG-2 BC, and MPEG-2 AAC are described. The basics of multichannel and binaural coding are also introduced in these chapters.

Chapter 14 is devoted to Dolby AC-3, the audio coder used in digital television standards and in DVDs.

Chapter 15 introduces the latest MPEG family of audio coding standards, MPEG-4, which allows for audio coding at very low bit rates and other advanced functionalities. MPEG-4 looks to be the coding candidate of choice for deployment in emerging wireless and wired network applications.

Acknowledgements

Audio coding is an area full of lore where you mostly learn via shared exploration with colleagues and the generous sharing of experience by previous explorers. This book is our attempt to pass on what we've learned to future trekkers. Some of the individuals we have been lucky enough to learn from and with during our personal explorations include: Louis Fielder and Grant Davidson from Dolby Laboratories; Karlheinz Brandenburg, Martin Dietz, and Jürgen Herre from the Fraunhofer Gesellschaft; Jim Johnston and Schulyer Quackenbush from AT&T; Leonardo Chariglione the esteemed MPEG Convener; Gerhard Stoll from IRT; and David Mears from the BBC. To all of the above (and the many others we've had the privilege to work with), we offer heartfelt thanks for their generosity of spirit and shared good times.

The course this book is based upon came into being due to the encouragement of John Chowning, Max Mathews, Chris Chafe, and Julius Smith at Stanford University. It was greatly improved by the nurturing efforts of its über-TA Craig Sapp (whose contributions permeate the course, especially the problem sets) and the feedback and good sportsmanship of its many students over the last 6 years. Thanks to Louis Fielder, Dan Slusser of DTS, and Baryn Futa of MPEG LA® for allowing MB to fit teaching into a full-time work schedule. Thanks also to Karlheinz Brandenburg and Louis Fielder for their guest lectures on MP3 and the Dolby coders, respectively, and to Louis for hosting the class at the Dolby facilities to carry out listening tests.

Not being able to find an appropriate textbook, the course made due for several years with extensive lecture notes. That would probably still be the case were it not for the intervention and encouragement of Joan L. Mitchell, IBM Fellow. Joan made the writing of a book seem possible and shared her hard-won insight into the process. You would not be holding this book in your hands were it not for Joan's kind but forceful encouragement.

Thanks to Joan Mitchell, Bernd Edler from Universität Hannover, Leonardo Chariglione, Louis Fielder, and Karlheinz Brandenburg for their careful review of early drafts of this book – their comments and feedback helped the clarity of presentation immensely. Thanks to Sarah Kane of the Brattle Group for her tireless yet cheerful administrative support during the writing process. Thanks also to Baryn Futa and Jamie Read from the Brattle Group for their support in ensuring that work demands didn't prevent finding the time for writing.

In spite of the generous help and support of many individuals, there are surely still some murky passages and possibly errors in the text. For any such blemishes, the authors accept full responsibility. We do sincerely hope, however, that you find enough things of novelty and beauty in the text that any such findings seem minor in comparison.

To Alex

PART I: AUDIO CODING METHODS

Chapter 1

Introduction

1.   REPRESENTATION OF AUDIO SIGNALS

We hear a sound and we want to store it for later replay – what information do we need to capture? Physicists tell us that sound is a pressure wave (i.e., vibration) in the air so we can measure this pressure wave with a mechanical device and then mechanically reproduce the pressure wave later. This is the principle used by Thomas Edison and other manufacturers of early gramophones (precursors to phonographs) in which a large cone concentrated the vibrations to a point where a needle scratched its vibrating path onto a spinning cylinder or disk. Later, a hand-cranked or other form of motor would turn the spinning cylinder or disk and the needle's forced movement along its prior path would cause the cone to recreate the pressure wave. The advent of electronic technology has allowed us to convert the pressure wave into a voltage reading that can be transferred onto a variety of storage media, for example as a changing degree of magnetization along a cassette tape. The basic idea in analogue technology, however, is still the same – to represent sound by the amplitude of its vibration over time. This tells us that one basic representation of sound is as a changing function of time t, which we denote x(t) as shown in *Figure 1*.

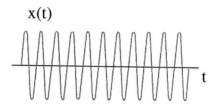

Figure 1. Time-domain representation of a sound tone

When we listen to sound, however, we hear clear distinctions between tonal content that tells us something about the sound. For example, some sounds seem high pitched like the squeaking of a sticky door while others are low pitched like the boom of a kettledrum. Tonal content is naturally described in terms of the frequencies contained in the sound. Since we perceive sounds in terms of their tonal content, in many instances it is more appropriate to describe audio signals as some function X(f) that shows how much of the frequency f is present in the signal (see *Figure 2*).

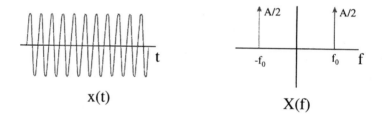

Figure 2. Time-domain versus frequency representation of a sound tone of amplitude A at frequency f_0

Audio coding, the main subject of this book, allows for the representation of sound in a very compact way without losing its perceptual characteristics.

2.   WHAT IS A DIGITAL AUDIO CODER?

What exactly is a digital audio coder? Any sound in nature has analogue characteristics. Since we live in a computer era, we would like to have this information in a digital form so that we can record, process, transmit and play it digitally. A typical digital audio coder, or codec for encoder-decoder, is a device that takes analogue audio signals as input and transforms them temporarily into a convenient digital representation. This transformation process takes place in the encoder stage of the coder. Once we have the signal represented as a series of numbers then we can store it, process it, or transmit it. At some point, we would like to be able to listen again to the sound. To do so we need to transform the signal from its digital representation back to an analogue signal so that the human ear can detect and enjoy it. This inverse transformation from digital back to analogue takes place in the decoder stage of the coder.

Figure 3. Digital audio coding chain

In general, an audio coder or codec is an apparatus which has as input an audio signal and as output a perceptually identical (or very close) delayed copy of the input signal. In *Figure 3*, the typical digital audio coding chain is shown. It is very important to emphasize that the very first stage of the audio coding chain is the source of the sound and the very last stage of the audio coding chain is the human ear. These two parts of the coding chain are important because they can play an important role in the design of the audio coder. If we can somehow develop a good understanding of the sound source then we can optimize the way we represent the audio signal, i.e. we can use a more compact description of the sound. Taking into consideration that the last stage is the human ear and applying models of the ear and its processing of acoustical stimuli, we can also reduce the amount of information contained in our representation of the audio signal that is irrelevant to our perception.

3.   AUDIO CODING GOALS

Once we decide that we would like to obtain a digital representation of the audio signal, a number of trade-offs come into play in order to carry out this transformation. In general, we would like to maximize the perceived quality but we also would like to minimize the amount of information needed to represent the signal. The challenge in designing an audio coding system is to balance these two conflicting goals while maintaining an acceptably low cost system.

Some of the most important factors we need to take into consideration when we are designing or assessing an audio coder are:

- Fidelity
- Data rate
- Complexity
- Delay

The balance between these factors will be determined by the application the technology is meant to support.

Fidelity addresses how perceptually equivalent the output of a codec is to the original input signal. The overall system quality is the most important attribute of any coding system. Depending on the application, however, we may have differing requirements for acceptable quality. So-called "telephony quality" is considered acceptable for applications that require intelligibility of the spoken words but not adequate for applications involving electronic distribution of music in which we would like to have "CD-like" quality audio signals, for example. Unfortunately, higher fidelity usually requires higher data rates, greater system complexity, and higher system delays.

The data rate of the audio coding system is linked to the throughput, the storage, and the bandwidth capacity of the overall system. Typically, we know the restrictions of the medium we are using for storage, transmission and playback of the audio signals under consideration. These restrictions, combined with the target quality of our application, are the parameters that determine our system data rate. Higher data rates typically imply higher costs in transmission and storage of the digital audio signals.

The complexity of carrying out the encode/decode process in a system translates into hardware and software costs in the encoder and decoder. Again, the target application will give guidance as to what trade-offs are acceptable here. For example, in a point-to-multipoint broadcast application, low cost and widely disseminated decoders are usually desired. In this case, we usually try to keep as much of the required processing complexity in the encoder to decrease the cost of the decoders. Moreover, by appropriately designing the encoder/decoder system, one can also maintain the ability to make some improvements to the coding process without having to alter (or replace) the installed base of decoders in the marketplace. In contrast, when we need to be able to encode/decode audio signals in real-time, like for example in desktop video-conferencing over the internet, then keeping the complexity low in both encoder and decoder is important. It should be noted that, with the current trend of decreasing memory costs and increasing computer horsepower, what was prohibitive in terms of complexity a few years ago, is currently considered acceptable. Some may even argue that complexity will soon become a non-issue. Implementation cost, however, is still a very important factor in the design of a coder.

Other important factors in the design of audio coders include coding delay (for example in telephony and teleconferencing), scalability (for example in internet broadcasts to users with very different connection speeds), and error robustness (for example in wireless transmission).

In general, we will assume that the design goal of any audio coding system is to provide high fidelity with low data rates, while maintaining the complexity of the system as low as possible.

4. THE SIMPLEST CODER – PCM

The simplest, best-understood, and most established audio coder is based on pulse code modulation, PCM. Block diagrams of the PCM encoder and decoder are shown in *Figure 4*. In the PCM encoder, the analogue audio signal is sampled at regular time intervals and then the signal amplitude of each sample is quantized into one of a limited set of digital codes, each representing a range of signal amplitude. While during the sampling process we don't lose any information if we sample often enough, the quantization process is inherently a lossy process and some of the information contained in the original signal is irrevocably lost.

In the decoder stage of the PCM scheme, the quantized codes are decoded and then the discrete time samples are interpolated to create an output analogue signal. The higher the number of discrete values employed in the quantization process, the more accurately the output signal will approximate the input signal.

PCM Encoder:

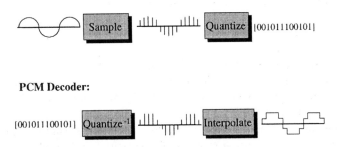

PCM Decoder:

Figure 4. The PCM coder

5.   THE COMPACT DISK

One widely employed application of the audio PCM scheme is the compact disk (CD) format. Introduced in the mid-eighties by Sony and Philips, the CD was the result of many years of research in storage, laser reader technology, and error correction. Its popularity exponentially increased over the years to the point where we now find millions of units in homes entertainment systems, cars, "boom boxes", and computer systems. The CD today has become a consumer standard for audio quality in the marketplace against which other audio coding systems are often compared.

In the CD format the audio signal is digitally represented as a stereo signal (i.e. two channel audio signal) sampled at time intervals of 0.023 ms or equivalently with a sampling frequency, F_s, of

$F_{sCD} = 44.1$ kHz.

where the time interval between adjacent time samples equals the reciprocal of the sampling frequency. This sampling frequency is adequate to preserve frequency content up to 22.05 kHz. From the psychoacoustics point of view, the CD sampling frequency is well selected since the average upper frequency limit for human hearing is around 20 kHz [Zwicker and Fastl 90]. It should be noted that the CD preserves a much wider range of frequency content than previously established analogue systems. For example, LPs typically only allow for frequency content below 10 kHz.

Some of you may wonder why such an "odd" number is employed rather than, for example, using a sampling rate of 40 kHz or 48 kHz which seem more natural choices. In fact, the 44.1 kHz sample rate is merely a historical artifact arising from the fact that VCR technology was used to store audio data in the early days of CD development [Watkinson 89].

The audio sample precision depends on the number of bits, R, employed to represent a sample. In the CD format R equals:

$R_{CD} = 16$ bits per sample.

This precision allows for up to $2^{16} = 65536$ discrete levels to represent the audio sample amplitudes and it can cover a nominal dynamic range of over 90 dB, which again widely exceeds the dynamic range of LPs which is typically less than 50 dB.

In the case of a digital system like the CD, one measure of its quality is given by the signal to noise ratio (SNR) measured in decibel, dB. Typical values for SNR_{CD} approach 90 dB. While in general this signal to noise ratio is quite good, psychoacoustics studies show that in the mid-range frequencies (between 2 and 5 kHz) it is not sufficient for all listeners. In this frequency range the human ear is very sensitive and more bits per sample are needed to transparently reproduce sounds. Ideally, 18-20 bits per sample are needed for describing audio samples in this frequency content.

The data rate or bitrate of a system, I, in bits per second, (or kb/s kilo, thousands, bits per second or Mb/s Mega, Millions, bits per second) per channel is given by the sampling frequency times the audio sample precision. In the case of the CD we have:

$$I_{CD} = F_{sCD} * R_{CD} = 705.6 \text{ kb/s per audio channel}$$

or

$$I_{CDTotal} = 706.5 * 2 = 1.4112 \text{ Mb/s.}$$

The maximum length of music that can be stored on the CD is about 75 minutes. The total amount of storage devoted to audio on the CD is less than 800 MBytes. The maximum CD duration again comes from the historical development of the CD and the fact that some of the storage area in the CD is devoted to error correction codes and control data [Immink 98].

6. POTENTIAL CODING ERRORS

Several types of errors can be introduced into the signal during any coding scheme, even one as simple as PCM. Errors can be introduced from inadequate sampling, poorly designed quantization, and from corruption during transmission or storage. The following are the main types of potential errors that can occur in an audio coder:

Sampling Errors – What happens if we sample the audio signal at time intervals too widely spaced? In doing so we irrevocably shift some of the signal's frequency content to where it doesn't belong. The signal frequencies above half the sampling frequency are mirrored to lower frequencies giving rise to a noticeable distortion called "aliasing". Aliasing can be avoided by either selecting an adequate sampling frequency or by passing the signal through a low-pass filter that eliminates the frequency content that would be aliased. (Although low-pass filtering can cause audible changes in a signal, the resulting changes are far less annoying than aliasing errors are.)

Quantization Errors – We encounter two types of quantization errors: overload errors and round-off errors. Overload errors occur when the input signal range exceeds the maximum value of the quantizer. This type of error is very annoying and needs to be carefully avoided. In contrast, round-off error is always present in the quantization process and the goal in audio coding is to reduce it to inaudible levels. A good portion of this book is devoted to understanding how to design coders that minimize the audible effects of round-off error.

Storage and transmission errors – Storage media and transmission channels can introduce errors in stored or transmitted signals. Additional bits can be included in a stored/transmitted signal to detect and even correct limited numbers of errors. Although error detection and correction is a very important topic in developing data systems to support audio coding, we consider it beyond the scope of this book. For the sake of simplicity, we will in general assume ideal transmission channels and storage media for the rest of this book.

7.   A MORE COMPLEX CODER

We noted while examining the CD format that, even at its high data rates, potentially audible round-off errors may be introduced in the mid-range frequencies of sound. If the application goal is to produce a perceptually transparent sound while operating at CD data rates or lower, "smarter" audio coding schemes are needed.

As an example of a smarter approach, consider an encoder that employs a transformation of the signal representation from the time domain (like in PCM) to the frequency domain so that it can dynamically allocate bits through the frequency spectrum based on the frequency content of the signal. In this manner, the coder can try to take bits from frequencies where our ear's dynamic range is lower and move them to the mid-range frequencies where the ear is very sensitive. In the decoder, the inverse bit allocation and transformation from the frequency to the time domain is applied. In *Figure 5* a block diagram of such a transform coder is shown.

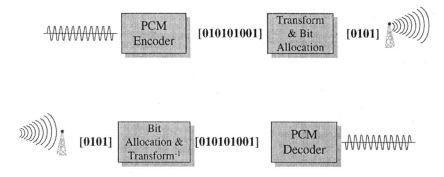

Figure 5. An example of a more complex coder than PCM

One may ask: "Do we really need such complex coders?" If we examine the CD representation of the audio signals we soon realize that a lot of repetitive information is stored when representing the signal. In other words, information that is not necessarily needed to uniquely reconstruct the signal is accumulated. For example, the PCM representation of a sine wave is a long series of time sampled values. If we were to instead describe the sine wave in the frequency domain, we would need only to store its frequency, amplitude and phase to completely characterize the signal (see also *Figure 2*). By doing so we greatly reduce the amount of data needed to represent the signal while not losing any information.

In general, although pure deterministic waves such as sine waves are improbable for sound, the statistical nature of audio signals is quasi-periodic. This important characteristic of sound implies that more often than not the PCM representation of sound contains a significant amount of redundant information. This redundancy can be reduced by simply applying a frequency transformation to the signal and appropriately allocating bits to the populated region of the spectrum. Other methods for reducing redundancies in the signal can also be used such as prediction methods and entropy coding (e.g., Huffman coding) that exploit symbol likelihood statistics.

The PCM representation of a signal often also contains a significant amount of irrelevant information, i.e. signal content which is inaudible. For example, information about sounds in the low frequency range which are too soft to be heard or normally audible sounds that are masked by louder sounds. This information does not need to be included in the coded signal. Perceptual audio coders reduce signal bit rate by reducing both redundancy and irrelevancy in the audio signal representation. In the next chapters of this book we will examine the basic principles and implementation choices used by state-of-the-art audio coders to carry out this bit rate reduction.

8.   REFERENCES

[Immink 98]: K. Immink, "The Compact Disc Story, " JAES vol. 46, number 5, pp. 458-462, May 1998.

[Watkinson 89]: J. Watkinson, *The Art of Digital Audio*, Focal Press, London, 1989.

[Zwicker and Fastl 90]: E. Zwicker and H. Fastl, *Psychoacoustics: Facts and Models*, Springer-Verlag, Berlin Heidelberg 1990.

9.   EXERCISES

a) Data rates:
Compute the data rates for the following audio signals:
1. A mono signal (i.e. single channel signal) sampled at 8 kHz using 8 bits per sample
2. A stereo signal sampled at 44.1 kHz using 16 bits per sample
3. Five channel (L, C, R, LS, RS) audio sampled at 44.1 kHz using 16 bits per sample
4. Five channel (L, C, R, LS, RS) audio sampled at 96 kHz using 24 bits per sample

b) The need for compression:
There has been recent discussion in the audio market about the need to move to higher sample representation precision and greater sample rates. In particular, a format using 24 bits per sample and a sample rate of 96 kHz has been discussed. Let's look at some implications for using such a format for passing 5-channel audio.
1. How much storage is needed for 2 hours of this type of audio signal?
2. If the CD format throughput for audio is equal to 1.411 Mb/s (Mega or Millions bits per second), what compression ratio is needed to pass this type of signal through a CD system?
3. If the DVD Video format throughput for audio is equal to 6.144 Mb/s, what compression ratio is needed to pass this type of signal through a DVD Video system?

Chapter 2

Quantization

1. INTRODUCTION

As we saw in the previous chapter, sound can be represented as a function of time, where both the sound amplitude and the time values are continuous in nature. Unfortunately, before we can represent an audio signal in digital format we need to convert continuous signal amplitude values into a discrete representation that is storable by a computer – an action which does cause loss of information. The reason for this conversion is that computers store numbers using finite numbers of bits so amplitude values can be stored with only finite precision. In this chapter, we address the quantization of continuous signal amplitudes into discrete amplitudes and determine how much distortion is caused by the process. Typically, quantization noise is the major cause of distortion in the coding process of audio signals. In later chapters, we address the perceptual impacts of this signal distortion and discuss the design trade-off between signal distortion and coder data rate. In this chapter, however, we focus on the basics of quantization.

In the following sections, we first review the binary representation of numbers. Computers store information in terms of binary digits ("bits") so an understanding of binary numbers is essential background to the quantization process. We also discuss some ways to manipulate the individual bits in a binary number. Next, we discuss different approaches to quantizing continuous signal amplitudes onto discrete values storable in a fixed number of bits. We look in detail at uniform and floating point quantization methods. Then we quantify the level of distortion introduced into the audio signal by quantizing signals to different numbers of bits for the different quantization approaches.   Finally, we discuss how entropy coding methods can be used to further reduce the bits needed to store the quantized signal amplitudes.

2.   BINARY NUMBERS

We normally work with numbers in what is called "decimal" or "base 10" notation. In this notation, we write out numbers using 10 symbols

$$(0,1,\ldots,9)$$

and we use the symbols to describe how we can group the number in groups of up to 9 of each possible power of ten.   In decimal notation the right-most digit tells us how many ones (10^0) there are in the number, the next digit to the left tells us how many tens (10^1), the next one how many hundreds (10^2), etc. For example, when we write out the number 1776 we are describing a number that is equal to

$$1*10^3 + 7*10^2 + 7*10^1 + 6*10^0 = 1000 + 700 + 70 + 6 = 1776$$

Computers and other digital technology physically store numbers using binary notation rather than decimal notation.   This reflects the underlying physical process of storing numbers by the physical presence or absence of a "mark" (e.g., voltage, magnetization, reflection of laser light) at a specific location.   Since the underlying physical process deals with presence or absence, we really have only two states to work with at a given storage point.

"Binary" or "base 2" notation is defined analogously to decimal notation but now we only work with 2 symbols (0,1) and we describe the number based on grouping it into groups of up to 1 of each possible power of 2. In binary notation, the rightmost column is the number of ones (2^0) in the number, the next column to the left is the number of twos (2^1), the next to the left the number of fours (2^2), etc. For example, the binary number

$$[0110\ 0100]$$

represents

$$0*2^7 + 1*2^6 + 1*2^5 + 0*2^4 + 0*2^3 + 1*2^2 + 0*2^1 + 0*2^0 = 64 + 32 + 4 = 100$$

Note that to minimize the confusion between which numbers are written in binary and which are in decimal, we try to always write binary numbers in square brackets. Therefore, the number 101 will have the normal decimal interpretation while the number [101] will be the binary number equal to five in decimal notation.

If we had to write down a decimal number and could only store two digits then we are limited to represent numbers only from 0 to 99. If we had three digits we could go all the way up to 999, etc. In other words, the number of digits we allow ourselves will determine how big a number we can represent and store. Similarly, the number of binary digits ("bits") limits how high we can count in binary notation. For example, *Table 1* shows all of the binary numbers that can be stored in only four bits counting up from [0000], 0, all the way to [1111], 15. Notice that each number is one higher than the one before it and, when we get to two in any column we need to carry it to the next column to the left just like we carry tens to the next column in normal decimal addition. In general, we can store numbers from 0 to 2^R-1 when we have R bits available. For example, with four bits we see in the table that we can store numbers from 0 to 2^4-1 = 16-1 = 15. If binary numbers are new to you, we recommend that you spend a little time studying this table before reading further in this section.

Table 1. Decimal numbers from 0 to 15 represented in 4-bit binary notation

Decimal	Binary (four bits)
0	[0000]
1	[0001]
2	[0010]
3	[0011]
4	[0100]
5	[0101]
6	[0110]
7	[0111]
8	[1000]
9	[1001]
10	[1010]
11	[1011]
12	[1100]
13	[1101]
14	[1110]
15	[1111]

2.1   Signed Binary Numbers

We sometimes want to write both positive and negative numbers in binary notation and so need to augment our definition to do this. Recall that in decimal notation we just add an additional symbol, the minus sign, to show what is negative. The whole point of binary notation is to get as far as we can keeping ourselves limited to just the two symbols 0, 1. There are two commonly used ways of expressing negative numbers in binary notation:

1) "folded binary" notation or "sign plus magnitude" notation

2) "two's complement" notation.

In either case, we end up using one bit's worth of information keeping track of the sign and so can only store numbers with absolute values up to roughly half as big as we can store when only positive numbers are considered.

In folded binary notation, we use the highest order bit (i.e., left-most bit) to keep track of the sign. You can consider this bit to be equivalent to a minus sign in decimal, in that the number is negative when it is set to 1 and positive when it is set to 0. For example, with four bits we would use the first bit as a sign bit and be able to store absolute values from 0 to 7 using the remaining three bits. In this notation, [1011] would now signify –3 rather than 11.

Two's complement notation stores the positive numbers the same as folded binary but, rather than being symmetric around zero (other than the sign bit), it starts counting the lowest negative number after the highest positive one, ending at –1 with all bits set to 1. For example, with four bits, we would interpret binary numbers [0000] up to [0111] as 0 to 7 as usual, but now [1000] would be –8 instead of the usual +8 and we would count up to [1111] being –1. In other words, we would be able to write out numbers from –8 to +7 using 4-bit two's complement notation. In contrast, folded binary only allows us to write out numbers from –7 to +7 and leaves us with an extra possible number of –0 being unused.

Computers typically work with two's complement notation in their internal systems but folded binary is easiest for humans to keep straight. Since we are more concerned with writing our own code to translate numbers to and from bits, we adopt the easier to understand notation and use folded binary notation whenever we need to represent negative numbers in this book.

2.2   Arithmetic Operations and Bit Manipulations

Binary numbers can be used to carry out normal arithmetic operations just as we do with normal decimal arithmetic, we just have to remember to carry twos rather than tens. As a few examples:

$3 + 4 = [11] + [100] = [111] = 7$

$5 + 1 = [101] + [1] = [110] = 6$

where in the last expression we carried the 2 to the next column,

$3 * 4 = [11] * [100] = (2^1 + 2^0) * 2^2 = (2^3 + 2^2) = [1100] = 12$

In addition, most computer programming languages provide support for some special binary operations that work bit by bit in binary numbers. These operators are the NOT, AND, OR, and XOR operators. The NOT operator flips each bit in a binary number so that all 1s become 0s and vice-versa. For example:

NOT [1100] = [0011]

The AND operator takes two binary numbers and returns a new number which has its bits set to 1 if both numbers had a 1 in that bit and sets them to 0 otherwise. For example:

[1100] AND [1010] = [1000]

Notice that only the left-most or "highest order" bit position had a one in both input numbers.

The OR operator also has two binary numbers as input but it returns a one in any bit where either number had a one. For example:

[1100] OR [1010] = [1110]

Notice that only the right-most or "lowest order" bit didn't have a one in either number.

Finally, the XOR (exclusive OR) function differs from the OR function in that it only returns 1 when one of the bits is one but not when both are one. For example:

[1100] XOR [1010] = [0110]

Notice that the highest order bit is now zero.

Having defined binary numbers, we would like to be able to manipulate them. The basic idea is to define storage locations as variables in a computer program, for example an array of integers or other data types, and to read and write coder bits to and from these variables. Then we can use standard programming (binary) read/write routines to transfer these variables, their values being equal to our stored bits, to and from data files or other output media. The binary digits themselves represent various pieces of data we need to store or transmit in our audio coder.

Suppose we have a chunk of bits that we want to read from or write to. For example, we could be writing a computer program and using 2-byte integer variables to store 16 bits in. Remember that a byte is equal to 8 bits so that a 2-byte integer variable gives us 16 bits with which to work. To read and write bits from this variable we need to know how to test and set individual bits. Our knowledge of binary notation provides us with the tools to do this. We can test and set bits using bit masks and the AND and XOR operations. Let's talk about some ways to do this.

A bit mask is a series of bits where specific bits are set to determined values. We know from binary notation that the number 2^n is represented in binary with the nth bit to the left of the right-most bit set equal to 1 and all others zero. For example:

$$2^3 = [1000], 2^2 = [0100], 2^1 = [0010], \text{ and } 2^0 = [0001]$$

Therefore we can easily create variables that have single bits set to one by using the programming language to set integer variables equal to powers of two. We call such a variable a "bit mask" and we will use it for setting and testing bits.

The AND operator lets us use a bit mask to read off single bits in a number. Remember that the AND operator only returns a one when both bits are equal to one and zero otherwise. If we AND together a bit mask with a number, the only possible bits that could be one in the result are the ones the bit mask has set to one. If the number has ones in those positions, the result will be exactly equal to the bit mask; if the number has zeros in those positions then the result will be zero. For example:

[0100] AND [abcd]

equals

[0100] for b = 1

or

[0000] for b =0

The XOR operator lets us use a bit mask to write a sequence of bits into a bit storage location. When we XOR a bit mask with a number, the bit values that are masked are flipped from one to zero and vice-versa. For example:

[0100] XOR [abcd]

equals

[a0cd] for b = 1

or

[a1cd] for b = 0

This means that we can take a number with zeros in a set of bit locations and use the XOR to flip specific bits to one.

If we aren't sure that the bit storage location was already set to all zeros, we can erase the values in that location before writing in new values. We can do this by first creating a number that has all ones in its bit location, for example 2^R-1 for unsigned variables and -1 for signed ones – remember computers use two's complement arithmetic. We then flip all the bits in the region we want to erase to zero by using XOR and bit masks. Finally, we AND this number with our bit storage location to erase the values. For example, to clear the right-most 2 bits in the 4-bit location [abcd], we create the number [1111], we flip the last 2 bits to get [1100], and then we AND this with the bit storage location to get [abcd] AND [1100] = [ab00]. Now we are ready to write bits into that location by using XOR to flip the bits we want equal to one.

Another set of operations that we sometimes find useful are shift operations. Shift operations move all bit values to the right or to the left a given number of columns. Some computer programs provide support for the bit-shift operators, denoted << n here for a left shift by n and denoted >> n for a right shift by n, but you can use integer multiplication and division to create the same effect. Basically, a multiplication by two is equivalent to a left bit-shift with n = 1; multiplying by 2^n is equivalent to a left shift by n, etc. Remember that when bits are left shifted any new position to the right is filled in with zeros. For example:

3 * 2 = [0011] << 1 = [0110] = 6

and

$3 * 2^2 = [0011] << 2 = [1100] = 12$

Similarly, a division by two is equivalent to a right bit shift by one; dividing by 2^n is equivalent to a right shift by n, etc. Remember that when bits are right shifted any new position to the left is filled in with zeros. For example:

$12 \div 2 = [1100] >> 1 = [0110] = 6$ and $12 \div 2^2 = [1100] >> 2 = [0011] = 3$

If we have a set of eight 4-bit numbers that we want to write into a 32-bit storage location, we can choose to write all eight into their correct locations. An alternative is to write the first (i.e., left-most) one into the first four bits and left-shift the storage location by four, write in the next one and left shift by four, etc. Likewise, we could read off the first four bits, right shift by four, etc. to extract the stored 4-bit numbers.

Computers store data with finite word lengths that also allow us to use shift operators to clear bits off the ends of data. Character variables are typically eight bits, short integers are usually 16 bits, and long integers are usually 32 bits in size. We clear off n left bits by shifting left by n and then shifting right by n. The way zeros are filled in on shifts means that we don't get back to our original number. For example:

$([1111\ 1111] << 2) >> 2 = [1111\ 1100] >> 2 = [0011\ 1111]$

Note that this is very different from normal arithmetic where multiplying and then dividing by four would get us back to the starting number. To clear off n right bits we shift right by n and then shift left by n. For example:

$([1111\ 1111] >> 2) << 2 = [0011\ 1111] << 2 = [1111\ 1100]$

Having learned how to work with binary numbers and bits, we now turn to the subject of translating audio signals into series of binary numbers, namely to quantization.

3.   QUANTIZATION

Quantization is the mapping of continuous amplitude values into codes that can be represented with a finite number of bits. In this section, we discuss the basics of quantization technology. In particular, we focus on instantaneous or scalar quantization, where the mapping of an amplitude value is not largely influenced by previous or following amplitude values. This is not the case, for example, in "vector quantization" systems. In vector quantization a group of consecutive amplitude values are quantized into a single code. As we shall see later in this chapter when we discuss Huffman coding, this can give coding gain when there are strong temporal correlations between consecutive amplitude values. For example, in speech coding certain phonemes follow other phonemes with high probability. If the reader is interested in this subject, good references are [Gray 84, Gersho and Gray 92]. While vector quantization is in general a highly efficient technique at very low data rates, i.e. much less than one bit per audio sample, it makes perceptual control of distortion difficult. In audio coding, vector quantization is employed for intermediate quality, very low data rates (see for example MPEG-4 Audio [ISO/IEC 14496-3]).

As we saw in the last section, R bits allow us to represent a maximum of 2^R different codes per sample, where each of these codes can represent a different signal amplitude. Dequantization is the mapping of the discrete R-bit codes onto a signal amplitude. The mapping from continuous input signal amplitudes onto quantized-dequantized output signal amplitudes depends on the characteristics of the quantizer used in the process.

Signal amplitudes can have both positive and negative values and so we have to define codes to describe both positive and negative amplitudes. We typically choose quantizers that are symmetric in that there are an equal number of levels (codes) for positive and negative numbers. In doing so, we can choose between using quantizers that are "midrise" (i.e., do not have a zero output level) or "midtread" (i.e., do pass a zero output). *Figure 1* illustrates the difference between these two choices. Notice that midrise has no zero level and quantizes the input signal into an even number of output steps. In contrast, midtread quantizers are able to pass a zero output and, due to the symmetry between how positive and negative signals are quantized, necessarily have an odd number of output steps. With R number of bits the midtread quantizer allows for 2^R-1 different codes versus the 2^R codes allowed by the midrise quantizer. In spite of the smaller number of codes allowed, in general, given the distribution of audio signal amplitudes, midtread quantizers yield better results.

Midtread   **Midrise**

Figure 1. Midtread versus midrise quantization

3.1   Uniform Quantization

We first examine the simplest type of quantizer: a uniform quantizer. Uniform quantization implies that equally sized ranges of input amplitude are mapped onto each code. In such a quantizer, the input ranges are numbered in binary notation and the code for an input signal is just the binary number of the range that the input falls into. To define the input ranges and hence the quantizer itself we need three pieces of information:

1) whether the quantizer is midtread or midrise,
2) the maximum non-overload input value x_{max} (i..e. a decision as to what range of input signals will be handled gracefully by the quantizer), and
3) the size of the input range per code Δ (which is equivalent information to the number of input ranges N once x_{max} is selected since $N = 2 * x_{max}/\Delta$).

The third data item defines the number of bits needed to describe the code since, as we learned in the last section, R bits allow us to represent 2^R different codes.

For a midrise quantizer, R bits allow us to set the input range equal to:

$$\Delta = 2 * x_{max}/2^R$$

Midtread quantizers, in contrast, require an odd number of steps so R bits are used to describe only 2^R-1 codes, and so a midtread uniform quantizer with R bits has the slightly larger input range size of:

$$\Delta = 2 * x_{max}/(2^R-1)$$

Since the input ranges collectively only span the overall input range from $-x_{max}$ to x_{max}, the question arises as what to do if the signal amplitude is outside of this range. This event is handled by mapping all input signals with amplitude higher than the highest range into the highest range, and mapping all input signals with amplitude lower (i.e., more negative) than the lowest range into that range. The term for this event is "clipping" or "overload", and it typically causes very audible artifacts. In this book we adopt the convention of defining units of amplitude such that $x_{max} = 1$ for our quantizers. In other words, we describe quantizers in terms of how they assign codes for input amplitudes between -1 and 1.

3.2   Midrise Quantizers

Figure 2 illustrates a two-bit uniform midrise quantizer. The left hand side of the figure represents the range of input amplitudes from –1 to 1. Since it is a 2-bit midrise quantizer, we can split the input range into 4 bins. Because we are discussing a uniform quantizer, these 4 bins are equally sized and divide the input range as shown in *Figure 2*. The bins are numbered using "folded binary" notation (recall from last section that this uses the first bit as a sign bit) and the middle of the figure shows codes for each bin: they are numbered consecutively from the bottom as [11], [10], [00], [01], literally, -1, -0, +0, +1.

Having quantized the input signal into 2-bit codes, we have to address how to convert the codes back into output signal amplitudes. We would like to do this in a manner that introduces the least possible error on average. Take, for example, bin [00] that spans input amplitudes from 0.0 up to 0.5. Assuming that the amplitude values are uniformly distributed within the intervals, the choice of output level that has the lowest expected error power would be to pick the exact center of the bin, namely 0.25. Analogously, the best output level for the other bins will be their centers, and so the quantizer maps codes [11], [10], [00], [01] onto output values of –0.75, -0.25, 0.25, 0.75, respectively.

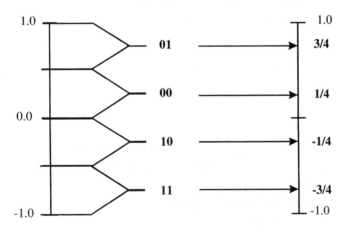

Figure 2. A two-bit uniform midrise quantizer

Uniform midrise quantizers with more than two bits can be described in similar terms. *Figure 3* describes a general procedure for mapping input signals onto R-bit uniform midrise quantizer codes and also for dequantizing these codes back onto signal amplitudes. To better understand this process, let's apply it to the two-bit quantizer we just described.

Consider an input amplitude equal to 0.6 which we can see from *Figure 2* should be quantized with code [01] and dequantized onto output amplitude 0.75. According to the procedure in *Figure 3*, the first bit of the code should represent the sign of the input amplitude leading to a zero. The second bit should be equal to

$$INT(2*0.6) = INT(1.2) = 1$$

leading to the correct code of [01] for an input of 0.6, where INT(x) returns the integer portion of the number x .

In dequantizing the code [01] the procedure of *Figure 3* tells us that the leading zero implies a positive number and second bit corresponds to an absolute value of

$$(1 + 0.5)/2 = 1.5/2 = 0.75$$

Putting together the sign and the absolute value gives us the correct output value of 0.75 for an input value of 0.6.

We recommend that you spend a little time trying to quantize and then dequantize other input values so you have a good feel for how the procedure works before continuing further in this chapter.

Quantize:

$code(number; R) = [s][|code|]$

where

$$s = \begin{cases} 0 & number \geq 0 \\ 1 & number < 0 \end{cases}$$

$$|code| = \begin{cases} 2^{R-1} - 1 & when\ |number| \geq 1 \\ INT(2^{R-1}|number|) & elsewhere \end{cases}$$

Dequantize:

$number(code; R) = sign*|number|$

where

$$sign = \begin{cases} 1 & if\ s = 0 \\ -1 & if\ s = 1 \end{cases}$$

$$|number| = (|code| + 0.5)/2^{R-1}$$

Figure 3. Quantization/dequantization procedure for an R-bit uniform midrise quantizer

3.3   Midtread Quantizers

Figure 4 illustrates a two-bit midtread uniform quantizer. Notice how the two bits are only used to describe three input amplitude bins. This means that we have divided the input range into thirds rather than quarters leading to a larger bin-size than in the midrise case. We have numbered the bins consecutively [11], [00], [01], literally, -1, 0, 1, and have chosen not to use the [10] code. The dequantized values are still the centers of the bins which are –2/3, 0, and +2/3, respectively, in this case. Notice that the zero value is passed. Audio signals often have quiet portions so quantizers that can represent a signal as having zero amplitude tend to sound better. For this reason, it is usually worth the cost of throwing away one possible code value and using midtread quantizers for audio coding.

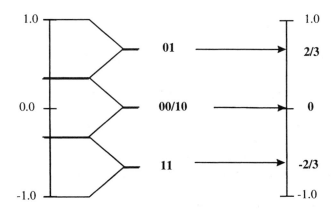

Figure 4. A two-bit uniform midtread quantizer

Figure 5 describes a procedure for implementing an R-bit uniform midtread quantizer. Let's again see how a value of 0.6 is quantized using this procedure. As before, the first bit is the sign bit, which should be zero for this input. The second bit should be equal to

INT((3*0.6+1)/2) = INT(2.8/2) = INT(1.4) = 1

leading to a code of [01]. In dequantizing we see that the first bit gives us a positive amplitude and the second bit gives an absolute value of 2 * 1/3 = 2/3. In other words, the procedure agrees with *Figure 4* and says the 0.6 should be mapped onto an output amplitude of +2/3. Again, we recommend that you try out a few more input amplitudes before moving on in this section.

Quantize:

$code(number; R) = [s][|code|]$

   where

$$s = \begin{cases} 0 & number \geq 0 \\ 1 & number < 0 \end{cases}$$

$$|code| = \begin{cases} 2^{R-1} - 1 & when\ |number| \geq 1 \\ INT(((2^R - 1)|number| + 1)/2) & elsewhere \end{cases}$$

Dequantize:

$number(code; R) = sign * |number|$

   where

$$sign = \begin{cases} 1 & if\ s = 0 \\ -1 & if\ s = 1 \end{cases}$$

$$|number| = 2.|code|/(2^R - 1)$$

Figure 5. Quantization/dequantization procedure for an R-bit uniform midtread quantizer

The uniform quantizer has a maximum round-off error equal to half of the bin width (i.e., $\Delta/2$) at any, non-overload input level. However, this error level could be huge relative to a very low amplitude signal. Since the perception of round-off distortion is more related to the relative error in amplitude than to the absolute size of the error, this means that uniform quantizers perform significantly worse on low power input signals than they do on higher power signals. This observation is the inspiration behind non-uniform quantization, which is described in the next sections.

3.4   Non-Uniform Quantization

In general, there is no requirement that the step sizes of a quantizer be of uniform size. Quantizers with step sizes that vary with input amplitude are called "non-uniform" quantizers. Although non-uniform quantization can be implemented using table lookups for the step sizes, it is more often implemented using the "companding" method, which we briefly discuss here.

In the companding method, an input x is passed through a monotonically increasing function

$y = c(x)$

prior to being uniformly quantized. Dequantization then is carried out by first dequantizing the uniformly quantized code into a value y' and then passing that value through the inverse function

$$x' = c^{-1}(y')$$

The function $c(x)$ is normally anti-symmetric around $x=0$ so that it maps negative values of x onto negative values. This implies that we can fully define $c(x)$ if we specify $c(|x|)$. If we consider both our input signal and our uniform quantizer to be normalized so that they run from -1.0 to 1.0, then the companding function $c(|x|)$ should map inputs from 0 to 1.0 onto the range from 0 to 1.0. The requirement that $c(x)$ be monotonically increasing is so that $c(x)$ is easily invertible.

To get a feel for how companding affects quantization, let's see how the size of the quantizer bins varies with input level x. We know that the quantizer bins are uniformly sized with regard to the level of y since y is uniformly quantized. If we do our quantization using a large number of bits so that the bins are small, then, over the size of a bin, the mapping $y = c(x)$ is approximately linear and we have that

$$y(x) \approx c(x_0) + dc/dx \ (x\text{-}x_0)$$

for some fixed x_0 in the bin. This tells us that the width in x of a quantizer bin is scaled down by a factor of dc/dx from the width in y (see *Figure 6*). For example, if we pick a function $c(x)$ that has a high slope for small values of x, then we have lower quantization noise in that region as compared to uniform quantization. However, since the function $c(|x|)$ must run monotonically from 0 to 1.0, a high slope for low values of x implies a lower slope and hence more quantization noise for higher values of x. Thus, we can use the companding function to move quantization noise from low amplitude inputs to higher amplitude inputs. Non-uniform quantization can be used to slow down the drop-off in signal to noise ratio as the signal amplitude decreases at the cost of lowering the peak signal to noise ratio of the quantizer.

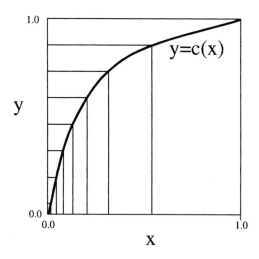

Figure 6. Effect of companding on quantizer bin widths for a four-bit midtread quantizer

3.4.1   Power-Law Companding

There are two common functional types used in companding: power law companding and logarithmic companding. In power law companding we use a function of the form:

$$c_{power}(|x|) = |x|^p$$

where choosing the parameter p so that it lies between 0 and 1.0 makes sure that the slope of c(x) is high for low values of x. We will see in Chapters 11 and 13 that power law companding is used in MPEG Layer III and MPEG AAC where p = 0.75.

3.4.2   Logarithmic Companding

In logarithmic companding we use a companding function based on the function log(x). Unfortunately, log(0) is equal to -∞ so this function cannot be used without adjustment. One common form of logarithmic companding is the so-called "μ law" (or "mu law") companding, which uses a function of the form:

$$c_\mu(|x|) = \frac{\log_b(1+\mu|x|)}{\log_b(1+\mu)}$$

where the base of the logarithm b doesn't affect the function (since a change of base would just multiply both numerator and denominator by a fixed factor) and the parameter μ determines the slope of dc/dx near zero. The slope of c(x) near x=0 is equal to $\mu/\ln(1+\mu)$ and so values of $\mu>>1$ are typically used. Notice how the 1 inside the logarithm avoids problems at x=0.

Another common form of logarithmic companding is the so-called "A law" companding which uses a function of the form:

$$c_A(|x|) = \begin{cases} \dfrac{1+\ln(A|x|)}{1+\ln(A)} & \text{for } |x| > 1/A \\ \dfrac{A}{1+\ln(A)}|x| & \text{for } |x| \le 1/A \end{cases}$$

Notice how the problem at small x is explicitly addressed by changing the functional form to linear for |x| smaller than 1/A. Again, having a high slope near x=0 implies that values of A>>1 are typically used.

Both μ law and A law companding have found significant application in the telecommunications industry and have been standardized by the CCITT (Telephone and Telegraph Consultative Committee, now known as the ITU-T, International Telecommunication Union Telecommunications Sector). The particular values of μ and A selected in the standards, $\mu=255$ and A=87.56, were so chosen so that the quantization characteristics, when used with an 8 bit uniform quantizer, can be reproduced digitally by manipulating the results of a longer uniform quantizer. In fact, the floating point quantization scheme that is presented in the next section is a variant of the digitally-companded A law scheme. The main difference is that the standardized 87.56 A law compander is carried out with a midrise quantizer rather than midtread. Nowadays, almost all logarithmic companding is carried out digitally through the use of various floating point quantizers. For more information on non-uniform quantization methods, the reader is encouraged to consult [Jayant and Noll 84].

3.5   Floating Point Quantization

The basic idea of floating point quantization is to scale the quantizer bin size to the size of the input signal: low input signals would use very small quantizer bins and high input signals will have larger bins. To implement this approach in a coder requires, however, that we know how big the bin sizes are when we dequantize the signal. This requires passing bits that describe the bin size. In other words, we split the bits in the code into two sets. Some bits are used to describe the bin size, where this set of bits represents the "scale factor" or "exponent" of the amplitude value. The rest of the bits are used to uniformly quantize the signal with those size bins, where this set of bits represents the "mantissa" of the amplitude value. Floating point quantization will give up accuracy for high-level signals since the scale factor bits that could have described bins are now used to tell the coder when the bins are smaller, but gains significant accuracy for low level signals, since the bins are now better sized for the signal. In general, the signal to noise ratio, SNR, will depend on the number of mantissa bits and will stay roughly constant over the whole range of input signal powers. This contrasts with uniform quantization where the SNR is highest at high powers (but low enough to avoid clipping) and decreases as the signal power decreases. We see examples of this in the next section. You can peek ahead to *Figure 10* to see a graph of this behavior.

We describe here a particular implementation of floating point quantization that is very similar to linearized A-law companding as was specified by the CCITT based on [Jayant and Noll 84]. The performance of the method depends on the number of scale factor bits R_s and mantissa bits R_m. Note that the total bits per sample R equals

$$R = R_s + R_m$$

At high level inputs this method is roughly equivalent to a uniform quantizer with

$$R = R_m + 1$$

In contrast, however, this quantizer performance will not degrade as the signal power is lowered until it reaches a level determined by the number of scale bits R_s. Once below this lower limit, the performance of this quantizer will also degrade but it will be comparable to how a uniform quantizer with

$$R = (2^{R_s}-1) + R_m$$

performs for such input signals. For example, eight bits divided into three scale bits and five mantissa bits perform for medium to high level signals about equivalently to a six-bit uniform quantizer sized to the signal power, and perform for low level signals like a 12-bit uniform quantizer.

To convert from a number into a scale-mantissa floating point code with Rs scale bits and Rm mantissa bits:

I.   Quantize the number as an R bit code where $R=2^{Rs}-1+Rm$.

II.   Count the number of leading zeros in |code|. If the number of leading zeros is less than $2^{Rs}-1$ then set the scale equal to $2^{Rs}-1$ minus the number of leading zeros; otherwise set the scale equal to zero.

III.   If scale equals zero then set the first mantissa bit equal to s and set the remaining Rm-1 bits equal to the bits following the $2^{Rs}-1$ leading zeros in |code|; otherwise set the first mantissa bit equal to s and set the remaining Rm-1 bits equal to the bits following the leading zeros omitting the leading one.

To convert from scale-mantissa floating point code with Rs scale bits and Rm mantissa bits into a number:

I.   Create an R bit code where $R=2^{Rs}-1+Rm$ from the mantissa and scale factor where s is the first mantissa bit and |code|

     A.   has $2^{Rs}-1$-scale leading zeros

     B.   followed by the remaining Rm-1 mantissa bits if scale is zero, otherwise followed by a one and then the remaining mantissa bits

     C.   followed by a one and as many trailing zeros as will fit if scale is greater than one.

II.   Dequantize the R bit code into the number.

Figure 7. Procedure for quantizing/dequantizing using a floating point quantizer

Figure 7 describes the floating point quantization and dequantization procedures. The basic idea is to first uniformly quantize the input signal using the highest number of bits for which the floating point quantizer is comparable. Then the scale bits are used to keep track of the number of leading zeros in the uniformly quantized code so one can strip them off the code. Finally, the mantissa bits are used to store the highest order bits in the remaining code, taking advantage of the fact that you know that the leading zeros were followed by a one. In order to dequantize, we apply the procedure in reverse. The scale factor tells us how many zeros to add to the front of the stripped off code (and leading 1, when appropriate) while we use the mantissa bits to recreate the rest of the code as accurately as possible. For code bits beyond what we stored in the last mantissa bit we pick the middle of the unknown range by following the last mantissa bit with a 1 and then zeros. For example, if we had three unknown trailing bits we would not know what they were ranging from [000] = 0 up to [111] = 7. We split the difference and use [100] = 4. Finally, we dequantize this code to get our output amplitude.

Figure 8 shows how floating point quantization is carried out for the case of $R_s = 3$ scale bits and $R_m = 5$ mantissa bits. This eight bit total floating point quantizer reaches an accuracy at low signal levels comparable to a 12-bit uniform quantizer so we first quantize our input signals using a 12-bit uniform quantizer. We use one of the five mantissa-bits to store the sign bit, using the remaining four mantissa-bits and the three scale-bits for storing the code. Notice that three scale bits can count up to seven leading zeros. We get to the 12-bit accuracy at low signal levels when one mantissa bit is the sign, the three scale bits tell us that the next seven bits are zero, and we use the remaining four mantissa bits to capture the next highest order bits in the code ($12 = 1 + 7 + 4$). Note that by convention the scale factor is set equal to zero when there are the maximum allowed seven leading zeros and the scale factor then counts up to seven as the number of leading zeros drops to none. The convention arises from the scale factor being equal to the number of left shifts that need be applied to the stripped off code (and leading one, when appropriate) described by the mantissa bits. This choice has no impact on accuracy of the coder and you could just as easily do it in the more natural approach where the scale factor is the number of leading zeros. We use this representation until the signal power reaches a level where the 12-bit quantization has less than seven leading zeros. In this case, we know that the first bit following those zeros is a one and does not need to be stored so we use the four mantissa-bits to store the four bits after the leading one. From this point all the way up to overload levels the quantization acts like a uniform quantizer with six bits corresponding to the sign bit, plus the leading one, plus the four other mantissa bits. As we see in the last line of the example, for the lower order bits beyond what we stored in the mantissa bits we split the difference and use a 1 followed by zeros when we recreate the 12-bit code. Once the 12-bit code is created we dequantize it back onto output signals as described in the previous section for uniform quantizers.

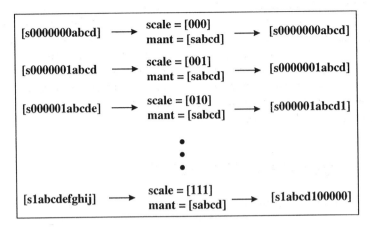

Figure 8. Applying floating point quantization with Rs=3 scale bits, Rm=5 mantissa bits

3.5.1   Block Floating Point

In building coders, we often use a variant of floating point quantization referred to as "block floating point" quantization or "block companding". Block floating point shares the bit-cost of the scale bits across several mantissa values. In other words, several floating point numbers are encoded using the same value of the scale factor. This cuts down on the number of scale bits used per number but, if the numbers are of different sizes, limits the ability of floating point quantization to adjust its bin sizes to the numbers being quantized. In this case, the signal SNR decreases for those values that are lower than the one used to set the exponent for the block. In general, to implement block floating point quantization, the scale factor for the group of numbers is set to the scale factor of the number with the largest absolute value. That single scale factor is then used to floating point quantize and dequantize all of the numbers in the group using floating point quantization but without assuming that there is a leading 1 in any mantissas. The reason that you cannot assume a leading 1 is that you won't know at dequantization time which number was the one that was used to set the scale factor. For example, any number with a magnitude more than a factor of two smaller than the magnitude of the number that set the scale factor will have a leading zero in the mantissa.

In summary, quantization is a critical step in the digitization process and in the design of audio coders. We can choose between midtread and midrise quantizers where midrise quantizers have slightly smaller round-off error but midtread quantizers accurately pass zero values. In general, we recommend the use of midtread quantizers for audio coders. Floating point quantizers allow you to allocate bits to match the quantizer bins to the size of the signal being quantized. Floating point quantization gives up some accuracy for high power signals but gains both much better accuracy for low power signals and a more consistent performance over a wide range of input signal powers. Block floating point quantization is a useful compromise method that uses fewer bits per number for scale factor storage by sharing scale factors across multiple signal values.

4. QUANTIZATION ERRORS

In the previous sections, we described how to implement quantizers and qualitatively discussed their performance. In this section, we take a more quantitative approach in describing quantization errors. Quantization errors are important because they are typically the major source of distortion in the audio coding process. Coder design requires trading off the bitrate of the coder against the fidelity of the decoded signal – more bits increases bitrate load but reduces the quantization error. Using some form of bit allocation to control the level of quantization error is a key feature of audio coders. We will discuss in later chapters how to optimize this trade-off; our goal for this section is the simpler task of describing the quantization error that results from using a given number of bits.

One way to characterize the quantization error is to compare the input signal, $x_{in}(t)$, with the output signal, $x_{out}(t)$, and measure the power in the difference or "error" signal q(t), where

$$q(t) = x_{out}(t) - x_{in}(t)$$

However, a more perceptually relevant measure would be to scale the error signal by the input signal to get the relative power of the error signal. We use this approach and describe quantization error in terms of its SNR measured in decibels (dB). The SNR in dB is defined as:

$$SNR = 10 \log_{10}(<x_{in}^2>/<q^2>)$$

Notice that in these units a low quantization error corresponds to a high SNR.

The SNR is certainly not a perfect (perceptual) measure of quality. In fact, many would say that it is a terrible one since it ignores many important perceptual effects including signal masking, different noise sensitivity at different frequencies, etc. SNR, however, is the general-purpose quality measurement most widely adopted. The reason is that it is objectively measurable and easily understood. Other objective measurements like, for example, perceptual objective measurements, rely on parameterized models of human perception and only in recent years have successfully addressed audio quality measurements (see also Chapter 10). In any event, for this section we stick with the SNR and return to the issue of human perception and perceptual measures of coder quality in later chapters.

4.1   Round-off Error

There are two types of quantization error and they sound very different: round-off error and overload or clipping error.

Round-off error comes from mapping ranges of input signal amplitudes onto a single code (and hence output level); the wider the range of input amplitude that maps onto a single code, the worse is the round-off error. We can estimate the relationship between round-off error and the number of bits in a uniform quantizer by assuming that the amplitude falls randomly into each quantization bin.

With such an assumption, the round-off error is equally likely to be any value between $-\Delta/2$ and $\Delta/2$. In other words the probability density of the error signal $q(t)$ at any time is approximately equal to $1/\Delta$ in the range between $-\Delta/2$ and $\Delta/2$ and zero elsewhere. Note that this assumption is well approximated when the quantizer has a large number of levels, but it is not true for quantizers with only a small number of levels, an extreme example being delta modulators.

Given the error probability distribution, we can calculate the expected error power for the quantizer:

$$< q^2 > = \int_{-\infty}^{\infty} q^2 p(q) dq = \int_{-\Delta/2}^{\Delta/2} q^2 \frac{1}{\Delta} dq = \frac{\Delta^2}{12}$$

In the case of a uniform quantizer with R bits we have that $\Delta \approx 2 * x_{max}/2^R$ leading to

$$< q^2 > \approx \frac{x_{max}^2}{3 * 2^{2R}}$$

If we feed this quantizer a signal with an input power equal to $<x_{in}^2>$ then we can expect the SNR (in dB) from the quantizer to be roughly equal to

$$SNR = 10 \log_{10}\left(< x_{in}^2 > / < q^2 >\right)$$

$$\approx 10 \log_{10}\left(< x_{in}^2 > \frac{3 * 2^{2R}}{x_{max}^2}\right)$$

$$\approx 10 \log_{10}\left(\frac{< x_{in}^2 >}{x_{max}^2}\right) + 20 * R * \log_{10}(2) + 10 * \log_{10}(3)$$

$$\approx 10 \log_{10}\left(\frac{< x_{in}^2 >}{x_{max}^2}\right) + 6.021 * R + 4.771$$

From this equation, we can see that the SNR increases as the signal power increases, at least until we start hitting overload, and that we improve the SNR by about 6 dB for every bit we add. The first term in this equation is the input signal power measured in dB relative to the quantizer x_{max}. This relative measure is the relevant measure of signal power for a uniform quantizer, which is why we can choose to set our quantizer x_{max} equal to 1 without any loss of generality. *Figure 9* shows both of these effects for a uniform quantizer that is fed a sine wave input signal with amplitude A less than x_{max}. In this case, the expected power of the input signal is $<x_{in}^2>$ = $A^2/2$. Notice that the curve for 16 bits is roughly 50 dB higher than the 8-bit curve in agreement with the 6-bit rule of thumb ($8 * 6 \approx 50$). Also, notice how the SNR increases with signal power.

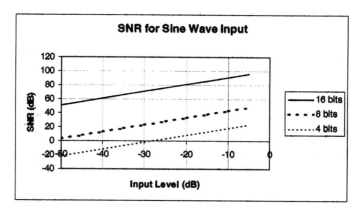

Figure 9. SNR for a sine wave input as a function of input level and number of bits

4.2 Overload error

Overload error comes from signal amplitudes that are too high in level for the quantizer. As we discussed in the previous section, these amplitudes are clipped to the highest and lowest quantizer steps. When the signal is well-sampled, overload error tends to present itself in bursts with quite audible effects. This error comes from input amplitudes $|x_{in}(t)|$ greater than the quantizer's maximum amplitude x_{max} (usually defined to be 1 in this book). If we can describe the probability distribution (or frequency distribution for known signals) of input signal amplitudes we can characterize the amount of overload error for a given quantizer x_{max}. We would like to set the quantizer's x_{max} high enough so that clipping doesn't occur, however, high x_{max} implies wide levels and hence large round-off error. Quantizer design requires a balance between the need to reduce both types of errors.

4.3 Error Effects

Figure 10 gives an example showing the effects of both types of quantization error for an input signal whose amplitude is random and uniformly distributed over an amplitude range between zero and A. In this case the expected power is related to the maximum amplitude A by

$$\left\langle x_{in}^2 \right\rangle = \int_0^A x_{in}^2 \frac{1}{A} dx_{in} = \frac{A^2}{3}$$

This figure shows the SNR of midtread uniform quantizers with various bit resolutions as well as showing the SNR of a 3-scale-bit, 5-mantissa-bit floating point quantizer. Notice that such a signal begins to overload the quantizer at input levels equal to –4.771 dB rather than at 0 dB. This is because 0 dB says that on average the signal has power equal to x_{max}^2, but by this point the highest amplitudes seen may be much larger than x_{max}. The maximum amplitude is just equal to x_{max} for uniformly distributed amplitudes when the input level is –4.771 dB (= $10*\log_{10}(1/3)$). Notice again the improvement in SNR with signal power for the uniform quantizers and the roughly 6 dB/bit improvement as R increases. Also notice how the 3-5 floating point quantizer is equal to the 6-bit uniform quantizer at high signal powers, holds on to its high SNR as the signal power drops, and finally drops just like the 12-bit uniform quantizer at very low level signals.

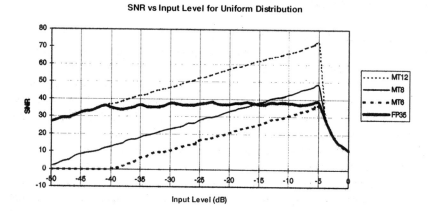

Figure 10. SNR for an input signal with random, uniformly distributed amplitudes for different types of quantizers (MT=midtread uniform quantization, FP = floating point) with R =12, 8, 6

5.   ENTROPY CODING

At this point in the book, we have learned how to represent amplitudes relative to an analogue audio signal with discrete codes that represent the quantized signal amplitudes. The codes we have discussed to this point have all been the result of scalar quantization with equal numbers of bits. We can often lower the data rate, however, by translating from these codes to a different symbol representation that uses a variable number of bits per code. The idea is to make common codes shorter so that the average bit rate will go down. Implementing this idea requires us to decide what codes are more common or, in other words, to estimate the probability of each possible code that might be seen. This requires us to describe something about the source of the signal. Having developed such probabilities, we can use "entropy coding" or "noiseless coding" methods to design variable bit length codes that reduce the overall number of bits needed to transmit the coded signal.

As an example, consider a 2-bit quantized signal that has the codes [00], [01], [10], [11]. Suppose we had a signal that we wanted to encode, whose frequency counts for each of these codes was 70%, 15%, 10%, 5%, respectively. Consider using the following code mapping instead: [00]→[0], [01]→[10], [10]→[110], and [11]→[111]. Notice that this new mapping is a "comma code" in that the zero symbol tells the decoder when the code terminates with less than 3 bits. This new code mapping has a lower bit rate on average for the signal since now the average bits per code has been reduced from two bits per code down to

$$<R> = 70\% * 1 + 15\% * 2 + 15\% * 3 = 1.45 \text{ bits/code}$$

We can translate our signal into the new mapping before storing or transmitting it and then translate back only when we are ready to decode the signal. Over a long signal, we will have managed to store or transmit the signal with less than ¾ of the bits needed for the original coding scheme. This example should make clear that exploiting information about code probabilities could allow us to squeeze bits out of a signal without any loss of information. Of course, to capture these savings we need to know how to decode the signal, which requires us to pass information. Any information we need to pass eats up some of the bit savings and needs to be taken into account when assessing the data rate of the system.

How might we develop code probabilities? First of all, some codes may be more common than others in certain classes of audio signals. If we knew that our coder was going to be passed a signal from that class, we would have some information about the code probabilities. Secondly, we could study the specific signal we wanted to encode and develop the probabilities from that analysis. In batch mode, we could study the whole signal before encoding it or, in streaming mode, we could run statistics on recent signal data that has gone by. Of course, this type of analysis would have implications as to the complexity and delay of the coder. In some cases, the bit reduction benefits might not be worth the complexity cost. Finally, we could exploit some type of prediction routine to predict the next code symbol and use codes to characterize the difference between the predicted value and the actual value – for such a system we would expect small values to be more common than large ones and could develop estimates of the probabilities of the difference codes.

Having developed code probabilities, we next should ask ourselves if the savings we can get from employing entropy coding is worth the time and effort of implementing it. "Entropy" is the measure that can answer this question for us. The entropy is a function of the probabilities p_n of the next code symbol being the nth code and is defined as

$$\text{Entropy} \equiv \sum_{n}^{codes} p_n \log_2(1/p_n)$$

When we are pretty sure what code will come out next, the entropy will be very low and when we have little idea as to which code will come out next, the entropy will be high. Shannon [Shannon 48] proved that the entropy as defined here exactly equals the lowest possible number of bits per sample any coder could produce for this signal.

To get a better feel for this measure called entropy, let's look at the case with only 2-code symbols. If the probability of the first symbol is equal to p then the probability of the other symbol is equal to 1-p and the entropy is equal to

$$\text{Entropy} = p \log_2(1/p) + (1-p) \log_2\big(1/(1-p)\big)$$

Figure 11 shows a graph of the 2-code entropy as a function of p. When p = 0 or p = 1 we know for sure what the next code will be and we find that the entropy is equal to zero. Since we know what the next symbol will be, we don't need to send any bits. Maximum lack of knowledge about the next symbol is when p = 50%. The entropy is equal to one for this value and we find that one bit is required to distinguish between the two outcomes. The interesting case is when p << 0.5 but not equal to zero (or p >> 0.5 but not equal to 1). Here Shannon's theorem tells us that there exist coding schemes that can encode a single-bit code using less than 1 bit per code symbol on average.

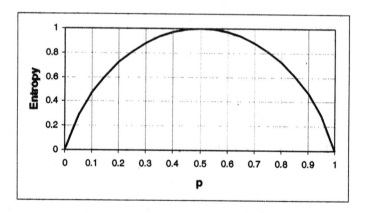

Figure 11. Entropy with 2-code symbols as a function of code probability p

Going back to the general definition of entropy for arbitrarily many code symbols

$$\text{Entropy} \equiv \sum_{n}^{\text{codes}} p_n \log_2(1/p_n)$$

we can see that there will not be any contribution to entropy for any symbol with probability zero (if it isn't going to show up, there is not need to allocate bits to it). Likewise, the overall entropy will be zero if any one symbol has probability 1 and all other symbols correspondingly have probability zero. In this case, we know what is coming beforehand so there is no need to send any information at all.

The maximum value of the entropy comes when all code symbols are equally likely. In this case, we won't get any savings from employing entropy coding methods. For example, consider the entropy of 2^R equally probable code symbols. Since they are equally probable, the probability of each symbol is $1/2^R$ and the entropy is equal to:

$$\text{Entropy} = 2^R * (1/2^R * \log_2(2^R)) = R \text{ bits}$$

In other words, using R bits for each of the 2^R equally probable codes is the best one can do. What we learn is that optimal coders only allocate bits to differentiate between symbols with near equal probabilities.

5.1 Huffman Coding

Huffman coding is a method for creating code symbols based on the probabilities of each symbol's occurrence. Huffman coding is a variable length code in that different symbols are given different length codes. In specific, Huffman coding gives short codes to more common symbols and longer codes to rarer ones. Shannon proved that the average number of bits per sample $<R>_{\text{Huffman}}$ in a Huffman code is within one bit of the entropy:

$$\text{Entropy} \leq < R >_{\text{Huffman}} \leq \text{Entropy} + 1$$

Huffman coding can reduce bits over fixed bit coding if the symbols are not evenly distributed.

It should also be noted that additional coding gain could often be achieved, at the cost of additional delay and complexity, by grouping consecutive symbols into a new set of longer symbols before creating the Huffman code. This additional coding gain from vector quantization comes about from two reasons: 1) it allows the Huffman code to exploit correlations between consecutive symbols in developing the codes, and 2) it allows the maximum difference from optimal coding of 1 bit per symbol to be spread out over more symbols. We will see examples of using vector Huffman coding in MPEG Layer III and in MPEG AAC (see also Chapters 11 and 13).

The Huffman code depends on the probabilities of each symbol. One creates Huffman codes by recursively allocating bits to distinguish between the lowest probability symbols until all codes are accounted for. To decode the Huffman code you need to know how the bits were allocated. Either you can recreate the allocation given the probabilities or you can pass the allocation with the data. The encoding algorithm goes as follows: Find the 2 lowest probability symbols and allocate a bit to distinguish between them. Consider the pair of symbols that you just distinguished as a single symbol with their combined probability and repeat the previous step until all symbols have been distinguished.

Figure 12 shows the development of a Huffman code for a case with 4 symbols with quite unequal probabilities. Notice that this is the example we examined at the very start of this section. Notice also that the two lowest probability symbols [10] and [11] get allocated a bit and then combined into a symbol with 15% probability. Now the two lowest probabilities are [01] and the combined [10]/[11] symbol and another bit is allocated to distinguish this pair. Finally, one last bit is needed to distinguish [00] from the combined [10]/[10]/[11] symbol. The final result is that [00], [01], [10], [11] get replaced with [0], [10], [110], [111], respectively. As we saw before, this new coding scheme reduces the average bit rate from 2 bits per sample down to 1.45 bits per sample.

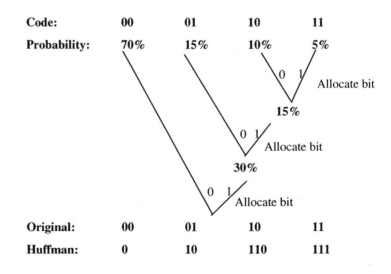

Figure 12. Huffman coding for four symbols with very unequal probabilities

Figure 13 shows the development of a Huffman code for the case with four symbols with equal probabilities. Since all four symbols have the same probability, we arbitrarily decide to first allocate a bit to the last two and combine them. Now the first two symbols are the lowest probability pair and so we allocate a bit and combine them. Finally, we need another bit to distinguish between the combined [00]/[01] and the combined [10]/[11] symbols. The result is that Huffman coding has just reproduced the initial coding scheme with no net gain. Again, we emphasize that Huffman coding only gives you gains when the symbols you are encoding have very different probabilities.

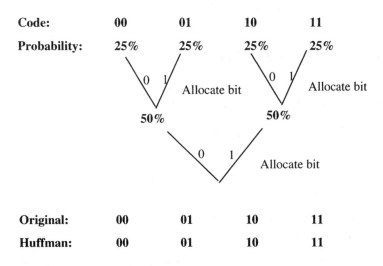

Figure 13. Huffman coding for four symbols with equal probabilities

In summary, we have learned that entropy is a measure of the minimum number of bits needed to represent a given signal. Entropy coding allows us to exploit redundancies in the signal representation in order to develop a new representation that requires fewer bits for the same information. These redundancies are identified and reduced based on the symbol probabilities of the various codes – if some symbols occur with much greater likelihood than others we can represent them with fewer bits. Huffman coding was presented as one commonly used implementation of these ideas.

6. SUMMARY

In summary for this chapter, we have shown that quantization can turn continuous amplitudes into discrete codes. We have seen that quantization produces errors and that different quantization schemes have different error effects. We have discussed how quantization error is the primary source of coding error in most perceptual coders and have analyzed the SNR produced using different numbers of bits in both uniform and floating point quantization. Finally, we have discussed how entropy coding techniques can be used to reduce the number of bits necessary to transmit or store quantized data. In the next chapters, we turn to the issues of frequency representation of audio signals and what types of errors can be heard by the human ear so that we will be in a position to decide what trade-offs should be made between fidelity and bit rate reduction.

7.   REFERENCES

[Gersho and Gray 92]: A. Gersho and R. M. Gray, *Vector Quantization and Signal Compression*, Kluwer Academic Press, 1992.

[Gray 84]: R. M. Gray, "Vector Quantization", IEEE ASSP Magazine, pp.4-29, April 1984.

[Jayant and Noll 84]: N. Jayant, P. Noll, *Digital Coding of Waveforms: Principles and Applications to Speech and Video*, Prentice-Hall, Englewood Cliffs, 1984.

[Shannon 48]: C. E. Shannon, "A Mathematical Theory of Communications", Bell Sys. Tech. J., Vol. 27, pp. 379-423, July 1948.

[ISO/IEC 14496-3]: ISO/IEC 14496-3, Information Technology, "Coding of audio-visual objects, Part 3: Audio", 1999-2001.

8.   EXERCISES

a) Working with bits:

Text is normally stored as 8 bit ASCII codes (of which only the lowest seven bits are used to store the basic character set). For this exercise, you will create a lossy text codec that stores text using only five bits per character. In doing so, you will gain experience in dealing with reading/writing binary coded files.

1. Write a function that takes the M lowest bits from an unsigned integer and writes them starting at the Nth bit location in an array of BYTES (unsigned character variables).

2. Define a mapping from the basic ASCII character set onto only five bits. (Obviously, you will need to sometimes map multiple ASCII

characters onto the same five-bit code. For example, you will need to map both capital and small letters onto the same code.)

3. Write a text encoder/decoder that allows you to read in ASCII text files (e.g., .txt files from Notepad), map the text into five bit codes, pack the coded text into arrays of BYTES, write the packed arrays into a coded file, read in your coded files, decode your coded files back into ASCII codes, and write out your decoded text file.

4. Test your five-bit text codec on several sample text files. Check file sizes to see what compression ratio you achieved. How readable is your decoded file?

5. Change your mapping to use two of your codes as control characters. Let one code signify that the next character is capitalized. Let the other code signify that the next character comes from a different set of mappings onto five-bit codes (i.e. include some important characters that weren't included in the basic mapping). How does this change impact compression ratio? How does this change impact readability?

b) Quantization and quantization noise:

In this exercise you will develop quantization routines that you will use when developing audio coders in later exercises.

1. Write functions that quantize floating point numbers from −1.0 up to 1.0 using R-bit midtread uniform quantization; R-bit midrise uniform quantization; and R_s scale factor bits, R_m mantissa bits midtread floating point quantization.

2. Create a version of *Figure 10* using 1.1 kHz sine waves sampled at 8 kHz as input.

c) A first audio coder:

In this exercise you will build a simple audio coder that allows you to test the effects of different quantization routines. You will also put in place the basic routines for reading and writing audio files that will be useful in later exercises.

1. Find a 16-bit PCM audio file format that 1) is well documented, 2) you can play on your computer, and 3) has sound samples you can find. (For example, information about the WAV file format is readily available on the internet. The Sound Recorder utility can be used to record and play PCM wave files.) Describe the file format. Get yourself a few good quality sound samples for testing codecs. Make sure you can play your samples.

2. Write an audio encoder/decoder that reads in 16 bit PCM audio files, dequantizes the audio samples to floating point values from −1.0 to 1.0, quantizes the samples using the quantization functions you

prepared for the prior exercise, packs the quantized samples into arrays of BYTES, writes the results into a coded file format you define, reads in your coded file, converts your data back into 16 bit PCM codes, and writes out your decoded audio data into an audio file you can play. Verify that your coder is bug-free by making sure that files coded using 16-bit midtread uniform quantization do not sound degraded when decoded.

3. Test your codec on some sound samples using 1) four-bit midtread uniform quantization, 2) four-bit midrise uniform quantization, 3) eight-bit midtread uniform quantization, and 4) three scale bits, five mantissa bits midtread floating point quantization. What compression ratios do you get? Describe the quantization noise you hear.

4. Estimate symbol probabilities for each of the 15 codes used in four-bit midtread uniform quantization using your sound samples. Use these probabilities to define a set of Huffman codes for your four-bit quantization codes. Modify your codec to read/write the coded file format using these Huffman codes when encoding using four-bit midtread uniform quantization. How much do the Huffman codes improve the compression on your sound samples?

Chapter 3

Representation of Audio Signals

1. INTRODUCTION

In many instances it is more appropriate to describe audio signals as some function of frequency rather than time, since we perceive sounds in terms of their tonal content and their frequency representation often offers a more compact representation. The Fourier Transform is the basic tool that allows us to transform from functions of time like x(t) into corresponding functions of frequency like X(f). In this chapter, we first review some basic math notation and the "Dirac delta function", since we will make use of its properties in many derivations. We then describe the Fourier Transform and its inverse to see how signals can be translated between their frequency and time domain representations. We also describe summary characteristics of signals and show how they can be calculated from either the time or the frequency-domain information. We discuss the Fourier series, which is a variation of the Fourier Transform that applies to periodic signals. In particular, we show how the Fourier series provides a more parsimonious description of time-limited signals than the full Fourier Transform without any loss of information. We show how we can apply the same insight in the frequency domain to prove the Sampling Theorem, which tells us that band-limited signals can be fully represented using only discrete time samples of the signal. Sampling allows us to convert continuous-time signals x(t) into discrete-time samples x[n] without loss of information if the sampling rate is high enough. Finally, we introduce prediction principles to represent a time series of audio samples in a more compact way than its PCM representation.

2. NOTATION

Before jumping straight into the Fourier Transform equations, we need to review some basic math notation and the "Dirac delta function", since we will make use of its properties in many derivations. In this book we try to use math notation symbols such as

$$x, y, t, f, \theta, \varphi \quad \in \Re$$

for variables that vary continuously over their allowed range (i.e. real numbers). In contrast, we try to use symbols such as

$$i, k, l, m, n, M, N \in \Im$$

for variables that are discrete over their allowed range (i.e. integers).

The Fourier Transform involves complex numbers and we use the symbol j to denote the square root of -1. Complex numbers can be represented as

$$z = x + jy$$

where x is called the "real part" of the complex number z and y is called the "imaginary part"[1]. Audio signals are usually real-valued, but it is sometimes convenient to consider complex-valued signals as well.

If z(t) is a complex-valued signal and x(t) and y(t) are real-valued signals that represent its real and imaginary parts, respectively, then we can also write

$$x(t) = Re \{ z(t) \} \quad and \quad y(t) = Im \{ z(t) \}$$

The "complex conjugate" of z is denoted as z^* and is equal to

$$z^* = x - jy$$

The "magnitude" (or "norm" or "modulus") of z is denoted as $|z|$ and is equal to

$$|z| = (x^2 + y^2)^{\frac{1}{2}}$$

---

[1] Sometimes the letter i is utilized instead of j to indicate the imaginary part of a complex number. In this book we adopt the "engineering notation" and employ j.

We can also write out the magnitude of z as

$$|z| = (z^*z)^{\frac{1}{2}} = (z\,z^*)^{\frac{1}{2}}$$

We often work with sinusoidal signals and make heavy use of the "Euler Identity" which defines an exponential of pure imaginary argument as a complex sum of a cosine and a sine:

$$e^{j\theta} = \cos(\theta) + j\sin(\theta)$$

We can use the definition of the magnitude to check that an exponential of pure imaginary argument has magnitude equal to 1. We call such a complex exponential a "pure phase" since the magnitude is 1 but there is still a phase angle θ that is specified. The "argument" (or "phase angle") of any complex number can be defined in terms of such an exponential by dividing the complex number by its magnitude and defining the argument as the angle θ in the expression $e^{j\theta}$ that has the same real and imaginary parts.

The Euler identity can also be used to convert sine and cosine into sums of complex exponentials:

$$\cos(\theta) = \frac{1}{2}\left(e^{j\theta} + e^{-j\theta}\right)$$
$$\sin(\theta) = \frac{1}{2j}\left(e^{j\theta} - e^{-j\theta}\right)$$

The reason for wanting to do this is that analytical calculations are typically much easier with exponentials than with sines and cosines.

3.   DIRAC DELTA

The Dirac delta function $\delta(t)$ is actually a "distribution", which is a generalized kind of a function, equal to zero everywhere except for at one point where its value is infinite:

$$\delta(t) \equiv \begin{cases} \infty & t = 0 \\ 0 & \text{elsewhere} \end{cases}$$

This would be a useless function were it not for a property of the function that it has finite integral equal to 1.

$$\int_{-\infty}^{\infty} \delta(t)\, dt = 1$$

Since $\delta(t)$ is zero everywhere except for at $t = 0$, an integral over a delta function only gets weight at $t = 0$ and we can pick off values of functions using the following:

$$\int_{-\infty}^{\infty} f(t)\delta(t - t_0)\, dt = f(t_0)$$

One way to derive the Dirac delta function is as the limit of a simple rectangular function:

$$\delta(t) = \lim_{A \to 0} f_A(t) \quad \text{where} \quad f_A(t) \equiv \begin{cases} \dfrac{1}{|A|} & -\dfrac{|A|}{2} \le t \le \dfrac{|A|}{2} \\[2mm] 0 & \text{elsewhere} \end{cases}$$

which is useful in trying to understand how the delta function behaves. For example, we can use this definition to derive how to rescale the argument of a Dirac delta function:

$$\delta(\alpha\, t) = \lim_{A \to 0} f_A(\alpha\, t) = \lim_{A \to 0} \frac{1}{|\alpha|} f_{A/\alpha}(t) = \frac{1}{|\alpha|} \lim_{B = A/\alpha \to 0} f_B(t) = \frac{1}{|\alpha|}\delta(t)$$

The limit of a rectangular function is not the only way to derive the Dirac delta function. Another derivation is as the limit of the sinc function as follows

$$\delta(t) = \lim_{A \to \infty}\{A\, \mathrm{sinc}(At)\} = \lim_{A \to \infty} \frac{\sin(\pi A t)}{\pi t}$$

where the sinc function is defined as

$$\mathrm{sinc}(x) \equiv \sin(\pi x)/\pi x$$

We can use the second definition of the Dirac delta function to derive a critical relationship that allows us to invert the Fourier Transform (see below):

$$\int_{-\infty}^{\infty} e^{\pm j2\pi ft} df = \lim_{F \to \infty} \left\{ \int_{-F/2}^{F/2} e^{\pm j2\pi ft} df \right\} = \lim_{F \to \infty} \left\{ \frac{\sin(\pi Ft)}{\pi t} \right\} = \delta(t)$$

The final property of the Dirac delta function that we find useful is the Poisson sum rule that relates an infinite sum of delta functions with a sum of discrete sinusoids:

$$\sum_{n=-\infty}^{\infty} \delta(\alpha - n) = \sum_{m=-\infty}^{\infty} e^{j2\pi m\alpha}$$

We can see that the right hand side is infinite (each term in the sum is equal to 1) when α is integer and it averages to zero for non-integer values – exactly the behavior described by the sum of Dirac delta functions on the left hand side. This relationship will be useful to us when we discuss the Fourier Transform of periodic or time-limited functions below.

4.   THE FOURIER TRANSFORM

The Fourier Transform is the basic tool for converting a signal from its representation in time $x(t)$ into a corresponding representation in frequency $X(f)$. The Fourier Transform is defined as:

$$X(f) \equiv \int_{-\infty}^{\infty} x(t) \, e^{-j2\pi ft} dt$$

and the inverse Fourier Transform which goes back from $X(f)$ to $x(t)$ is equal to:

$$x(t) = \int_{-\infty}^{\infty} X(f) \, e^{j2\pi ft} df$$

We can check that the inverse Fourier Transform applied to $X(f)$ does indeed reconstruct the signal $x(t)$:

$$\int_{-\infty}^{\infty} X(f) e^{j2\pi ft} df = \int_{-\infty}^{\infty} \left[\int_{-\infty}^{\infty} x(s) e^{-j2\pi fs} ds \right] e^{j2\pi ft} df$$

$$= \int_{-\infty}^{\infty} \left[\int_{-\infty}^{\infty} e^{-j2\pi f(s-t)} df \right] x(s) ds$$

$$= \int_{-\infty}^{\infty} \delta(s-t) x(s) ds$$

$$= x(t)$$

The inverse transform shows us that knowledge of $X(f)$ allows us to build $x(t)$ as a sum of terms, each of which is a complex sinusoid with frequency f. When we derive Parseval's theorem later in this chapter, we will see that $X(f)$ represents strength of the signal at frequency f. The Fourier Transform therefore is a way to pick off a specific frequency component of $x(t)$ and to calculate the coefficient describing the strength of the signal at that frequency. The Fourier Transform allows us to analyze a time signal $x(t)$ in terms of its frequency content $X(f)$.

Note that, although we deal with real-valued audio signals, the Fourier Transform is calculated using complex exponentials and so is complex-valued. In fact, for real valued signals we have that

$$X(f)^* = X(-f)$$

which implies that the real part of $X(f)$ is equal to the average of $X(f)$ and $X(-f)$, and the imaginary part is equal to their difference divided by 2j. The Euler identity tells us that $\cos(2\pi ft)$ has real-valued, equal coefficients at positive and negative frequencies while $\sin(2\pi ft)$ has purely imaginary coefficients that differ in sign. Likewise, any sinusoidal signal components differing in phase from a pure cosine will end up with imaginary components in their Fourier Transforms. We are stuck with working with complex numbers when we work with Fourier Transforms!

We can get some experience with the Fourier Transform and verify our intuition as to how it behaves by looking at the Fourier Transform of a pure sinusoid. Consider the Fourier Transform of the following time-varying signal:

$$x(t) = A \cos(2\pi f_0 t + \phi)$$

Notice that this signal is just a pure sinusoid with frequency f_0 and is identical to a pure cosine when the phase term equals zero, $\phi = 0$, and identical to a pure sine when $\phi = -\pi/2$.

We can calculate the Fourier Transform of this function to find that:

$$X(f) = \int_{-\infty}^{\infty} x(t) e^{-j2\pi ft} dt$$

$$= \int_{-\infty}^{\infty} A \cos(2\pi f_0 t + \phi) e^{-j2\pi ft} dt$$

$$= \int_{-\infty}^{\infty} A \tfrac{1}{2} \left(e^{j2\pi f_0 t + j\phi} + e^{-j2\pi f_0 t - j\phi} \right) e^{-j2\pi ft} dt$$

$$= \tfrac{A}{2} e^{j\phi} \int_{-\infty}^{\infty} e^{-j2\pi(f-f_0)t} dt + \tfrac{A}{2} e^{-j\phi} \int_{-\infty}^{\infty} e^{-j2\pi(f+f_0)t} dt$$

$$= \tfrac{A}{2} e^{j\phi} \delta(f - f_0) + \tfrac{A}{2} e^{-j\phi} \delta(f + f_0)$$

Notice that the Fourier Transform has components only at frequencies equal to positive and negative f_0. The period of this time-varying signal is $T_0 = 1/f_0$ and we can see that it only has frequency components at frequencies equal to integer multiples of $1/T_0$. We shall see shortly that this is a general property of periodic functions. Due to the presence of the phase terms $e^{\pm j\phi}$ – these phase terms come from the phase of the sinusoid- we can see that the Fourier Transform is in general complex valued. By inspection one can also see that $X(f)^* = X(-f)$ as required for real signals $x(t)$. Finally, notice the great deal of data reduction associated with representing this signal with the three parameters: A, f_0, and ϕ, as opposed to having to store its value at every point in time $x(t)$. The fact that most audio signals are highly tonal makes the Fourier Transform an important part of the audio coder's toolkit! For further reading on the Fourier Transform and its properties, we recommend [Brigham 74].

5.   SUMMARY PROPERTIES OF AUDIO SIGNALS

We often wish to summarize the general properties of an audio signal so that we can define coders that work well for broad classes of similar signals. Some of the most important properties include the bias $<x>$, the energy E, the average power P, and the standard deviation σ. In defining these quantities, we consider a finite-extent signal that is non-zero only between times $-T/2$ and $T/2$ for some time scale T. You may often see these definitions written taking the limit as $T \to \infty$, but, in reality, we always do our calculations over some finite time scale T.

The mean or bias of a signal is a measure of the average signal amplitude and is typically equal to zero for audio signals:

$$< x > \equiv \tfrac{1}{T} \int_{-T/2}^{T/2} x(t) dt$$

The average power of the signal is a measure of the rate of energy in the acoustic wave and is defined as

$$P \equiv \tfrac{1}{T} \int_{-T/2}^{T/2} x(t)^2 dt$$

Notice that power is defined as an average over time of the instantaneous power $P(t) = x(t)^2$. If we think of $x(t)$ as a voltage, this should hearken back to the electrical definition of power as Voltage^2/Resistance.

The energy of a signal is the integral of the instantaneous power over time and is equal to

$$E \equiv \int_{-T/2}^{T/2} x(t)^2 dt = P\,T$$

The standard deviation σ is a measure of the average power in the signal after any bias has been removed, and is defined by:

$$\sigma^2 \equiv \tfrac{1}{T} \int_{-T/2}^{T/2} (x(t) - < x >)^2 dt = P - < x >^2$$

Notice that the power and the square of the standard deviation are equal when there is no bias, as is usually the case for audio signals.

To get a feel for these signal properties let's calculate them for a pure sinusoid with frequency f_0. We look over a time interval T much longer than the period of the sinusoid T_0. Since the average of a sinusoid over an integral number of periods is equal to zero, we can see that the bias of this function is approximately equal to zero and we have that

$$<x> \approx 0$$

We know that the average of cosine squared over an integral number of periods is equal to ½, so we can see that

$$P \approx \frac{1}{2} A^2$$

Since the bias is about zero we can then write that

$$\sigma \approx \sqrt{P} = \frac{\sqrt{2}}{2} A$$

Finally, we can use the relationship between energy and average power to see that

$$E = P T$$

Notice that all of these properties are well-behaved in the limit as $T \to \infty$ except for the signal energy which becomes infinite for the sinusoid. In contrast, the bias, power, and standard deviation are all driven to zero for a finite extent signal as $T \to \infty$ since all the zero signal values outside the signal extent dominate the time averages. Again, in practice, all of these properties are usually calculated over a finite extent and taking the limit as $T \to \infty$ is an unnecessary but surprisingly common operation.

The Fourier Transform of a signal can also be used to calculate the above signal properties. For example, the bias is related to the Fourier Transform with frequency equal to zero:

$$< x >= \frac{1}{T} X(0)$$

where, if the true signal is not time limited, we limit the signal to the time window from $-T/2$ to $T/2$ consistently with our definitions of the signal summary properties before taking the Fourier Transform. In other words, we calculate the Fourier Transform using the time-limited signal:

$$x(t)' = \begin{cases} x(t) & T/2 < t \leq T/2 \\ 0 & \text{elsewhere} \end{cases}$$

instead of the true signal $x(t)$. The statement that audio signals tend to have zero bias is equivalent to the statement that the Fourier component at zero frequency is usually extremely small.

As another example, Parseval's theorem tells us that the energy in a signal can be written as an integral over the square of the frequency domain signal $X(f)$:

$$E = \int_{-\infty}^{\infty} |X(f)|^2 \, df$$

This theorem shows us that we can consider $X(f)$ to be a measure of the contribution to the signal energy from a particular location in the frequency domain. We often call the quantity $|X(f)|^2$ the "power spectral density", psd, in recognition of this fact.

We can prove Parseval's theorem as follows:

$$E = \int_{-\infty}^{\infty} x(t)^2 \, dt$$

$$= \int_{-\infty}^{\infty} \left(\int_{-\infty}^{\infty} X(f) e^{j2\pi ft} \, df \right) x(t) \, dt$$

$$= \int_{-\infty}^{\infty} X(f) \left(\int_{-\infty}^{\infty} x(t) e^{j2\pi ft} \, dt \right) df$$

$$= \int_{-\infty}^{\infty} X(f) X(f)^* \, df$$

$$= \int_{-\infty}^{\infty} |X(f)|^2 \, df$$

Given the energy and the bias, the other signal properties can be quickly derived.

Let's now calculate the signal properties for our sinusoidal example using the frequency domain information in the Fourier Transform instead of the time domain information. If we take the Fourier Transform of the time-limited sinusoid (see *Figure 1*), we find the following Fourier Transform:

$$X(f) = \int_{-\infty}^{\infty} x(t)e^{-j2\pi ft}dt$$

$$= \int_{-T/2}^{T/2} A\cos(2\pi f_0 t + \phi)\, e^{-j2\pi ft}dt$$

$$= \int_{-T/2}^{T/2} A\frac{1}{2}\left(e^{j2\pi f_0 t + j\phi} + e^{-j2\pi f_0 t - j\phi}\right)e^{-j2\pi ft}dt$$

$$= \frac{A}{2}e^{j\phi}\int_{-T/2}^{T/2} e^{-j2\pi(f-f_0)t}dt + \frac{A}{2}e^{-j\phi}\int_{-T/2}^{T/2} e^{-j2\pi(f+f_0)t}dt$$

$$= \frac{A}{2}e^{j\phi}\delta_T(f-f_0) + \frac{A}{2}e^{-j\phi}\delta_T(f+f_0)$$

where

$$\delta_T(f) = \frac{\sin(\pi Tf)}{\pi f}$$

which, we can recall from our definition of the Dirac delta function, approaches a Dirac delta function for large T. We find that the Fourier Transform of this time-limited sinusoid looks like the Fourier Transform of the infinite sinusoid other than the replacement of Dirac delta functions with similar functions of finite frequency width.

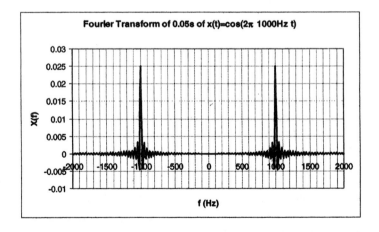

Figure 1. Fourier Transform of a time-limited sinusoid

To the degree that $T \gg T_0$, neither term will give much contribution to the Fourier Transform at $f = 0$ (since they are centered at $f = f_0$ and $f = -f_0$) so we again have that

$$<x> = F(0)/T \approx 0$$

To the degree that the $\delta_T(f \pm f_0)$ functions are narrow enough to not overlap significantly, we can approximate the energy as:

$$E = \int\limits_{-\infty}^{\infty} |X(f)|^2 \, df \approx \frac{A^2}{4} \int\limits_{-\infty}^{\infty} \delta_T (f - f_0)^2 df + \frac{A^2}{4} \int\limits_{-\infty}^{\infty} \delta_T (f + f_0)^2 df$$

$$\approx \frac{A^2}{4} T + \frac{A^2}{4} T \approx \frac{A^2}{2} T$$

where the integrals over the squared $\delta_T(f \pm f_0)$ functions can be integrated to find that they are each equal to T. This result also leads to estimates of average power and standard deviation that agree with our direct calculation in the time domain. In summary, one can see that the Fourier Transform information can also be used to describe the signal properties.

When we used the Fourier Transform to calculate the signal properties in the time interval from $-T/2$ to $T/2$, we time-limited the signal. One way to look at this time-limiting is to say that we multiplied the original signal $x(t)$ by a window function $w_T(t)$ equal to 1 in the time interval from $-T/2$ to $T/2$, and equal to zero elsewhere:

$$w_T(t) = \begin{cases} 1 & T/2 < t \le T/2 \\ 0 & \text{elsewhere} \end{cases}$$

When we looked at the resulting signal in the frequency domain, we found that it looked like the Fourier Transform of the original signal other than the fact that the Dirac delta functions were spread out a bit. This result is a specific example of the convolution theorem of the Fourier Transform.

The convolution theorem states that multiplication by a function in the time domain (e.g., windowing) corresponds to convolution (i.e. spreading) by that function in the frequency domain. In addition, the convolution theorem also states that convolution in the time domain corresponds to multiplication in the frequency domain. Defining the convolution of two functions $x_1(t)$ and $x_2(t)$ as follows:

$$y(t) = \int\limits_{-\infty}^{\infty} x_1(\tau) x_2(t - \tau) d\tau = x_1(t) \circ x_2(t)$$

we can then prove the convolution theorem:

$$
x(t)w_T(t) = \left(\int_{-\infty}^{\infty} X(f)e^{j2\pi ft}df \right)\left(\int_{-\infty}^{\infty} W_T(g)e^{j2\pi gt}dg \right) = \int_{-\infty}^{\infty} X(f)e^{j2\pi ft}\left(\int_{-\infty}^{\infty} W_T(g)e^{j2\pi gt}dg \right)df
$$

$$
= \int_{-\infty}^{\infty} X(f)e^{j2\pi ft}\left(\int_{-\infty}^{\infty} W_T(g-f)e^{j2\pi (g-f)t}dg \right)df = \int_{-\infty}^{\infty} X(f)\left(\int_{-\infty}^{\infty} W_T(g-f)e^{j2\pi gt}dg \right)df
$$

$$
= \int_{-\infty}^{\infty} \left(\int_{-\infty}^{\infty} X(f)W_T(g-f)df \right)e^{j2\pi gt}dg = \int_{-\infty}^{\infty} (X(g)\circ W_T(g))e^{j2\pi gt}dg
$$

where the last expression represents the inverse Fourier Transform of the convolution of $X(f)$ and $W_T(f)$. Notice how the Fourier Transform of the product is equal to the convolution of the Fourier Transforms. The reverse can also be proven as follows:

$$
\int_{-\infty}^{\infty} x(t-s)h(s)ds = \int_{-\infty}^{\infty} \left(\int_{-\infty}^{\infty} X(f)e^{j2\pi f(t-s)}df \right)h(s)ds = \int_{-\infty}^{\infty} X(f)\left(\int_{-\infty}^{\infty} h(s)e^{-j2\pi fs}ds \right)e^{j2\pi ft}df
$$

$$
= \int_{-\infty}^{\infty} X(f)H(f)e^{j2\pi ft}df
$$

where $x(t)$ represents the audio signal convolved with a filter having impulse response $h(t)$. Notice how this form of the theorem shows that filtering multiplies the frequency content of a signal with the frequency content of the filter. In Chapters 4 and 5 when we discuss filter banks, we will make frequent use of this result for discrete-time functions.

6. THE FOURIER SERIES

Suppose we only care about a signal over a finite time interval of time T which we define as being from time $-T/2$ up to time $T/2$. We saw in the previous section that the summary properties of the signal can be calculated either from the signal values in that time interval or from the Fourier Transform of the signal windowed to that time interval. In fact, the Fourier Transform of the windowed signal

$$
X(f) = \int_{-\infty}^{\infty} x(t)w_T(t)\, e^{-j2\pi ft}dt = \int_{-T/2}^{T/2} x(t)e^{-j2\pi ft}dt
$$

has enough information to perfectly reconstruct the signal in the time interval since the Fourier Transform can be inverted to recover the windowed signal.

We shall see in this section that we do not even need all of the Fourier Transform data to fully reconstruct the signal in the time interval. Rather, we can fully reconstruct the time-limited signal using only the values of $X(f)$ at discrete frequency points $f = k/T$ where k is an integer. We carry out this data reduction by replacing the windowed signal with a signal equal to it in the time interval from $-T/2$ to $T/2$ but repeated periodically outside of that interval. As long as we only care about signal values in the time interval from $-T/2$ to $T/2$, making this change will not affect the results.

Let's calculate the Fourier Transform of such a signal that is periodic with period T so that $x(t + nT) = x(t)$ for all t and any integer n. In this case, we have that:

$$X(f) = \int_{-\infty}^{\infty} x(t)e^{-j2\pi ft}dt = \sum_{n=-\infty}^{\infty} \int_{-T/2}^{T/2} x(t+nT)e^{-j2\pi f(t+nT)}dt$$

$$= \sum_{n=-\infty}^{\infty} \int_{-T/2}^{T/2} x(t)e^{-j2\pi f(t+nT)}dt = \int_{-T/2}^{T/2} x(t)e^{-j2\pi ft}\left(\sum_{n=-\infty}^{\infty} e^{-j2\pi nfT}\right)dt$$

$$= \int_{-T/2}^{T/2} x(t)e^{-j2\pi ft}\left(\frac{1}{T}\sum_{k=-\infty}^{\infty} \delta(f - k/T)\right)dt$$

$$= \frac{1}{T}\sum_{k=-\infty}^{\infty} \delta(f - k/T) \int_{-T/2}^{T/2} x(t)e^{-j2\pi kt/T}dt$$

$$= \frac{1}{T}\sum_{k=-\infty}^{\infty} \delta(f - k/T) X[k]$$

where we have used the Poisson sum rule that relates an infinite sum of Dirac delta functions to an infinite sum of complex exponentials and have defined the quantities $X[k]$ as

$$X[k] \equiv \int_{-T/2}^{T/2} x(t)e^{-j2\pi kt/T}dt$$

Notice that the Fourier Transform of this period signal is non-zero only at the discrete set of frequencies $f = k/T$. The inverse Fourier Transform for this periodic signal then becomes

$$x(t) = \int\limits_{-\infty}^{\infty} X(f) e^{j2\pi ft} df = \frac{1}{T} \sum_{k=-\infty}^{\infty} X[k] e^{j2\pi kt/T}$$

Noting that $X[k]$ is just the Fourier Transform of our time-limited signal, this result shows us that a periodic version of our time-limited signal can be exactly recovered using only the Fourier Transform values at the discrete frequency points $f = k/T$ where k is integer. This transformation of a periodic signal $x(t)$ to and from the discrete frequency values $X[k]$ is known as the "Fourier Series", which we have just seen is a special case of the Fourier Transform.

We can apply the Fourier Series to non-periodic signals if we are only interested in the signal values in a limited time interval. If we window our signal to the time interval of interest then we can fully represent the signal in that interval by its content at only the set of discrete frequencies. However, as we saw from the convolution theorem, the windowed signal has frequency content that is a blurred version of the original signal. Moreover, we find that many frequency components are needed to accurately reproduce the signal if there is too sharp a discontinuity in the windowed signal at the edge of the time interval. For this reason, we typically use smooth windows to transition the signal to zero near the edges of the time interval before taking transforms. In Chapter 5 we will discuss how careful selection of window shape and length can keep the frequency domain blurring to a minimum while also limiting the creation of high frequency content from edge effects.

7.   THE SAMPLING THEOREM

The Sampling Theorem [Shannon 48] tells us that continuous-time signals can be fully represented with discrete-time samples of the signal if we sample the signal often enough. Moreover, the theorem specifies exactly what sample rate is needed for a signal.

Suppose we have a signal whose frequency content is entirely contained in the frequency range from $-F_{max}$ to F_{max}. Since the frequency content is zero outside of this frequency range, the signal can be fully recovered given only the frequency content in this region by taking the appropriate inverse Fourier Transform. If we choose some frequency interval $F_s > 2*F_{max}$, we can periodically continue the signal's frequency spectrum outside of the range from $-F_s/2$ to $F_s/2$ without corrupting any of the frequency content needed to recover the original signal. The same reasoning that tells us that a time-periodic function only has discrete frequency components can be employed to show that a frequency-periodic function only has discrete-time components.

Repeating the line of reasoning used formerly to derive the Fourier Series leads us to the following representation of the spectrum that has been periodically continued in frequency with "period" F_s:

$$X(f) = \frac{1}{F_s} \sum_{n=-\infty}^{\infty} x[n] e^{-j2\pi nf / F_s} \quad \text{where} \quad x[n] \equiv \int_{-F_s/2}^{F_s/2} X(f) e^{j2\pi nf / F_s} df$$

We can draw several immediate conclusions from this result.

Firstly, if we define $T \equiv 1/F_s$ then we find that $x[n]$ is exactly equal to $x(nT)$. They are equal because the true Fourier Transform of the frequency-limited signal is equal to its periodically-continued version in the frequency range from $-F_s/2$ to $F_s/2$ and zero elsewhere. This implies that the inverse Fourier Transform of the real $X(f)$ at time $t = nT$ is exactly what we defined $x[n]$ to be. In other words, the periodic $X(f)$ is fully defined in terms of the true signal $x(t)$ sampled with sample time T (i.e. sample rate F_s).

Secondly, since the periodic $X(f)$ is fully defined from the signal samples then the continuous-time signal $x(t)$ must also be fully defined by the samples. This can be shown by throwing out the frequency content of the periodic $X(f)$ outside of the frequency range from $-F_s/2$ to $F_s/2$ to recover the signal's true Fourier Transform $X(f)$. We can then take the inverse Fourier Transform of the true $X(f)$ to recover the full signal:

$$x(t) = \int_{-\infty}^{\infty} X(f) e^{j2\pi ft} df = \int_{-F_s/2}^{F_s/2} X(f) e^{j2\pi ft} df$$

$$= \int_{-F_s/2}^{F_s/2} \left(\frac{1}{F_s} \sum_{n=-\infty}^{\infty} x[n] e^{-j2\pi nfT} \right) e^{j2\pi ft} df$$

$$= \sum_{n=-\infty}^{\infty} x[n] \left(\frac{1}{F_s} \int_{-F_s/2}^{F_s/2} e^{j2\pi f(t-nT)} df \right)$$

$$= \sum_{n=-\infty}^{\infty} x[n] \left(\frac{\sin(\pi F_s (t - nT))}{\pi F_s (t - nT)} \right)$$

$$= \sum_{n=-\infty}^{\infty} x[n] \left(\frac{\sin(\pi (F_s t - n))}{\pi (F_s t - n)} \right)$$

$$= \sum_{n=-\infty}^{\infty} x[n] \, \text{sinc}(F_s t - n)$$

In other words, the signal x(t) can be fully recreated from only the samples x(nT) by sinc function interpolation, where sinc (x) = [sin(πx)]/ πx.

Having drawn the conclusion that the signal x(t) can be fully represented by its samples x(nT), we need to remember the assumptions leading to that result. The critical assumption was that the periodicity frequency F_s was greater than twice the highest frequency component in the signal, i.e. $F_s \geq 2$ F_{max}. Without this assumption, we would not have been able to recover the true spectrum from the periodically continued one. This constraint relates the sampling rate F_s (T = 1/ F_s) to the frequency content of the signal. We call this minimum sample rate 2 F_{max} the "Nyquist frequency" of the signal. The Sampling Theorem tells us that we can do all our work with discrete-time samples of a signal without losing any information if the sampling rate is larger than the Nyquist frequency.

In addition, the Sampling Theorem tells us that the frequency spectrum of sampled data is periodic with period F_s. If we sample a signal with frequency components greater than $F_s/2$ (i.e. F_s is smaller than the Nyquist frequency) then the Sampling Theorem tells us that the frequency spectrum will be corrupted. In particular, high frequency content will be irretrievably mixed in with the frequency content from lower than $F_s/2$. This frequency mix up is called "aliasing" and it produces a very unpleasant distortion in the original signal. Normally, we low-pass filter any input signal that might have frequency components above $F_s/2$ before we sample it to prevent aliasing from occurring. For example, telephone service is typically sampled at 8 kHz so it is low-pass filtered down to below 4 kHz before sampling. (Yes, you do sound different on the phone!)

In the next section we discuss how prediction can be used to represent a time series of audio samples while reducing the number of bits needed to encode the sample.

8.   PREDICTION

We often find that we can predict a quantized audio sample with reasonable accuracy based on the values of prior samples. The basic idea is that, if the difference between the actual quantized sample and the predicted sample is typically much smaller in magnitude than the range of sample values, we should be able to quantize the differences using fewer bits than were needed to quantize the actual samples without increasing the quantization noise present.

What do we mean by prediction? Prediction means recognizing a pattern in the input data and exploiting that pattern to make a reasonably accurate guess as to the next data sample prior to seeing that sample. For example, if we were trying to quantize a slowly varying parameter describing our sound signal, we might be able to predict the next value based on simple linear extrapolation of the two prior samples:

$$y_{pred}[n] = y[n-1] + (y[n-1] - y[n-2]) = 2y[n-1] - y[n-2]$$

As another example, sound in an excited resonant cavity, for example, a struck bell, decays away at a predictable rate given the resonance frequency and decay time:

$$y_{pred}[n] = 2\cos(2\pi f_0 / F_s) e^{-1/(F_s \tau)} y[n-1] \quad - \quad e^{-2/(F_s \tau)} y[n-2]$$

where f_0 is the resonance frequency, τ is the decay time, and F_s is the sample rate. Notice how both of these examples predict the next value as a weighted sum of the prior values:

$$y_{pred}[n] = \sum_{k=1}^{N} a_k \, y[n-k]$$

where N is known as the "order" of the prediction. (In the two examples presented N was equal to 2 so they would be known as second order predictions.) Such a weighted sum of prior values is known as an "all-poles filter" and is a common method of prediction in low bit rate speech coding.

Suppose we had a way to predict samples with reasonable accuracy, how would we use it to save bits?   One way to do this is to quantize the prediction error rather than the signal itself. If the prediction works reasonably well then the error signal $e[n] = y[n] - y_{pred}[n]$ should be small. For example, *Figure 2* compares an input signal y[n] (from a resonant cavity with a resonance frequency of 2 kHz and a decay time of 10 ms that is being excited by both noise and periodic pulses) with the error signal from using the second order predictor described above. Notice that, although the input signal covers most of the range from −1.0 to +1.0, other than the occasional spike (coming from the periodic pulses) the error signal is mostly contained in the region from −0.01 to +0.01. In other words, this example shows a typical amplitude reduction of roughly a factor of 100 in the scale of the prediction error versus that of the input signal. If we know how much quantization noise was allowed in the signal, we would need about 6 fewer bits to quantize the error signal to the same level of quantization noise as is needed to quantize the input signal.

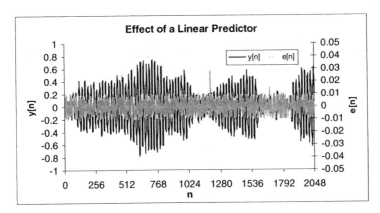

Figure 2. Prediction leads to an error signal with much lower amplitude than the original input signal

As a very stylized example of how we can achieve this bit rate reduction, let's consider a case where we can predict the next 16 bit quantized sample to within three quantizer spacings 99% of the time. (Although a predictor that can get within three quantizer spacings of a 16 bit quantized sample is quite unlikely, this exaggerated example should make the mechanics of bit rate reduction by prediction more clear.) In this case, we could code the difference signal using 3 bits according to the following pattern:

Error	Code
-3	[111]
-2	[110]
-1	[101]
0	[000]
1	[001]
2	[010]
3	[011]
beyond	[100]

where any sample whose predicted value was more than 3 quantizer spacings away would have the [100] code followed by the full quantized code of the input sample. If the original samples were quantized at 16 bits then this encoding of the prediction errors would have an average bit rate of 3.16 bits per sample (3 bits to code the prediction error plus another 16 bits 1% of the time when the predicted value is beyond 3 quantizer spacings away from the input signal).

If we also knew that the prediction error was clustered around low values we could supplement prediction with an entropy coding routine to further reduce the required bit rate. For example, if the prediction error in the prior example had the following probability distribution:

Error	Prob
-3	1%
-2	3.5%
-1	15%
0	60%
1	15%
2	3.5%
3	1%
beyond	1%

we could encode it using the following Huffman code table:

Error	Prob	Code
-3	1%	[1111110]
-2	3.5%	[11110]
-1	15%	[110]
0	60%	[0]
1	15%	[10]
2	3.5%	[1110]
3	1%	[111110]
beyond	1%	[1111111]

to get an average bit rate of 2.03 bits per sample.

In implementing prediction in a coder there are a number of issues that need to be confronted. First of all, a decision needs to be made as to the form of the prediction. This depends a lot on the source of the data being predicted. The all-poles filter approach has been used in low bit rate speech coding, often implemented with 10th order prediction. The all-poles filter approach is attractive for predicting speech samples since we know that speech is formed by passing noise-like (e.g., the hiss in a sibilant) or pulsed (e.g., glottal voicing) excitation through the resonant cavities of the vocal tract and sinuses, but the appropriate prediction routine for other types of information could very well take very different forms.

Secondly, the parameters describing the prediction function must be determined. In predictive speech coding, the filter coefficients (the a_k in the all-pole filter expression above) are usually set to minimize the variance of the error signal. This is carried out on a block-by-block basis where the block length is chosen to be shorter than the typical phoneme time scale. The resulting matrix equation for the a_k depends on the autocorrelation of the signal over the block (averages of y[n-k]*y[n-p] over all block samples n for various values of k and p) and has been studied sufficiently that very high speed solutions are known. For other forms of prediction equation, corresponding parameter fitting routines need to be defined.

Thirdly, information about the predictor form and coefficients needs to be passed to the decoder. Such information requires additional bits and therefore removes some of the performance enhancement from prediction. This loss is kept to a minimum by using a set of predictor coefficients as long as is possible without causing significant degradation of the prediction. For example, in low bit rate speech coding each set of predictor coefficients is typically used for a passage of 20-30 ms.

Fourthly, to limit the growth of quantization errors over time, prediction is almost always implemented in "backwards prediction" form where quantized samples are used as past input values in the prediction equation rather than using the signal itself. The reason is that the quantization errors produced during backwards prediction only arise from the coarseness of the quantizer while the errors in "forward prediction" form (i.e., doing the prediction using the prior input samples and not their quantized versions) can add up over time to much larger values.

Finally, a coding scheme must be selected to encode the prediction errors. Quantizing the error signal with a lower x_{max} and fewer bits than are used for the input signal is the basic idea behind the "differential pulse code modulation" (DPCM) approach to coding. Choosing to use a quantizer where x_{max} changes over time based on the scale of the error signal is the idea behind "adaptive differential pulse code modulation" (ADPCM). (For more information about DPCM and ADPCM coding the interested reader can consult [Jayant and Noll 84].) In low bit rate speech coding several very different approaches have been used. For example, in "model excited linear prediction" (MELP) speech coders the error signal is modeled as a weighted sum of noise and a pulse train. In this case, the error signal is fit to a 3-parameter model (the relative power of noise to pulses, the pulse frequency, and the overall error power) and only those 3 parameters are encoded rather than the error signal itself. As another example, in "code excited linear prediction" (CELP) speech coders the error signal is mapped onto the best matching of a sequence of pre-defined error signals and the error signal is encoded as a gain factor and a codebook entry describing the shape of the error signal over the block. (For more information about predictive speech coding the interested reader can consult [Shenoi 95]. Also, see Chapter 15 to learn more about the role of CELP and other speech coders in the MPEG-4 Audio standard.)

9.   SUMMARY

In this chapter, we discussed the representation of audio signals in both the time and frequency domains. We used the Fourier Transform and its inverse as a means for transforming signals back and forth between the time and frequency domains. We learned that we only need to keep track of frequency content at discrete frequencies if we only care about signal values in a finite time interval. We also learned that we can fully recover a signal from its discrete-time samples if the sampling rate is high enough. Having learned that we can work with only discrete-time samples of a signal, we learned how to represent a series of quantized audio in a more compact way by predicting samples from previous ones.

In the next chapters, we address the issue of time-to-frequency mapping discrete quantized samples and learn how we can use the computer to transform finite blocks of signal samples into equivalent information in the frequency domain. Once in the frequency domain, we have greater ability to use the tonal properties of the input signal and the limits of human hearing to remove redundant and irrelevant data from how we store and transmit audio signals.

10.   APPENDIX – EXACT RECONSTRUCTION OF A BAND-LIMITED, PERIODIC SIGNAL FROM SAMPLES WITHIN ONE PERIOD

Let's consider a band-limited, periodic signal, $x(t)$, with a maximum frequency F_{max} and period T_0. We can recover the exact input signal from its samples if we can sample it with a sample rate $F_s = 1/T \geq 2F_{max}$ using the reconstruction formula

$$x(t) = \sum_{n=-\infty}^{n=\infty} x[n] \left(\frac{\sin\left(\pi F_s \left(t - \frac{n}{F_s}\right)\right)}{\pi F_s \left(t - \frac{n}{F_s}\right)} \right)$$

All samples contribute to $x(t)$ when $t \neq n/F_s$ with a contribution that drops slowly with distance in time according to the function $\sin[\pi(t\text{-}t')]/\pi(t\text{-}t')$.

In the particular case of a periodic signal, we can choose to sample an integer number of times per period, i.e., $T = 1/F_s = T_0/M \leq 1/2F_{max}$, so for each period the sample values are the same. In this case defining $n = m + kM$ and noting that $x[n + kM] = x[n]$, we have:

$$x(t) = \sum_{k=-\infty}^{k=\infty} \sum_{m=0}^{m=M-1} x[m] \left(\frac{\sin(\pi(tF_s - m - kM))}{\pi(tF_s - m - kM)} \right)$$

$$= \sum_{k=-\infty}^{k=\infty} \sum_{m=0}^{m=M-1} (-1)^{kM} x[m] \left(\frac{\sin(\pi(tF_s - m))}{\pi(tF_s - m - kM)} \right)$$

$$= \sum_{m=0}^{m=M-1} x[m] \sin(\pi(tF_s - m)) \sum_{k=-\infty}^{k=\infty} \left(\frac{(-1)^{kM}}{\pi(tF_s - m - kM)} \right)$$

Combining positive and negative k terms with equal |k|, we obtain:

$$x(t) = \sum_{m=0}^{m=M-1} x[m] \sin(\pi(tF_s - m)) \left\{ \frac{1}{\pi(tF_s - m)} + \sum_{k=1}^{k=\infty} \left(\frac{2(-1)^{kM}(tF_s - m)}{\pi[(tF_s - m)^2 - (kM)^2]} \right) \right\}$$

By using [Dwight 61]:

$$\frac{\cos(ax)}{\sin(a\pi)} = \left\{ \frac{1}{\pi a} + \sum_{k=1}^{k=\infty} \left(\frac{2\cos(kx)(-1)^{k+1} a}{\pi[k^2 - a^2]} \right) \right\}$$

with $a = (tF_s - m)/M$, $x = 0$ for M odd, and $x = \pi$ for M even, we obtain:

$$x(t) = \sum_{m=0}^{m=M-1} x[m] \frac{\sin(\pi(tF_s - m))}{M\sin(\frac{\pi}{M}(tF_s - m))} \qquad \text{for M odd}$$

$$x(t) = \sum_{m=0}^{m=M-1} x[m] \frac{\sin(\pi(tF_s - m))\cos(\frac{\pi}{M}(tF_s - m))}{M\sin(\frac{\pi}{M}(tF_s - m))} \qquad \text{for M even}$$

You can recognize that these equations allow us to reconstruct the full signal x(t) from a set of samples in one period of the periodic function. For M odd, the function multiplying the sample values is referred to as the "digital sinc" function in analogy with the sinc function interpolation formula derived in the discussion of the Sampling Theorem.

11. REFERENCES

[Brigham 74]: E. O. Brigham, *The Fast Fourier Transform*, Prentice Hall Englewood Cliffs, N. J. 1974.

[Dwight 61]: H. B. Dwight, *Tables of Integrals and other related Mathematical Data*, MacMillan Publishing Co., Inc., New York 1961.

[Jayant and Noll 84]: N. Jayant, P. Noll, *Digital Coding of Waveforms: Principles and Applications to Speech and Video*, Prentice-Hall, Englewood Cliffs, 1984.

[Shannon 48]: C. E. Shannon, "A Mathematical Theory of Communications", Bell Sys. Tech. J., Vol. 27, pp. 379-423, July 1948.

[Shannon 49]: C. E. Shannon, "Communication in the Presence of Noise", Proc. IRE, Vol. 37, pp. 10-31, January 1949 (reproduced in Proc. of IEEE, Vol. 86, no. 2, pp. 447-457, February 1998).

[Shenoi 95]: K. Shenoi, *Digital Signal Processing in Telecommunications*, Prentice-Hall PTR, 1995.

12.   EXERCISES

a) Signal Representation and Summary Properties:
Consider the following signal:

$$x(t) = \begin{cases} \sin(2000\pi t)\sin(4\pi\, t) & \text{for } 0 \le t \le \frac{1}{4} \\ 0 & \text{elsewhere} \end{cases}$$

which represents a 1 kHz sine wave windowed with a sine window to a duration of ¼ second. Do the following:
1. Graph the signal
2. Compute the signal summary properties from the time domain description of the signal.
3. Compute and graph the Fourier Transform of this signal.
4. Compute the signal summary properties from the frequency domain description of the signal.
5. Sample this signal at an 8 kHz sample rate.
6. Use sinc function interpolation to estimate the original signal from its samples, and compare with the original signal. Explain any differences.

b) Prediction:
Consider the signal

$$y(n) = \begin{cases} e^{-an}\cos n\omega_0 & n \geq 0 \\ 0 & n < 0 \end{cases}$$

where $a = 0.05$ and $\omega_0 = 0.3\ \pi$. Consider also a (rectangular) windowed version of this signal

$$\hat{y}_M^R(n) = \begin{cases} y(n) & -M \leq n \leq M \\ 0 & \text{elsewhere} \end{cases}$$

where $M = 128$ (i.e. rectangular window of length $2M+1$ centered at $n = 0$). Finally, consider a 2-term LPC predictor $A(z) \equiv a_1 z^{-1} + a_2 z^{-2}$ so that we try to predict a signal $y(n)$ as $y_{predict}(n) = a_1 y(n-1) + a_2 y(n-2)$.

1. Define the excitation (i.e. prediction error) $x(n)$ as $x(n) \equiv y(n) - y_{predict}(n)$. Show for arbitrary signals $y(n)$ that the transfer function relating $x(n)$ to $y(n)$ is equal to $1/(1 - a_1 z^{-1} - a_2 z^{-2})$. This transfer function can in general have two complex poles (i.e. the denominator is quadratic in z^{-1}). Assume that the poles are complex conjugates of each other (i.e. one pole is $c = re^{j\theta}$ and the other is $c^* = re^{-j\theta}$) and determine a_1 and a_2 in terms of r, θ.

2. Take the z-transform (see Chapter 5) of our specific $y(n)$ and relate r, θ to a, ω_0 by identifying the pole locations of $y(n)$. Given r, θ we also know a_1, a_2 (from part 1), so calculate the prediction error $x(n)$ for our two-point predictor applied to our $y(n)$ using its true pole locations.

3. Use the z-transform of $y(n)$ to calculate the Fourier Transform of $y(n)$ (i.e. $Y(\omega)$ is the z-transform evaluated at $z = e^{j\omega}$) and graph $|Y(\omega)|^2$. Take a close-up look at the positive frequency pole and fit a Lorentzian to the peak of the form

$$\frac{K}{A^2 + (\omega - W)^2}$$

where A, W, K are constants to fit by eye, analytical calculation, or numerical fitting. What is your best fit for A, W, K? Show a close-up graph at your fit vs. $|Y(\omega)|^2$ at the peak. How do you think A, W relate to a, ω_0?

4. Define the prediction error energy as

$$E = \sum_{n=-\infty}^{\infty}(y(n) - y_{predict}(n))^2$$

Substitute $y_{predict}(n) = a_1 y(n-1) + a_2 y(n-2)$ and write E as a quadratic function of a_1 and a_2. An alternative approach to that of part 2 for finding the predictor coefficients is to minimize $E(a_1, a_2)$ w.r.t. a_1 and a_2. Do this by setting both

$$\frac{\partial E}{\partial a_1} \text{ and } \frac{\partial E}{\partial a_2}$$

equal to zero. Write the result as a matrix equation for

$$\vec{x} = \begin{pmatrix} a_1 \\ a_2 \end{pmatrix}$$

in terms of the matrix elements

$$\phi(k,l) \equiv \sum_{n=-\infty}^{\infty} y(n-k)y(n-l)$$

5. Obviously we can't actually calculate $\phi(k,l)$ values since they require us to use infinite length signals. We can, however, calculate a windowed approximation to $\phi(k,l)$ using our windowed (and zero padded) signal

$$\hat{y}_M^R(n)$$

Define

$$\hat{\phi}_M^R(k,l) \equiv \sum_{n=-\infty}^{\infty} \hat{y}_M^R(n-k)\hat{y}_M^R(n-l)$$

and show that it can be calculated as a sum of a finite number of terms. Also show that it only depends on $|k-l|$ terms in the form

$$\hat{\phi}_M^R(k,l) = g_M^R(|k-l|)$$

Re-write the matrix equation from part 4 using the windowed approximation and show that we only need three numbers to solve for x. Namely, show that we only need to know

$$g_M^R(0), g_M^R(1), g_M^R(2)$$

to fill in the matrix elements in the equation for x.

6. Using M = 128, calculate

$$g_M^R(0), g_M^R(1), g_M^R(2)$$

for our signal and solve the matrix equation to find a_1 and a_2. Use the relations from parts 1, 2 to calculate the pole estimates a, ω_0 corresponding to this estimate of a_1 and a_2. Also use this estimate of a_1 and a_2 to calculate the prediction error $x(n) \equiv y(n) - y_{predict}(n)$. Graph this prediction error vs. that calculated on part 2. Calculate the prediction error energy in both cases: which is lower? Which do you think is a "better" predictor? Explain.

7. Repeat part 6 with one of the other windows discussed in Chapter 5 (e.g. Hanning, Kaiser-Bessel) in place of the rectangular window. In other words, for some other window type W, use

$$\hat{y}_M^W(n) = \begin{cases} W(n)y(n) & -M \leq n \leq M \\ 0 & \text{elsewhere} \end{cases}$$

To estimate

$$g_M^W(0), g_M^W(1), g_M^W(2)$$

and solve for the LPC predictor pole location. Compare the rectangular window's pole estimates a, ω_0 with these for your window for the cases M = 32, 64, 128, 256.

Chapter 4

Time to Frequency Mapping Part I: The PQMF

1.   INTRODUCTION

In this and the following chapter, we discuss common techniques used in mapping audio signals from the time domain into the frequency domain. The basic idea is that we can often reduce the redundancy in an audio signal by subdividing its content into its frequency components and then appropriately allocating the bit pool available. Highly tonal signals have frequency components that are slowly changing in time. The data necessary to fully describe these signals can be significantly less than that involved in directly describing the signal's shape as time passes.

Frequency domain coding techniques have the advantage over time domain techniques like, for example, predictive coding schemes such as ADPCM (see also Chapter 3 and [Jayant and Noll 84]), in that the number of bits used to encode each frequency component can be adaptable. Allocating different numbers of bits to different frequency components allows us to control the level of quantization noise in each component to ensure that we have the highest coding accuracy in the frequency components that most need it. In this sense, the frequency-domain signal representation provides an ideal framework for exploiting irrelevancies in the signal. This issue is intimately related to the main topic of Chapters 6 and 7, where we discuss how studies of human hearing allow us to determine which frequency components can accept significant quantization noise without producing audible artifacts.

The basic technique of time to frequency mapping is to pass the signal through a bank of filters that parse the signal into K different bands of frequencies. The signal from each frequency band is then quantized with a limited number of bits, putting most of the quantization noise in frequency bands where it is least audible. The quantized signal is then sent to a decoder where the coded signal in each band is dequantized and the bands are combined to restore the full frequency content of the signal. In most cases, an additional filter bank is needed in the decoder to make sure that each band's signal is limited to its appropriate band before we add the bands back together to create the decoded audio signal. An immediate issue with such an approach is the fact that, by splitting the signal into K parallel bands, we have multiplied our data rate by a factor of K. To avoid raising the data rate when passing the signal through the encoder filter bank, we throw away all but one out of every K samples or in other words we "down sample" by a factor of K. In *Figure 1*, a general overview of the time to frequency mapping process is shown. Remarkably, we shall see that we can cleverly design filter banks such that the original signal is fully recoverable from the down-sampled data.

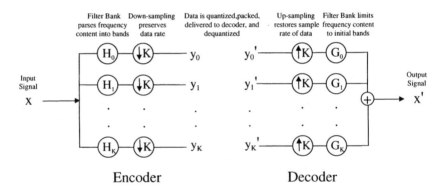

Figure 1. Overview of the time to frequency mapping process

In this chapter we discuss the constraints on the design of filter banks for parsing signals into their frequency domain content and meet some of the more commonly used filter banks in audio coding. We first introduce the discrete time generalization of the Fourier transform, the Z transform. The Z transform is a basic technique used in filter design for sampled data and is the easiest way to derive the basic filter bank coding techniques. We then introduce two-channel perfect reconstruction filter banks to get a better understanding of how filter design constraints allow us to recover the original signal from down-sampled frequency bands. We discuss how to create filter banks that generalize the two-channel frequency parsing to higher numbers of bands (e.g., 32 bands). We then present in detail a particular filter bank, the "pseudo quadrature mirror filter" PQMF, that has had a major impact in audio coding. As an example of its applications, the 32-band PQMF used in the MPEG audio coders is described in detail.

2. THE Z TRANSFORM

In Chapter 3 we saw that band-limited signals can be recovered from sampled data provided that the sample rate $F_s \equiv 1/T_s$ is higher than twice the highest frequency component in the signal. We saw in our derivation of the Sampling Theorem (you can derive this by replacing $2*F_{max}$ with F_s) that we can represent the frequency content of such a signal in the frequency range from $-F_s/2$ to $F_s/2$ using the following Fourier series:

$$X(f) = \frac{1}{F_s} \sum_{n=-\infty}^{\infty} x(nT_s) e^{-j2\pi \frac{nf}{F_s}}$$

The time domain content of such a band-limited signal can then be recovered using the inverse Fourier Transform to find that:

$$x(t) = \int_{-F_s/2}^{F_s/2} X(f) e^{j2\pi ft} df$$

However, we need to be careful about making sure that the signal is band-limited to within the frequency range from $-F_s/2$ to $F_s/2$ since sampling will "alias" any spectral components outside this frequency range into this frequency range. Moreover, since it is based on the Fourier series, any attempt to use the above formula for X(f) outside of this frequency range will find the frequency content in the range periodically continued throughout all possible frequencies. We can use this pairing of sampled time-domain data and frequency domain content in its own right to define a frequency representation of any sampled time-domain data as long as we recognize the limitations of this pairing in describing the true frequency content of a signal. Such a pairing is called the "discrete-time Fourier Transform".

The Z transform is a generalization of the discrete-time Fourier Transform. Define the mapping of the frequency f onto the complex number

$$z(f) = e^{j2\pi f/Fs}$$

Notice that z(f) is a complex number with values on the unit circle and periodic in frequency with period F_s. In terms of z, we can write the forward transform of the discrete time Fourier Transform as:

$$X(f) = \frac{1}{F_s} \sum_{n=-\infty}^{\infty} x(nT_s) \, z(f)^{-n}$$

The Z transform generalizes this forward transform to arbitrary complex values of frequency, i.e. values of z off the unit circle.

Given a data series x[n] and a complex number z, the Z transform of x[n] is defined as:

$$X(z) = \sum_{n=-\infty}^{\infty} x[n] \, z^{-n}$$

Notice that we can immediately associate a Z transform of a sampled data series with the discrete-time Fourier Transform of that data using

$x[n] = x(nT_s)$, $z = z(f) = e^{j2\pi f/Fs}$, and defining $X(f) = X(z(f))/F_s$

Like the discrete-time Fourier Transform, the Z transform has an inverse transform requiring an integration in the complex plane (see for example [Rabiner and Gold 75]), however, we will not need to use the inverse Z transform in this book. In this book, we mostly use the Z transform as a convenient way to derive analytical expressions for Fourier Transforms of sampled data.

2.1   Important Properties

Three extremely important properties of the Z Transform are its linearity, the convolution theorem, and the delay theorem. Linearity of the Z transform says that, given two data series $x_1[n]$ and $x_2[n]$ then the Z transform of any linear combination of the series:

$y[n] = A \, x_1[n] + B \, x_2[n]$

is just the linear combination of the Z transforms:

$Y(z) = A \, X_1(z) + B \, X_2(z)$

This is derived readily from the definition of the Z transform:

$$Y(z) = \sum_{n=-\infty}^{\infty} y[n] \, z^{-n} = \sum_{n=-\infty}^{\infty} (Ax_1[n] + Bx_2[n]) \, z^{-n}$$

$$= A \sum_{n=-\infty}^{\infty} x_1[n] \, z^{-n} + B \sum_{n=-\infty}^{\infty} x_2[n] \, z^{-n}$$

$$= AX_1(z) + BX_2(z)$$

The convolution theorem states that the Z transform of the convolution of two data series is equal to the product of the Z transforms. In other words, if we define the convolution of two data series $x_1[n]$ and $x_2[n]$ as:

$$y[n] = x_1[n] \circ x_2[n] \equiv \sum_{m=-\infty}^{\infty} x_1[n-m]x_2[m]$$

then we have that

$$Y(z) = X_1(z) \, X_2(z)$$

This can be shown as follows:

$$Y(z) = \sum_{n=-\infty}^{\infty} y[n]z^{-n} = \sum_{n=-\infty}^{\infty} \sum_{m=-\infty}^{\infty} x_1[n-m]x_2[m]z^{-n}$$

$$= \sum_{n=-\infty}^{\infty} \sum_{m=-\infty}^{\infty} x_1[n-m]z^{-(n-m)}x_2[m]z^{-m} = \sum_{p=-\infty}^{\infty} \sum_{m=-\infty}^{\infty} x_1[p]z^{-p}x_2[m]z^{-m}$$

$$= \left(\sum_{p=-\infty}^{\infty} x_1[p]z^{-p} \right)\left(\sum_{m=-\infty}^{\infty} x_2[m]z^{-m} \right) = X_1(z)X_2(z)$$

Since passing a signal through a linear, time-invariant filter is equivalent to convolving the signal with the filter's impulse response function, the convolution theorem tells us that the Z transform of a filtered signal is just the product of the original signal's Z transform and the Z transform of the filter's impulse response function.

The delay theorem states that the Z transform of a signal delayed D time samples is equal to z^D times the signal's Z transform. This can be readily seen from the Z transform of a data series $y[n] = x[n-D]$:

$$Y(z) = \sum_{n=-\infty}^{\infty} y[n] \, z^{-n} = \sum_{n=-\infty}^{\infty} x[n-D] \, z^{-n} = \sum_{m=-\infty}^{\infty} x[m] \, z^{-m+D} = X(z)z^D$$

These three properties together tell us how to calculate the Z transform of a signal passed through a chain of filters: for all filters in series multiply the Z transforms, whenever parallel paths are summed add the Z transforms, and whenever delay lines are used multiply by z^D.

2.2   Down-Sampling

Before proceeding to the design of perfect reconstruction filter banks, we need to establish two more properties of the Z transform: the effects of down-sampling and up-sampling on a signal's Z transform. We need these properties because, as shown in *Figure 1*, we typically down-sample the data coming out of the filter bank to keep the data rate constant and then we typically up-sample the data (i.e. intersperse zeroes between data points) prior to recombining the sub-band signals to space it out back to the original data rate. We first derive the effect of down-sampling and then derive the effect of up-sampling.

Prior to deriving the effects of down-sampling, we take a brief digression to discuss a useful result that makes the derivation easier. This result concerns the properties of the Kth roots of 1.

The K different K^{th} roots of 1 are symmetrically located on the unit circle and can be enumerated as

$$\{x_r = e^{j2\pi r/K} \text{ for } r = 0,..., K-1\}$$

Due to their symmetric location on the unit circle, the sum of the roots is equal to zero. An interesting property of the roots of 1 is that the set of each of the roots raised to the mth power:

$$\{x_r^m = e^{j2\pi rm/K} \text{ for } r=0,...,K-1\}$$

is just another enumeration of the full set of roots provided that m is not a multiple of K. For example, consider the cube roots of 1 (see also *Figure 2*):

$$\{1, e^{j2\pi/3}, e^{j4\pi/3}\}$$

The set of cube roots to the first power is trivially the set itself. The set of cube roots to the second power is again the set itself but in a different order:

$$\{1, e^{j4\pi/3}, e^{j8\pi/3} = e^{j2\pi/3}\}$$

However, the set of cube roots to the 3rd power is just 1 repeated:

$\{1, e^{j6\pi/3} = 1, e^{j12\pi/3} = 1\}$

And so on…

This property of the Kth roots of 1 allows us to establish the following sum rule for powers of the roots:

$$\frac{1}{K}\sum_{r=0}^{K-1} e^{j2\pi r m/K} = \begin{cases} 1 \text{ if } m = nK \\ 0 \text{ otherwise} \end{cases}$$

We make use of this rule in our derivation of the Z transform for down-sampled data.

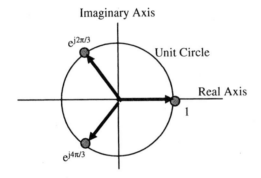

Figure 2. The cube roots of 1

We now derive the Z transform of down-sampled data. Consider the data series y[n] = x[nK] which represents data down-sampled by a factor of K. The Z transform of y[n] is equal to:

$$Y(z) = \sum_{n=-\infty}^{+\infty} y[n]z^{-n} = \sum_{n=-\infty}^{+\infty} x[nK]z^{-n} = \sum_{m=-\infty}^{+\infty} x[m]z^{-m/K}\delta_{m,nK \text{ for some } n} =$$

$$= \sum_{m=-\infty}^{+\infty} x[m](z^{1/K})^{-m}\left(\frac{1}{K}\sum_{r=0}^{K-1} e^{j2\pi r m/K}\right) = \frac{1}{K}\sum_{r=0}^{K-1}\sum_{m=-\infty}^{+\infty} x[m](z^{1/K}e^{-j2\pi r/K})^{-m} =$$

$$= \frac{1}{K}\sum_{r=0}^{K-1} X(z^{1/K}e^{-j2\pi r/K})$$

Notice that the Z transform of down-sampled data is the sum of K terms. What does this mean?

Let's think about it in the frequency domain. Suppose we have data sampled with a sample rate F_s. If we down-sample this data by K, then each data point is spaced farther out in time and the new sample rate is equivalent to F_s/K. This means that any spectral content of the original signal outside of the frequency range of $-F_s/2K$ to $F_s/2K$ will be aliased. The K-1 extra terms in the down-sample Z transform are the aliasing terms of this spectral content. How do we see this in the frequency domain? First of all, we must realize that we consider all data series to be sampled with the same underlying sample rate. This means that, although the down-sampled data is really sampled with an effective sample rate of F_s/K, the data is viewed in the time domain as a new data series sampled at the usual sample rate F_s. In other words, we can directly use the above associations between the Z transform and the discrete time Fourier transform, using F_s for the sample rate, to relate the discrete time Fourier Transform of y[n] to that of x[n].

We find that:

$$Y(f) = \frac{1}{K} \sum_{r=0}^{K-1} X(f/K - r\, Fs/K)$$

We can see from this that the effect of down-sampling, other than signal power reduction due to the factor 1/K, is to:

1) spread out the spectral bandwidth by a factor of K (r = 0 term) and
2) to alias any part of the spectrum pushed outside of the range $-F_s/2K$ to $F_s/2K$ (r ≠ 0 terms).

For example, the contribution of Y(f) at f = 0 comes not only from X(f) at f = 0 but also at values of X(f) at f = r F_s/K for r = 1,..., K-1. When we try to develop perfect reconstruction filter banks, considerable effort will go into making sure that we can undo the aliasing caused by down-sampling. In *Figure 3*, the effects of aliasing caused by the down-sampling process are shown.

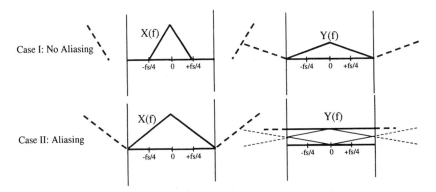

Figure 3. Aliasing effects in the down-sampling process

2.3   Up-Sampling

The effect of up-sampling data by a factor of K (i.e. intersperse K-1 zeroes between each data point) is just to replace z in the Z transform by z^K. This result is quickly derived by considering the Z transform of the data series

$$y[n] = \begin{cases} x[m] \text{ if } n = mK \\ 0 \text{ otherwise} \end{cases}$$

which is equal to

$$Y(z) = \sum_{n=-\infty}^{\infty} y[n]\, z^{-n} = \sum_{m=-\infty}^{\infty} x[m]\, z^{-mK} = \sum_{m=-\infty}^{\infty} x[m]\, (z^K)^{-m} = X(z^K)$$

In the frequency domain this becomes $Y(f) = X(Kf)$ so we can see that up-sampling shrinks the spectrum bandwidth by a factor of K. Note, however, that $X(f)$ is periodic with period F_s and so up-sampling will bring images of these copies of $X(f)$ into the $Y(f)$ spectrum. When we develop perfect reconstruction filter banks, we filter out these extra images with a band-pass filter corresponding to the frequency band of that signal component. In *Figure 4*, the effects of imaging caused by the up-sampling process are shown. For an in-depth description of multirate systems and filter banks the reader can consult [Vaidyanathan 93].

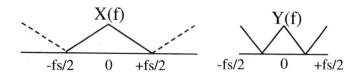

Figure 4. Imaging effects in the up-sampling process

3. TWO-CHANNEL PERFECT RECONSTRUCTION FILTER BANKS

Having developed all of the pieces we need, let's move ahead and describe how to design a two-channel perfect reconstruction filter bank. We then discuss how the two-channel filter bank can be extended to create multi-channel perfect reconstruction filter banks.

In a two-channel perfect reconstruction filter bank (see *Figure 5*), we pass the signal x[n] through two parallel filters with responses $h_0[n]$ and $h_1[n]$. Ideally, these two filters split the frequency spectrum between them so that different signal components will be isolated in each data stream. This will give us twice the data rate so we need to down-sample the data by a factor of two. In a real coder we would then quantize the two data streams, pack them together and send the packed data stream to the decoder, and then unpack and dequantize the two data streams before the next steps in the chain below, introducing quantization noise on the way. In this chapter we ignore the effect of quantization noise and try to characterize the steps by which the two uncorrupted intermediate data streams $y_0[n]$ and $y_1[n]$ can be combined into a new signal x'[n] which is exactly equal to the original signal x[n] other than possibly being delayed. What we will do to the intermediate data streams is to up-sample them (so they really reflect the original sample rate again), pass them through filters with responses $g_0[n]$ and $g_1[n]$, and then add them together. The challenge to us is to define filters $h_0[n]$, $h_1[n]$, $g_0[n]$, and $g_1[n]$ that allow perfect reconstruction of the input data x[n].

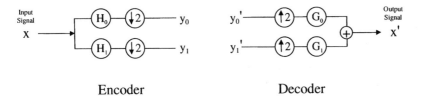

Encoder   Decoder

Figure 5. Two-channel perfect reconstruction filter bank

To begin our analysis we calculate the Z transform of signal x'[n] in terms of the input signal and the four filter responses. At each step, we use the properties described in the previous section for applying filters, adding signals, down-sampling, and up-sampling. First we write down the Z transform of x'[n] in terms of the Z transforms of $y_0[n]$, $y_1[n]$, $g_0[n]$, $g_1[n]$:

$$X'(z) = Y_0(z^2)G_0(z) + Y_1(z^2)G_1(z)$$

Then we write down the Z transforms of $y_0[n]$, $y_1[n]$ in terms of x[n] and the filters $h_0[n]$ and $h_1[n]$:

$$Y_i(z) = \tfrac{1}{2}\left(H_i(z^{1/2})X(z^{1/2}) + H_i(-z^{1/2})X(-z^{1/2})\right) \quad i = 0,1$$

Finally, we put it all together to get the final result:

$$X'(z) = \tfrac{1}{2}\left(H_0(z)G_0(z) + H_1(z)G_1(z)\right)X(z) +$$
$$\tfrac{1}{2}\left(H_0(-z)G_0(z) + H_1(-z)G_1(z)\right)X(-z)$$

Notice that the equation described above has one term proportional to X(z) and another proportional to X(-z) – in the frequency domain the X(-z) term represents the aliasing of spectral components at frequency f - F_s/2 onto frequency components at frequency f. The first thing we need to do in our filter design is to make sure that the coefficient of the aliasing X(-z) term is zero.

3.1   Aliasing Cancellation

One way to set up the filter banks so that there is no aliasing of the signal is to define the synthesis filters $g_0[n]$, $g_1[n]$ in terms of the analysis filters $h_0[n]$, $h_1[n]$ such that their Z transforms satisfy:

$$G_0(z) = -H_1(-z)$$
$$G_1(z) = H_0(-z)$$

Notice how this choice of filters eliminates the $X(-z)$ term and we are only left with the relationship that

$$X'(z) = \tfrac{1}{2}\left(-H_0(z)H_1(-z) + H_1(z)H_0(-z)\right)X(z)$$

The Z transform relationships between the analysis and synthesis filters can be written in the time domain as:

$$g_0[n] = -(-1)^n h_1[n]$$
$$g_1[n] = (-1)^n h_0[n]$$

These relationships can be quickly derived by comparing the Z transform relationship term by term for different powers of z.

3.2   Perfect Reconstruction: the QMF Solution

Having defined synthesis filters to eliminate aliasing, we still need to pick analysis filters that lead to perfect reconstruction. There are multiple ways known to do this but we focus here on one of the most common types: the Quadrature Mirror Filters, QMF [Croisier, Esteban and Galand 76]. The QMF solution defines the filter $h_1[n]$ in terms of the filter $h_0[n]$ as:

$$H_1(z) = -H_0(-z) \qquad h_1[n] = -(-1)^n h_0[n]$$

Notice that, if $h_0[n]$ is a low-pass filter then $h_1[n]$ will be high-pass (see *Figure 6*) . We can see this by going to the frequency domain where we find that

$$H_1(f) = -H_0(-F_s/2 + f)$$

If the filter $h_0[n]$ is low-pass filter then it has frequency components near $f = 0$ but not near $f = F_s/2$. The frequency domain relationship then shows us that the filter $h_1[n]$ has response near $f = 0$ like that of $h_0[n]$ near $F_s/2$, not much pass-through, while it has response near $F_s/2$ like that of $h_0[n]$ near zero, lots of pass-through. In other words, $h_1[n]$ will be high-pass and have its highest frequency response magnitude near $F_s/2$ and correspondingly near

–Fs/2. We can rewrite the definition of all 3 of the other filters in terms of $h_0[n]$:

$$H_1(z) = -H_0(-z) \qquad h_1[n] = -(-1)^n h_0[n]$$
$$G_0(z) = H_0(z) \qquad g_0[n] = h_0[n]$$
$$G_1(z) = H_0(-z) \qquad g_1[n] = (-1)^n h_0[n]$$

Figure 6. Qualitative relationship of frequency content in the two-channel QMF analysis filters h_0 and h_1

Having defined all of the other filters now in terms of $h_0[n]$, we can rewrite the Z transform of the output signal as:

$$X'(z) = \tfrac{1}{2}\left(H_0(z)^2 - H_0(-z)^2\right)X(z)$$

The output signal is just a delayed but perfectly reconstructed copy of the input signal if we can construct a filter $h_0[n]$ which satisfies the following:

$$H_0(z)^2 - H_0(-z)^2 = 2z^{-D}$$

3.2.1   An Example: the Haar Filter

To get a feel for the QMF solution, let's examine the exact 2-tap solution: the Haar filter. The Haar filter $h_0[n]$ has impulse response

$$h_0[n] = \{ \tfrac{1}{\sqrt{2}}, \tfrac{1}{\sqrt{2}}, 0, 0, \ldots \}$$

The Z transform of this filter $H_0(z)$ is equal to

$$H_0(z) = \tfrac{1}{\sqrt{2}}(1 + z^{-1})$$

It satisfies the perfect reconstruction condition since

$$H_0(z)^2 - H_0(-z)^2 = 2z^{-1}$$

which shows us that if we build a QMF filter bank using it, the output should equal the input signal delayed by 1 sample.

Let's check. The impulse responses of the other filters $h_1[n]$, $g_0[n]$, and $g_1[n]$ are defined in terms of the filter $h_0[n]$, and in the Haar case are:

$$h_1[n]=\{ -\frac{1}{\sqrt{2}}, \frac{1}{\sqrt{2}}, 0, 0, \ldots \}$$

$$g_0[n]=\{ \frac{1}{\sqrt{2}}, \frac{1}{\sqrt{2}}, 0, 0, \ldots \}$$

$$g_1[n]=\{ \frac{1}{\sqrt{2}}, -\frac{1}{\sqrt{2}}, 0, 0, \ldots \}$$

If we start with an input signal $x[n] = \{\ldots, 0, 0, x[0], x[1], x[2], \ldots\}$ then the signals $y_0[n]$, $y_1[n]$ will be equal to

$$y_0[n]=\{\ldots, 0, \frac{1}{\sqrt{2}}(x[0]+x[1]), \frac{1}{\sqrt{2}}(x[2]+x[3]), \ldots\}$$

$$y_1[n]=\{\ldots, 0, \frac{1}{\sqrt{2}}(x[0]-x[1]), \frac{1}{\sqrt{2}}(x[2]-x[3]), \ldots\}$$

After up-sampling and filtering with the synthesis filters these 2 series become

$$\{\ldots, 0, 0, \tfrac{1}{2}(x[0]+x[1]), \tfrac{1}{2}(x[0]+x[1]), \tfrac{1}{2}(x[2]+x[3]), \tfrac{1}{2}(x[2]+x[3]), \ldots\}$$

$$\{\ldots, 0, 0, \tfrac{1}{2}(x[0]-x[1]), -\tfrac{1}{2}(x[0]-x[1]), \tfrac{1}{2}(x[2]-x[3]), -\tfrac{1}{2}(x[2]-x[3]), \ldots\}$$

respectively. Finally, when added together to get the output signal, we find that $x[n] = \{\ldots, 0, 0, x[0], x[1], x[2], \ldots\}$ as expected.

Although Haar filters can be used to create a two-channel perfect reconstruction filter bank, the shortness of the filter impulse response makes the frequency localization quite poor for the two channels. The reason for using longer filters is to get a much shorter transition region between the pass-bands of the two analysis filters (see *Figure 7*). Unfortunately, no finite order FIR filter with more than 2 taps has been found to solve the QMF perfect reconstruction condition. Although no exact solution has been found, filter design techniques have been developed to find longer FIR filters that approximate the QMF perfect reconstruction conditions extremely well.

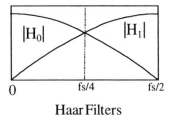
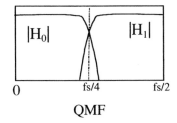

Haar Filters   QMF

Figure 7. Qualitative comparison of frequency response of the Haar filter QMF solution with that of longer approximate solutions to the QMF perfect reconstruction condition

3.3   Perfect Reconstruction: the CQF Solution

The QMF solution is not the only possible solution to the 2 channel perfect reconstruction equations. Another solution has been found that is better adapted to implementation using FIR filters: the Conjugate Quadrature Filter (CQF) solution [Smith and Barnwell 86]. For this solution the synthesis filters $g_0[n]$, $g_1[n]$ are just the time reverses of the analysis filters $h_0[n]$, $h_1[n]$. For FIR filters of length N this implies that:

$$g_0[n] = h_0[N-1-n]$$
$$g_1[n] = h_1[N-1-n]$$

Like in the QMF solution, the analysis filter $h_1[n]$ is a version of the low-pass filter $h_0[n]$ modulated by $(-1)^n$ to make it high-pass, but in this solution it is the time reverse of $h_0[n]$ that is modulated rather than $h_0[n]$ itself. In specific, the relationship between the two analysis filters for even length N is:

$$h_1[n] = -(-1)^n h_0[N-1-n]$$

If we rewrite these CQF relationships in the z domain we find that:

$$G_0(z) = z^{-(N-1)} H_0(z^{-1})$$
$$G_1(z) = z^{-(N-1)} H_1(z^{-1})$$
$$H_1(z) = z^{-(N-1)} H_0(-z^{-1})$$

A quick substitution into the 2-channel alias cancellation condition shows that these choices ensure alias cancellation for even length N. Substitution into the 2-channel perfect reconstruction condition shows that the base filter $h_0[n]$ must be designed to satisfy

$$H_0(z)H_0(z^{-1}) + H_0(-z)H_0(-z^{-1}) = 2$$

where the delay D = N-1. We can look at this condition in the frequency domain and we see that our two analysis filters must satisfy the "power complementarity" condition that:

$$|H_0(f)|^2 + |H_0(-F_s/2+f)|^2 = |H_0(f)|^2 + |H_1(f)|^2 = 2/F_s^2$$

Several standard methods exist for developing FIR filters that exactly satisfy this condition, see for example [Vetterli and Kovačević 95] for a summary of design methods. The value of this CQF solution is that it is the basis for expanding the two channel results to multiple channel filter banks.

4.   THE PSEUDO-QMF FILTER BANK, PQMF

Having shown with the two channel case that careful design of the analysis and synthesis filters can lead to perfect reconstruction filter banks, for practical applications we need many more channels than two. For example, in Chapter 6 we will see that the human ear's frequency response naturally divides into 20-30 "critical bands". How can we create perfect reconstruction filter banks with closer to that number of channels?

Early work on multi-channel filter banks tried to cascade QMF filter pairs to subdivide the spectrum into multiple channels. This tree structure approach has the disadvantage of long impulse responses and high computational complexity. A more efficient, parallel multi-band approach that represents an approximate generalization of the two channel CQF solution was developed variously called the "pseudo-QMF" [Nussbaumer 81], PQMF, and "polyphase quadrature" [Rothweiler 83] filter bank. We refer to it as the PQMF filter bank. The basic idea is to take a narrow low-pass filter and modulate copies of it to span the frequency domain. The filter is then defined so that it decays fast enough so that there is negligible overlap between next-near-neighbor filters and so that near-neighbor filters cancel aliasing and satisfy a CQF-type perfect reconstruction equation.

The PQMF solution to developing near-perfect reconstruction filter banks was extremely important historically. The Layer I and Layer II coders in MPEG-1 and MPEG-2 use this approach to time to frequency mapping. The ability of the PQMF filter banks to reconstruct input signals with extremely high accuracy and their efficient implementation allowed the development of early perceptual audio coders.

4.1   Basic Structure

The PQMF filter bank consists of K channels, each of which is a low-pass filter h[n] modulated by a cosine. The exact form of the analysis and synthesis filters is:

$$h_k[n] = h[n]\cos\left(\pi\left(\frac{k + \frac{1}{2}}{K}\right)\left(n - \frac{(N-1)}{2}\right) + \phi_k\right) \quad \text{for k=0,...,K-1}$$

$$g_k[n] = h_k[N - 1 - n]$$

where N is the length of h[n]. The phase ϕ_k is determined by an anti-aliasing conditions between adjacent bands and satisfies the relationship:

$$\phi_k - \phi_{k-1} = \tfrac{\pi}{2}(2r + 1)$$

where r is an integer. Note that the synthesis filters are just the time reverses of the analysis filters as in the CQF solution.

Recall that cosine has a delta-function frequency response at positive and negative frequencies. The convolution theorem, i.e. products in the time domain lead to convolutions in the frequency domain, tells us that $h_k[n]$ has a frequency response equal to that of H(f) shifted to both frequencies

$$f_k = \pm\frac{(k + \frac{1}{2})}{K}F_s / 2$$

The K channels therefore lay down 2K copies of H(f) to divide up the frequency spectrum between $-F_s/2$ and $F_s/2$. This means that the full-width of the low-pass filter H(f) should be equal to $F_s/2K$, i.e. the pass-band should be for frequencies up to $|f| \sim F_s/4K$.

The perfect reconstruction requirements are that we design h[n] so that its frequency components beyond $|f| = F_s/2K$ are negligible and that for lower frequencies we satisfy the PQMF power complementarity equation

$$|H(f)|^2 + |H(-F_s/2K+f)|^2 = 2/F_s^2 \quad \text{for } 0 \le |f| \le F_s/4K$$

Notice the similarity of this requirement to the power complementarity condition of the CQF solution.

4.2   The MPEG PQMF

MPEG-1 and 2 Layers I, II, (and also the hybrid filter of Layer III, see Chapter 11) use 32 channel PQMF filter banks where the base filter h[n] has 511 taps [ISO/IEC 11172-3 and ISO/IEC 13818-3]. The PQMF filter bank used in these coders employs the following analysis and synthesis filters, $h_k[n]$ and $g_k[n]$ respectively:

$$h_k[n] = h[n] \cos\left[\left(k + \frac{1}{2}\right)(n - 16)\frac{\pi}{32}\right]$$

$$g_k[n] = 32\, h[n] \cos\left[\left(k + \frac{1}{2}\right)(n + 16)\frac{\pi}{32}\right]$$

$$k = 0,1,...,31$$

$$n = 0,1,...,511$$

where k is the frequency index and n is the time index. The filter coefficients describing the prototype filter h[n] are shown in *Table 1*. (The filter coefficients for n > 256 can be found from the symmetry relation h[256+n] = h[256-n]). A good closed form approximation of the standard coefficients can be found in [Searing 91]. The filter length N is equal to 513, of which the first and last coefficients are zero.

Table 1. MPEG-1 Audio PQMF prototype filter coefficients h[n] [ISO/IEC 11172-3]

h[0] = 0.000000000	h[32] = -0.000013828	h[64] = -0.000101566	h[96] = 0.000218868	h[128] = 0.000971317	h[160] = -0.002457142	h[192] = -0.003134727	h[224] = 0.017876148
h[1] = -0.000000477	h[33] = -0.000014782	h[65] = -0.000103951	h[97] = 0.000247478	h[129] = 0.000953674	h[161] = -0.002630711	h[193] = -0.002841473	h[225] = 0.018756866
h[2] = -0.000000477	h[34] = -0.000016689	h[66] = -0.000105858	h[98] = 0.000277042	h[130] = 0.000930786	h[162] = -0.002803326	h[194] = -0.002521515	h[226] = 0.019634247
h[3] = -0.000000477	h[35] = -0.000018120	h[67] = -0.000107288	h[99] = 0.000307560	h[131] = 0.000902653	h[163] = -0.002974033	h[195] = -0.002174854	h[227] = 0.020506859
h[4] = -0.000000477	h[36] = -0.000019550	h[68] = -0.000108242	h[100] = 0.000339031	h[132] = 0.000868797	h[164] = -0.003141880	h[196] = -0.001800537	h[228] = 0.021372318
h[5] = -0.000000477	h[37] = -0.000021458	h[69] = -0.000108719	h[101] = 0.000371456	h[133] = 0.000829220	h[165] = -0.003306866	h[197] = -0.001399517	h[229] = 0.022228718
h[6] = -0.000000477	h[38] = -0.000023365	h[70] = -0.000108719	h[102] = 0.000404358	h[134] = 0.000783920	h[166] = -0.003467083	h[198] = -0.000971317	h[230] = 0.023074150
h[7] = -0.000000954	h[39] = -0.000025272	h[71] = -0.000108242	h[103] = 0.000438213	h[135] = 0.000731945	h[167] = -0.003622532	h[199] = -0.000519938	h[231] = 0.023907185
h[8] = -0.000000954	h[40] = -0.000027657	h[72] = -0.000106812	h[104] = 0.000472546	h[136] = 0.000674248	h[168] = -0.003771782	h[200] = -0.000033379	h[232] = 0.024725437
h[9] = -0.000000954	h[41] = -0.000030041	h[73] = -0.000105381	h[105] = 0.000507355	h[137] = 0.000610352	h[169] = -0.003914356	h[201] = 0.000475883	h[233] = 0.025527000
h[10] = -0.000000954	h[42] = -0.000032425	h[74] = -0.000102520	h[106] = 0.000542614	h[138] = 0.000539303	h[170] = -0.004048824	h[202] = 0.001011848	h[234] = 0.026310921
h[11] = -0.000001431	h[43] = -0.000034809	h[75] = -0.000099182	h[107] = 0.000576973	h[139] = 0.000462532	h[171] = -0.004174709	h[203] = 0.001573563	h[235] = 0.027073860
h[12] = -0.000001431	h[44] = -0.000037670	h[76] = -0.000095367	h[108] = 0.000611782	h[140] = 0.000378609	h[172] = -0.004290581	h[204] = 0.002161503	h[236] = 0.027815342
h[13] = -0.000001907	h[45] = -0.000040531	h[77] = -0.000090122	h[109] = 0.000646591	h[141] = 0.000288486	h[173] = -0.004395962	h[205] = 0.002774239	h[237] = 0.028532982
h[14] = -0.000001907	h[46] = -0.000043392	h[78] = -0.000084400	h[110] = 0.000680933	h[142] = 0.000191689	h[174] = -0.004489899	h[206] = 0.003411293	h[238] = 0.029224873
h[15] = -0.000002384	h[47] = -0.000046253	h[79] = -0.000077724	h[111] = 0.000714302	h[143] = 0.000088215	h[175] = -0.004570484	h[207] = 0.004072189	h[239] = 0.029890060
h[16] = -0.000002384	h[48] = -0.000049591	h[80] = -0.000069618	h[112] = 0.000747204	h[144] = -0.000021458	h[176] = -0.004638195	h[208] = 0.004756451	h[240] = 0.030526638
h[17] = -0.000002861	h[49] = -0.000052929	h[81] = -0.000060558	h[113] = 0.000779152	h[145] = -0.000137329	h[177] = -0.004691124	h[209] = 0.005462170	h[241] = 0.031132698
h[18] = -0.000003338	h[50] = -0.000055790	h[82] = -0.000050545	h[114] = 0.000809669	h[146] = -0.000259876	h[178] = -0.004728317	h[210] = 0.006189346	h[242] = 0.031706810
h[19] = -0.000003338	h[51] = -0.000059605	h[83] = -0.000039577	h[115] = 0.000838757	h[147] = -0.000388145	h[179] = -0.004748821	h[211] = 0.006937027	h[243] = 0.032248020
h[20] = -0.000003815	h[52] = -0.000062943	h[84] = -0.000027180	h[116] = 0.000866413	h[148] = -0.000522137	h[180] = -0.004752159	h[212] = 0.007703304	h[244] = 0.032754898
h[21] = -0.000004292	h[53] = -0.000066280	h[85] = -0.000013828	h[117] = 0.000891685	h[149] = -0.000661850	h[181] = -0.004737377	h[213] = 0.008487225	h[245] = 0.033225536
h[22] = -0.000004768	h[54] = -0.000070095	h[86] = 0.000000954	h[118] = 0.000915051	h[150] = -0.000806808	h[182] = -0.004703045	h[214] = 0.009287834	h[246] = 0.033659935
h[23] = -0.000005245	h[55] = -0.000073433	h[87] = 0.000017166	h[119] = 0.000935555	h[151] = -0.000956535	h[183] = -0.004649162	h[215] = 0.010103703	h[247] = 0.034055710
h[24] = -0.000006199	h[56] = -0.000076771	h[88] = 0.000034332	h[120] = 0.000954151	h[152] = -0.001111031	h[184] = -0.004573822	h[216] = 0.010933399	h[248] = 0.034412861
h[25] = -0.000006676	h[57] = -0.000080585	h[89] = 0.000052929	h[121] = 0.000968933	h[153] = -0.001269817	h[185] = -0.004477024	h[217] = 0.011775017	h[249] = 0.034730434
h[26] = -0.000007629	h[58] = -0.000083923	h[90] = 0.000072956	h[122] = 0.000980854	h[154] = -0.001432419	h[186] = -0.004357815	h[218] = 0.012627602	h[250] = 0.035007000
h[27] = -0.000008106	h[59] = -0.000087261	h[91] = 0.000093937	h[123] = 0.000989437	h[155] = -0.001597881	h[187] = -0.004215240	h[219] = 0.013489246	h[251] = 0.035242081
h[28] = -0.000009060	h[60] = -0.000090599	h[92] = 0.000116348	h[124] = 0.000994205	h[156] = -0.001766682	h[188] = -0.004049301	h[220] = 0.014358521	h[252] = 0.035435200
h[29] = -0.000010014	h[61] = -0.000093460	h[93] = 0.000140190	h[125] = 0.000995159	h[157] = -0.001937389	h[189] = -0.003859043	h[221] = 0.015233517	h[253] = 0.035586357
h[30] = -0.000011444	h[62] = -0.000096321	h[94] = 0.000165462	h[126] = 0.000991821	h[158] = -0.002110004	h[190] = -0.003643036	h[222] = 0.016112804	h[254] = 0.035694122
h[31] = -0.000012398	h[63] = -0.000099182	h[95] = 0.000191212	h[127] = 0.000983715	h[159] = -0.002283096	h[191] = -0.003401756	h[223] = 0.016994476	h[255] = 0.035758972
							h[256] = 0.035780907

We can relate the above filter bank description to our general form of the PQMF by noting that the phase ϕ_k in the MPEG PQMF is equal to:

$$\phi_k = \frac{\pi}{2}\left(\frac{N-1-K}{K}\right)\left(k + \tfrac{1}{2}\right)$$

which satisfies the near-neighbor alias-cancellation requirement that $\phi_k - \phi_{k-1}$ is equal to an odd multiple of $\pi/2$ for these values of N = 513 and K = 32. In addition, the encoder gain has been set to preserve the amplitude of input sinusoids on encoding to provide a gain reference for the psychoacoustic thresholds. The impulse response of the prototype low-pass filter h[n] compared with $h_k[n]$ for k = 0, 1 is shown in *Figure 8*. The frequency response of the prototype is shown in *Figure 9*.

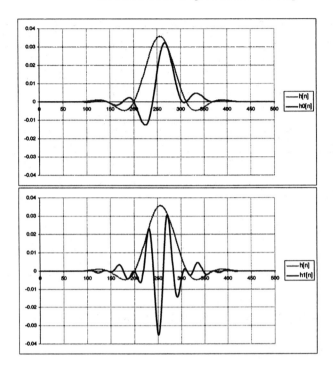

Figure 8. MPEG Audio PQMF prototype filter impulse response h[n] and h_k[n]for (a) k = 0
and (b) k = 1

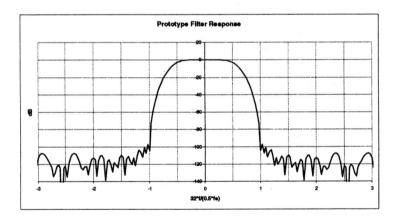

Figure 9. MPEG Audio PQMF prototype filter frequency response in units of F_s/64.

Each filter $h_k[n]$ is a modulated version of the prototype $h[n]$. The positive and negative frequency components of the modulating cosine lead to a frequency response with the frequency response of the prototype filter appearing in both positive and negative frequency locations with center frequencies f_k given by

$$f_k = \pm \frac{F_s}{2K} \left(k + \frac{1}{2} \right)$$

for $k = 0, 1, 2,..., 31$ and with each copy having a nominal bandwidth of $F_s/64$. In *Figure 10*, the frequency response of the first 4 filters of the MPEG Audio PQMF is shown. Notice how each successive set of filter pairs is shifted outward in frequency by $F_s/64$ from the prior pair.

As shown in *Figure 10*, the prototype filter does not have sharp cut-off at its nominal bandwidth of $F_s/64$. Because of this transition region, the frequency content of adjacent bands shows a certain amount of overlapping (see *Figure 10*). The phase shifts φ_k, in absence of quantization, ensure complete cancellation of the aliasing terms between neighbor bands in the synthesis stage of the decoder. Although the PQMF is not a perfect reconstruction filter bank, the MPEG prototype filter design guarantees a ripple of less than 0.07 dB for the composite frequency response of the analysis and synthesis filter banks [Noll and Pan 97].

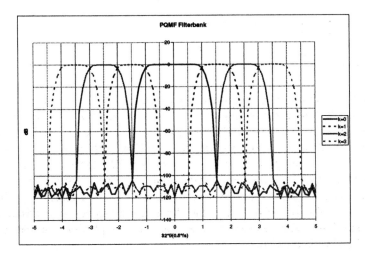

Figure 10. Frequency response of the first four bands of the MPEG Audio PQMF in units of $F_s/64$

4.3   Implementing the MPEG PQMF

In the standard implementation of the PQMF analysis stage the input buffer contains a time sequence x[n] of 512 samples that is multiplied by the filter coefficients to get the current output for each channel. Since the last filter coefficient is equal to zero, the filter bank is implemented as if the filter length is N = 512. Remembering that the PQMF is critically sampled, the filter bank can be implemented as a block transform on a 512-sample block that takes in 32 new samples in each pass.

A direct implementation of the PQMF filter bank results in 512 *32 = 16384 multiplications and 511*32 = 16352 additions for each set of 32 new samples (or about 512 multiplications and additions per sample). In the standard specifications (see *Figure 11*), a description of a medium complexity implementation, which involves about 80 multiplications and additions per sample is given as follows:

$$y_m[k] = \sum_{r=0}^{63} M[k,r] * \sum_{p=0}^{7} \left[C[r+64p] * x_m[r+64p] \right] \quad \text{for all m and k=0,...,31}$$

where

$$M[k,r] = \cos\left(\left(k + \frac{1}{2} \right) (r-16) \frac{\pi}{32} \right)$$

$$C[n] = (-1)^{int(\frac{n}{64})} h[n]$$

In these equations, k is the frequency index, $y_m[k]$ is the output of the kth analysis filter after processing the mth block of 32 new input samples and $x_m[n]$ represents a block of 512 audio input samples time-reversed with $x_m[n]$ equal to x[32*(m+1)-1-n]. A comparison of C[n] with the impulse response of the filter prototype h[n] is given in *Figure 12*. Other efficient PQMF implementations involve the utilization of fast algorithms for the computation of the discrete cosine transform [Kostantinides 94] with an additional reduction of over a factor of six in the number of multiplications and additions with respect to the medium complexity implementation described in the standard.

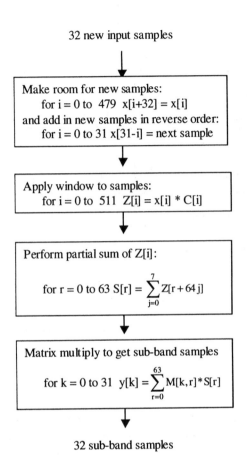

32 new input samples

Make room for new samples:
for i = 0 to 479 x[i+32] = x[i]
and add in new samples in reverse order:
for i = 0 to 31 x[31-i] = next sample

Apply window to samples:
for i = 0 to 511 Z[i] = x[i] * C[i]

Perform partial sum of Z[i]:

for r = 0 to 63 $S[r] = \sum_{j=0}^{7} Z[r+64j]$

Matrix multiply to get sub-band samples

for k = 0 to 31 $y[k] = \sum_{r=0}^{63} M[k,r]*S[r]$

32 sub-band samples

Figure 11. Flow chart of the MPEG PQMF analysis filter bank from [ISO/IEC 11172-3]

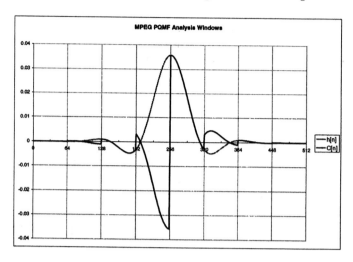

Figure 12. Comparison of C[n] with the impulse response of the filter prototype h[n]

The standard also specifies a medium complexity, efficient implementation for the synthesis filter bank. This specification is shown in flow-chart form in *Figure 13*. In this specification, the synthesis window D[i] is equal to 32*C[i] for i =0,1, … 511 and the matrix N[k,r] is given by

$$N[k,r] = \cos\left(\left(k + \frac{1}{2}\right)(r + 16)\frac{\pi}{32}\right)$$

The complexity for the synthesis filter is again greatly reduced from naïve implementation down to about 80 multiplications and additions per sample.

32 new sub-band samples y[k]

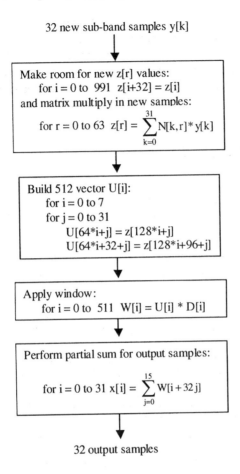

Make room for new z[r] values:
  for i = 0 to 991 z[i+32] = z[i]
and matrix multiply in new samples:

  for r = 0 to 63 $z[r] = \sum_{k=0}^{31} N[k,r] * y[k]$

Build 512 vector U[i]:
  for i = 0 to 7
  for j = 0 to 31
    U[64*i+j] = z[128*i+j]
    U[64*i+32+j] = z[128*i+96+j]

Apply window:
  for i = 0 to 511 W[i] = U[i] * D[i]

Perform partial sum for output samples:

  for i = 0 to 31 $x[i] = \sum_{j=0}^{15} W[i + 32j]$

32 output samples

Figure 13. Flow chart of the MPEG PQMF synthesis filter bank from [ISO/IEC 11172-3]

5.   SUMMARY

In this chapter we have learned that we can create filter banks that parse a signal into its high and low frequency components without any loss of information or increase in that data rate. We have then seen how this technology can be generalized into the creation of near-perfect reconstruction filter banks that can subdivide audio signals into reasonably narrow frequency bands. These PQMF methods of creating large numbers of frequency channels rely on modulating a prototype low-pass filter into the appropriate frequency band location. The PQMF filter banks developed in this manner were very important to the historical development of audio coding but become hard to work with as the number of desired frequency channels becomes very large. In the next chapter we turn towards transform coding methods as a means to create filter banks that can be used more efficiently to create filter banks with large numbers of frequency bands.

6.   REFERENCES

[Croisier, Esteban and Galand 76]: A. Croisier, D. Esteban, and C. Galand, "Perfect channel splitting by use of interpolation, decimation, and tree decomposition techniques", Proc. Intl. Conf. Inform. Sci. Syst., Patras Greece, pp. 443-446, August 1976.

[ISO/IEC 11172-3]: ISO/IEC 11172-3, Information Technology, "Coding of moving pictures and associated audio for digital storage media at up to about 1.5 Mbit/s, Part 3: Audio", 1993.

[ISO/IEC 13818-3]: ISO/IEC 13818-3, Information Technology, "Generic coding of moving pictures and associated audio, Part 3: Audio ", 1994-1997.

[Jayant and Noll 84]: N. Jayant, P. Noll, *Digital Coding of Waveforms: Principles and Applications to Speech and Video*, Prentice-Hall, Englewood Cliffs, 1984.

[Kostantinides 94]: K. Kostantinides, "Fast Sub-Band Filtering in MPEG Audio Coding ", IEEE Signal Processing Lett., pp. 26-28, 1994.

[Noll and Pan 97]: P. Noll and D. Pan, "ISO/MPEG Audio Coding" in N. Jayant (ed.), *Signal Compression- Coding of Speech, Audio, Text, Image and Video*, pp. 69-118, World Scientific 1997.

[Nussbaumer 81]: H. J. Nussbaumer, "Pseudo-QMF Filter Bank", IBM Tech. Disclosure Bull., vol. 24, pp. 3081-3087, November 1981.

[Rabiner and Gold 75]: L. R. Rabiner and B. Gold, *Theory and Applications of Digital Signal Processing*, Prentice-Hall, Englewood Cliffs, 1975.

[Rothweiler 83]: J. H. Rothweiler, "Polyphase Quadrature Filters - A new Subband Coding Technique", International Conference IEEE ASSP, Boston, pp. 1280-1283, 1983.

[Searing 91]: S. Searing, "Suggested Formulas for Audio Analysis and Synthesis Windows", ISO/IEC JTC1/SC29/WG11 MPEG 91/328, November 1991.

[Smith and Barnwell 86]: M. J. T. Smith and T. P. Barnwell III, "Exact reconstruction for tree-structured sub-band coders", IEEE Transactions on Acoustics, Speech, and Signal Processing, vol. ASSP-34, no. 3, pp. 431 – 441, June 1986.

[Vaidyanathan 93]: P. P. Vaidyanathan, *Multirate Systems and Filter Banks*, Prentice Hall, Englewood Cliffs, 1993.

[Vetterli and Kovačević 95]: M. Vetterli and J. Kovačević, *Wavelets and Subband Coding*, Prentice Hall, Englewood Cliffs, 1995.

7. EXERCISES

CQF filterbanks:

In this exercise you will design an N = 4 CQF filterbank. Recall that implementing the CQF 2-channel filterbank requires finding a low pass prototype filter $h_0[n]$ that satisfies the power complementarity condition $H_0(z)H_0(1/z) + H_0(-z)H_0(-1/z) = 2$. One way to create a CQF prototype filter $h_0[n]$ is to create a half-band filter $p[n]$ that has non-negative Fourier Transform and to factor its z Transform into $H_0(z)H_0(1/z)$. A candidate half-band filter $p[n]$ can be created by windowing the ideal half-band filter sinc[n/2] with a window centered on n=0.

1. Prove that the power complementarity condition is satisfied if $P(z) \equiv H_0(z)H_0(1/z)$ has only odd powers of z other than the constant term (z to the zero[th] power) which is equal to 1.

2. If $H_0(z) = a + b z^{-1} + c z^{-2} + d z^{-3}$ then show that P(z) satisfies the power complementarity condition if
$$a = -\sin(\theta)\cos(\phi) \quad c = \cos(\theta)\sin(\phi)$$
$$b = \sin(\theta)\sin(\phi) \quad d = \cos(\theta)\cos(\phi)$$
Write out the form of P(z) in terms of the angles θ and ϕ.

3. If $h_0[n]$ is 4 taps long then p[n] will be seven taps long and will run from n = -3 to n = 3. Create a candidate half-band filter p[n] using a 7-tap sine window $w_s[n]$: $p[n] = w_s[n+3]*sinc[n/2]$. Notice that this candidate filter satisfies the power complementarity condition. However, if the Fourier Transform of the candidate filter has negative values it will not be possible to factor it to determine $h_0[n]$. Tune the angles θ and ϕ so that P(z) best fits the z Transform of this candidate half-band filter. What are the filter coefficients $h_0[n]$ corresponding to these best fit angles θ and ϕ?

4. Use the other CQF filter conditions to determine $h_1[n]$, $g_0[n]$, and $g_1[n]$. Graph $|H_0(f)|$ and $|H_1(f)|$ in dB relative to $|H_0(0)|$ and compare with a similar graph for the Haar filter.

5. Use these CQF filters to implement a 2-channel perfect reconstruction filterbank. Test that the output exactly matches the input using the signal $x[n] = 0.3 \sin[3\pi n/4] + 0.6 \sin[\pi n/4]$. Graph the input signal against the outputs of the 2 channels. Now make another graph using the signal $x[n] = 0.3 \sin[3\pi n/4] \, w_S[n] + 0.6 \sin[\pi n/4] \, w_S[n - 50]$ where $w_S[n]$ is a sine window of length $N = 100$. How are the 2 sub-band signals related to the input signal $x[n]$?

Chapter 5

Time to Frequency Mapping Part II: The MDCT

1. INTRODUCTION

The PQMF solution to developing near-perfect reconstruction filter banks (see Chapter 4) was extremely important. Another approach to the time to frequency mapping of audio signals is historically connected to the development of transform coding. In this approach, block transform methods were used to take a block of sampled data and transform it into a different representation. For example, K data samples in the time domain could be transformed into K data samples in the frequency domain using the Discrete Fourier Transform, DFT. Moreover, exceedingly fast algorithms such as the Fast Fourier Transform, FFT, were developed for carrying out these transforms for large block sizes. Researchers discovered early on that they had to be very careful about how blocks of samples are analyzed/synthesized due to edge effects across blocks. This led to active research into what type of smooth windows and overlapping of data should be used to not distort the frequency content of the data. This line of research focused on windows, transforms, and overlap-and-add techniques of coding.

Although sub-band coding and the transform coding grew out of different areas using different building blocks, it became clear that they are just different views of the same underlying methodology. The windows used in transform coding are related to the low-pass filters that generate sub-band filters. The main differences between the techniques has to do with the number of bands that are used to parse the signal. In the current view of things, coders with a small number of frequency channels (e.g., MPEG Layers I and II [ISO/IEC 11172-3]) are still sometimes referred to as sub-band coders, and coders with a larger number of frequency channels (e.g.,

AT&T/Lucent PAC [Sinha, Johnston, Dorward and Quackenbush 98], Dolby
AC-2 and AC-3 [Fielder et al. 96], and MPEG AAC [ISO/IEC 13818-7]) are
sometimes referred to as transform coders. From the mathematical point of
view, however, there is no distinction between sub-band and transform
coding.

In this chapter we will learn about the DFT and how it can be used to
create a perfect reconstruction transform coder at the cost of enhanced data
rate. Then we will learn about the Modified Discrete Cosine Transform,
MDCT, developed by Princen and Bradley in 1986-87 and used in state-of-
the-art audio coding schemes such as MPEG AAC. The MDCT can be used
in a perfect reconstruction transform coder with a very high number of
frequency bands without requiring an increase in the coder data rate. As we
shall see, the MDCT can also be seen as a perfect reconstruction PQMF
filter bank showing the linkages between the material in this and the
previous chapter.

2.   THE DISCRETE FOURIER TRANSFORM

In Chapter 3 we discussed the Fourier Transform and its use in mapping
time signals x(t) into the frequency domain X(f). We learned that band-
limited signals, i.e. signals with frequency content only up to a finite upper
frequency F_{max}, can be fully represented with discrete time samples $x[n] \equiv
x(n*T_s)$ provided that the sampling time T_s is no longer than $1/(2*F_{max})$ or,
equivalently, the sample rate $F_s \equiv 1/T_s$ is at least as large as $2*F_{max}$.
Moreover, we learned that a time-limited signal, i.e. a signal with non-zero
values only in a finite time interval, can be fully represented with discrete
frequency samples $X[k] \equiv X(k/T)$ where T is the length of the time interval.
What we would really like to be able to do is to work with signals that are
both time and frequency limited so that we could work with finite blocks of
time-sampled data in the time domain and convert them into a finite number
of discrete samples in the frequency domain. In other words, we would like
to work with a finite extent of sampled data and be able to map it into
discrete frequencies in a finite range without any loss of information. Can
we do this? The answer turns out to be not exactly, but, with a careful
choice of windows, accurately enough. We find that we can window finite
blocks of our signal so that they remain essentially band-limited. This will
allow us to define a finite block version of the Fourier Transform, called the
Discrete Fourier Transform, that maps these blocks of time-samples into a
finite and discrete frequency-domain representation. Moreover, we shall see
that this transform can be carried out exceptionally fast for large block
lengths allowing for much greater frequency-domain resolution than is

typically available using PQMF filter banks. Let's see how this comes about.

2.1 Windowing the Signal in the Time Domain

Suppose we start with a band-limited signal, possibly band-limited from being passed through a low-pass filter, that we would like to sample at a given sample rate F_s. If the signal is band-limited so that $F_{max} \leq F_s/2$, we can work with samples x[n] and not lose any information. However, suppose we want to work with only a finite block of samples so that we can start making calculations without waiting for the signal to finish. In this case, we only consider signal values in the interval from t = 0 to t = T. One way to think of this time-limiting is that we are multiplying our original signal x(t) by a rectangular window function $w_R(t)$ equal to 1 from t = 0 to t = T and equal to zero elsewhere (see *Figure 1*). We need to ask ourselves if this time-limited signal is still band-limited enough to continue working with only the samples x[n].

2.1.1 The Rectangular Window

What happens to the frequency content of the signal after windowing? In Chapter 3 we learned about the convolution theorem, which tells us that windowing in the time domain is equivalent to a convolution in the frequency domain. We can quickly calculate the Fourier Transform of our rectangular window to find:

$$W_R(f) = \int_{-\infty}^{\infty} w_R(t) e^{-j2\pi ft} dt = \int_0^T e^{-j2\pi ft} dt = e^{-j\pi fT} \frac{\sin(\pi fT)}{\pi f}$$

Notice that this function has a main lobe centered on f = 0 whose width is proportional to 1/T and it has side lobes that drop off in amplitude like $1/|f|$ (see *Figure 2*). In general, we find that the main lobe of any window's Fourier Transform will get narrower as the window length T increases.

The Fourier Transform of our time-limited signal equals the original signal's Fourier Transform convolved with (i.e. spread out by) the function $W_R(f)$. The Fourier Transform of this window, however, drops off very slowly with frequency implying that the Fourier Transform of the time-limited signal is unlikely to remain band-limited enough to work with the sampled data x[n]. If we choose to go ahead and work with x[n] anyhow, we ·risk contaminating our analysis with aliasing. Does this mean we are out of luck? No, it just means we need to choose better windows than the rectangular window!

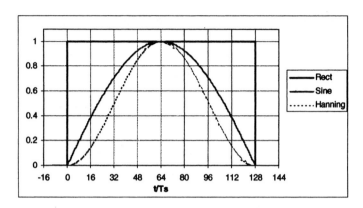

Figure 1. Time domain comparison of the rectangular, sine and Hanning windows for T=128 * T$_s$

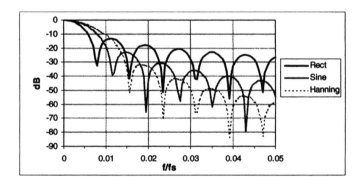

Figure 2. Frequency domain comparison of the rectangular, sine and Hanning windows for T=128*T$_s$ (Note: Windows are normalized to have integral equal to 1 prior to graphing.)

2.1.2   The Sine Window

The problem with the rectangular window is that its cut-off is very sharp at the edge of the window. Sharp changes in the value of a function lead to high frequency components in the Fourier Transform. A better selection for the window in the time domain would taper down slowly to the edges so that there is no sharp discontinuity in window value there. For example, we can consider the sine window $w_s(t)$ which is equal to

$w_s(t) = \sin(\pi t/T)$

when t is between 0 and T and is equal to zero elsewhere (see *Figure 1*). Note that when applied to discrete time signals over N samples this window is implemented as

$w_s[n] = \sin[\pi(n+\frac{1}{2})/N]$ for n=0,...,N-1

We can calculate the Fourier Transform of this window and we find that

$$W_S(f) = \int_{-\infty}^{\infty} w_S(t)e^{-j2\pi ft}dt = \int_{0}^{T} \sin(\frac{\pi t}{T})e^{-j2\pi ft}dt$$

$$= e^{-j\pi fT} \cos(\pi fT)\left(\frac{\frac{2T}{\pi}}{1-(2fT)^2} \right)$$

Although the main lobe around f = 0 is wider with respect to the rectangular window, the frequency domain amplitude of this window drops off much faster than the rectangular window (see *Figure 2*). Unlike the rectangular window, the sine window can be used to time-limit a reasonably-sampled signal without expecting it to spread out the frequency content enough to cause substantial aliasing. Notice also that the width of the main lobe is again proportional to 1/T showing that longer windows give better frequency resolution.

2.1.3   The Hanning Window

One might conclude from this discussion that it might be even better to use a window such as the Hanning window that doesn't have the sudden change in derivative at the edges that the sine window has. The Hanning window $w_H(t)$ is equal to

$w_H(t) = \frac{1}{2}(1-\cos(2\pi t/T))$

for times between zero and T and is equal to zero elsewhere (see *Figure 2*). Note that when applied to discrete time signals over N samples this window is implemented as

$w_H[n] = \frac{1}{2}(1-\cos[2\pi(n+\frac{1}{2})/N])$ for n=0,...,N-1

Again, we can compute its Fourier Transform and we find that

$$W_H(f) = \int_{-\infty}^{\infty} w_H(t) e^{-j2\pi ft} dt = \int_0^T \frac{1}{2}(1 - \cos(2\pi t/T)) e^{-j2\pi ft} dt$$

$$= e^{-j\pi fT} \frac{\sin(\pi fT)}{\pi f} \left(\frac{1/2}{1 - (fT)^2} \right)$$

If we compare this Fourier Transform with that of the sine window (see *Figure* 2), we find that the drop-off is indeed much faster for the Hanning window (good to avoid aliasing) but the width of its main lobe is much larger (bad for accurate frequency identification). In other words, we start to face trade-offs in window design: low side lobes energy (linked to the importance of spurious frequency components) versus width of the main lobe (linked to the frequency resolution of the window).

2.1.4   The Kaiser-Bessel Window

The Kaiser-Bessel window allows for different trade-offs between the main lobe energy and side lobes energy simply by changing a parameter α in its description. The Kaiser-Bessel window $w_{KB}(t)$ is equal to

$$w_{KB}(t) = \frac{I_0 \left(\pi\alpha\sqrt{1.0 - \left(\frac{t - T/2}{T/2}\right)^2} \right)}{I_0(\pi\alpha)}$$

for times between zero and T and is equal to zero elsewhere. $I_0(x)$ is the 0th modified Bessel function

$$I_0(x) = \sum_{k=0}^{\infty} \left(\frac{(x/2)^k}{k!} \right)^2$$

Note that when applied to discrete time signals over N+1 samples this window is implemented as

$$w_{KB}[n] = \frac{I_0 \left[\pi\alpha\sqrt{1.0 - \left(\frac{n - N/2}{N/2}\right)^2} \right]}{I_0[\pi\alpha]} \quad \text{for n=0,...,N}$$

There is no closed-form analytical expression for the Fourier Transform of the Kaiser-Bessel window, but we can approximate it as [Harris 78]

$$W_{KB}(f) = \frac{T}{I_0(\pi\alpha)} \frac{\sinh\left[\sqrt{\pi^2\alpha^2 - (T2\pi f / 2)^2}\right]}{\sqrt{\pi^2\alpha^2 - (T2\pi f / 2)^2}}$$

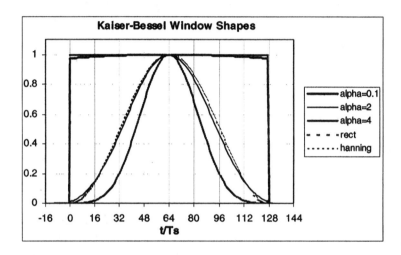

Figure 3. Time domain comparison of different shapes of the Kaiser-Bessel window for different values of the parameter α and the rectangular and Hanning windows for $T=128 * T_s$

The Kaiser-Bessel parameter α controls the trade-off between main lobe width and side lobe energy. For example, for $\alpha = 0$ the window is just equal to the rectangular window which we saw has a very narrow main lobe, but very high side lobes. For $\alpha = 2$ the window is very similar in shape to the Hanning window and likewise has low side lobe energy, but a wider main lobe than the rectangular window. As α gets larger, the side lobe energy continues to decrease at the cost of wider main lobe. *Figure 3* shows the time-domain shapes of a Kaiser-Bessel window with a very low α ($\alpha = 0.1$), with $\alpha = 2$, and with $\alpha = 4$ and their comparison with the rectangular and Hanning windows. *Figure 4* shows the frequency response of a Kaiser-Bessel window with a very low α ($\alpha = 0.1$), with $\alpha = 2$, and with $\alpha = 4$. Notice the clear trade-off between the main lobe width and the side lobe energy in the figure.

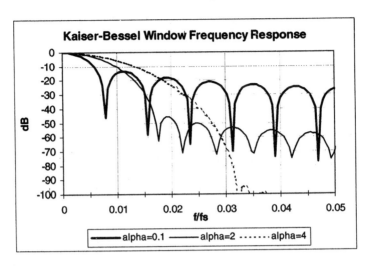

Figure 4. Frequency responses of the Kaiser-Bessel window with different values of α for T=128*T$_s$ (Note: Windows are normalized to have integral equal to 1 prior to graphing)

There are many windows defined that can achieve different points in the trade-off between main lobe width versus side lobe roll-off, and the best choice of window is application-dependent. The important conclusion is that, depending on the design requirements, we can achieve a good enough roll-off rate with non-rectangular windows that we can window a signal to finite length and still have a frequency spectrum that is reasonably well band-limited. For the reader interested in a complete review of different windows and window properties we recommend [Harris 78].

2.2   The DFT

Suppose we had a signal x(t) that was windowed to finite length and the windowed signal is also (reasonably) band-limited. Suppose further that we select an adequate sample rate $F_s = 1/T_s$ and that the signal duration is T = N*T$_s$. Since the windowed signal is finite length, we can work only with discrete frequency components

$$X[k] \equiv X(k/T) = X(k*F_s/N) \qquad \text{for } k = 0,..., N-1$$

and still recover the full windowed signal x(t). (Note that we have decided to use the frequency range from 0 to F_s as our set of independent Fourier series components rather than the range from $-F_s/2$ to $F_s/2$.) Since the windowed signal is (reasonably) band limited we can work only with a set of sampled values $x[n] \equiv x(n*T_s)$ for n=0,..., N-1. The Fourier series (see Chapter 3) tells us that we can get signal values as a sum over these frequency components; so we can write

$$x[n] \equiv x(nT_s) = \frac{1}{T}\sum_{k=0}^{N-1} X[k] e^{j2\pi(kF_s/N)(nT_s)} = \frac{1}{T}\sum_{k=0}^{N-1} X[k] e^{j2\pi kn/N}$$

for n = 0,..., N-1. Likewise, we can write the frequency components as a Fourier series sum over time samples as:

$$X[k] \equiv X(kF_s/N) = T_s \sum_{n=0}^{N-1} x[n] e^{-j2\pi(kF_s/N)(nT_s)} = T_s \sum_{n=0}^{N-1} x[n] e^{-j2\pi kn/N}$$

for k = 0,..., N-1. This transform pair is known as the "Discrete Fourier Transform" or DFT, and is the basis for all applied transform coding. The DFT pair is usually written in dimensionless form by absorbing a factor of F_s into the definition of X[k] as follows:

$$x[n] \equiv x(nT_s) = \frac{1}{N}\sum_{k=-0}^{N-1} X[k] e^{j2\pi kn/N} \qquad n = 0,...,N-1$$

$$X[k] \equiv F_s X(kF_s/N) = \sum_{n=0}^{N-1} x[n] e^{-j2\pi kn/N} \qquad k = 0,...,N-1$$

Notice the factor of F_s that now appears in the definition of X[k].

2.3   The FFT

One of the main reasons that the DFT became so important to applied coding is that it has a fast implementation called the "Fast Fourier Transform" (FFT). The forward DFT can be seen as a matrix multiplication between a vector of N time samples x[n] and an NxN matrix of phase terms (i.e. complex exponentials) leading to a new vector of N frequency samples X[k]. The inverse DFT is similarly a matrix multiplication between a vector of N frequency samples X[k] and an NxN matrix where now the matrix is the inverse of the matrix used in the forward transform. In either case, such a matrix multiplication would usually take N^2 complex multiplications and additions to carry out. Remarkably, the FFT allows us to carry out the exact same calculation in roughly $N*\log_2(N)$ complex multiplication/additions. A dramatic savings for large value of N! For example, a DFT 1024 ($=2^{10}$) samples long would require roughly 1,000,000 multiplication/additions in a straightforward calculation while the FFT would carry it out in roughly 10,000 multiplication/additions, i.e., only about 1% of the calculation time.

The trick that led to the development of the FFT is the observation that we can take an N-point DFT and turn it into the sum of two N/2-point DFTs as follows:

$$X[k] = \sum_{n=0}^{N-1} x[n]e^{-j2\pi kn/N}$$

$$= \sum_{n=0}^{N/2-1} x[2n]e^{-j2\pi k2n/N} + \sum_{n=0}^{N/2-1} x[2n+1]e^{-j2\pi k(2n+1)/N}$$

$$= \left(\sum_{n=0}^{N/2-1} x[2n]e^{-j2\pi kn/(N/2)} \right) + \left(\sum_{n=0}^{N/2-1} x[2n+1]e^{-j2\pi kn/(N/2)} \right) e^{-j2\pi k/N}$$

Notice that this equation says that the N-point DFT evaluated at the kth frequency sample is equal to the N/2-point DFT of the even samples at k plus a k-dependent complex constant times the N/2-point DFT of the odd samples at k. For each of the N values of k, this equation requires one addition and one multiplication. Let's see how this result leads to the operation count of the FFT. If we had the N/2-point DFT results ready at our disposal, we can calculate the N-point DFT at each value of k using only one addition and one multiplication. How do we get the N/2-point DFT results? We can recursively repeat this process B times for $N = 2^B$, until we only need to calculate a length two DFT. The final 2-point DFT can be directly carried out for each k with one multiplication and one addition, which is done in the straightforward way. For each of the B intermediate stages, only one addition and one multiplication are required for each of the N different values of k. If we add up all of the operations, we find that it costs us N multiplications and additions for each of the B stages. In other words, a total of roughly $N*B = N*\log_2(N)$ complex additions and multiplications to carry out the entire transform.

Normally recursive procedures like the one described above require large buffers of memory. By exploiting the symmetries of the NxN phase terms in the DFT multiplication matrix, Cooley and Tukey [Cooley and Tukey 65] developed a very elegant and efficient method for computing the FFT in place. In the Cooley-Tukey FFT algorithm, pairs of input data are processed in parallel with the so-called "butterfly" operation in a series of $\log_2(N)$ stages (see the flow diagram of *Figure 5* for an example). At the end of this process, the output values then need to be unscrambled by swapping values associated with bit-reversed indices. Note that the Cooley-Tukey FFT algorithm is defined for numbers of input samples equal to powers of 2 (i.e., $N = 2^B$ for some B). This algorithm is also known as a "radix-2" decimation in time FFT algorithm since the n components (time components) are separated into the butterfly kernels.

If the number of input samples for a DFT is even but not a power of two then a sequence decomposition to power-of-two sub-sequences can be carried out with radix-2 FFTs done for each subsequence and then merged into the final result (for more details see for example [Oppenheim and Schafer 75]). Note that for analysis purposes, i.e. if the data rate of the system is not an issue in the design of the system, one can instead choose to zero-pad non-power-of-two length sequences up to the next power of two length and then simply perform a radix-2 FFT.

Note also that you can also carry out an N-point FFT for a real function in less than $N*\log_2(N)$ operations. One can create a new input sequence by creating an N/2-point complex series for which the even terms become the real parts and the odd terms become the imaginary parts of the new sequence. One can then perform an N/2-point (instead of an N-point) FFT on the complex series and unwind the results to recover the N-point FFT values for the original real-valued series (for more details see for example [Brigham 74] and for efficient implementations [Duhamel 86]).

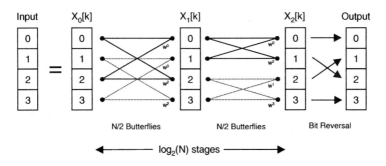

Figure 5. Block diagram of the Cooley-Tukey FFT algorithm for $N = 4$ $(w \equiv e^{-j2\pi/N})$.

3.   THE OVERLAP-AND-ADD TECHNIQUE

Having discovered that we can window a band-limited time signal in such a way that we can reasonably use the DFT or its fast implementation, the FFT, to transform the data into a discrete frequency-domain representation, how do we use this information to create an audio coder?

The first reason we discussed going into the frequency domain is that we can then easily remove redundancy from tonal signals. This reason suggests that we expect the frequency domain content to be relatively static over time (at least compared to the time-domain data) so that we have a more concise description of the signal to store or transmit. Secondly, as we'll discuss further in the next chapters, we can exploit frequency-domain masking to eliminate irrelevant signal components. We do this by completely throwing away inaudible frequency components and by allocating the available bit pool in such a way that the added quantization noise falls in areas of the spectrum where it will not be detectable. Having windowed the original signal to be able to carry out the DFT without creating significant aliasing effects, however, we need to ask ourselves how we recover the original signal from the transmitted/stored frequency domain data. We can carry out the inverse transform of the frequency-domain data to get an approximation of the windowed input signal but we still need to get the windowing effects out of the data.

The first idea one may have about getting the window effects out of the data is just to divide the output of the inverse DFT by the window coefficients. After all, we know what the window function is at each data point since we applied the window in the first place. The problem with this approach is that the quantization/dequantization process has typically created small errors in the signal. These errors may be inaudible but dividing the output of the inverse DFT by the window function could amplify the errors near the edges of the block of data since the window function is designed to go smoothly to zero in that region. If we take our dequantized data and divide them by the small values of the window function near the edges of the block, we are going to magnify the errors greatly. We need another approach.

The way we solve the window problem is to have our windowed blocks of input signal overlap each other and design our windows so that we can overlap and add the output signals in such a way that the original input signal (other than differences due to the presumably small quantization noise) is exactly recovered. We then put requirements on the window function so that the overlap-and-added output signal equals the (albeit delayed) original input signal in the absence of any quantization noise.

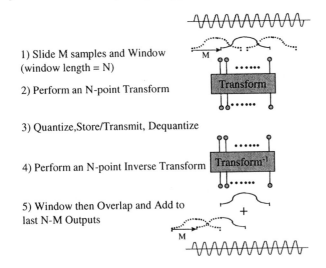

1) Slide M samples and Window
(window length = N)

2) Perform an N-point Transform

3) Quantize,Store/Transmit, Dequantize

4) Perform an N-point Inverse Transform

5) Window then Overlap and Add to
last N-M Outputs

Figure 6. Schematic of the window and overlap-and-add approach utilized to encode-decode
audio data

The overlap-and-add approach proceeds as shown in *Figure 6*. For an N-point DFT we decide on an overlap amount N-M. For simplicity, we require here that N-M be no larger than N/2 so that only adjacent blocks overlap. (For extended overlapping between more than two consecutive block the reader can consult [Malvar 92]). An overlap of N-M samples implies that each successive block starts M samples after the start of the prior block of data and includes M new time samples. In the encoder, we window the N data points of a particular block, perform a DFT, and quantize the N DFT frequency components. We can then transmit or store the encoded frequency-domain data from each block until we are ready to decode it. In the decoder, we dequantize each block's N frequency components, perform an inverse DFT to create N time samples, window again with a synthesis window, and transfer the first M samples of the result to an output buffer and the remaining N-M samples to a storage buffer. We add the N-M samples from the prior block's storage buffer to the first N-M samples of the current block's output buffer and send the M output buffer samples to the decoder output stream.

The reason we choose to window again after decoding is twofold. First, we need to make sure that quantization noise in the frequency domain remains small near the edges of the inverse-transformed block. Second, the analysis and synthesis stages can then be easily carried out symmetrically (see for example [Portnoff 80]). If the reader were to choose instead to only use analysis windows and not use synthesis windows, the conditions for the analysis windows can be derived from the following results by simply setting the synthesis window values to one in each equation.

In general, each N-sample block overlaps with N-M samples of the prior block and N-M samples on the forthcoming block. If, other than differences due to quantization noise, this overlap-and-add process is to recover the original signal then we must require certain conditions on the analysis and synthesis windows $w_a[n]$ and $w_s[n]$, respectively (see *Figure 7*).

In any block region without overlap, where n = N-M, ..., M-1, we require that the signal windowed with both the analysis and synthesis windows be equal to the original signal. Since we already saw that the DFT is invertible, this condition, in terms of the window functions, is equivalent to:

$$w_a^i[n] * w_s^i[n] = 1 \quad \text{for } n = N - M, ..., M - 1$$

where the superscript i on the window functions indicates the current block index. In the overlap regions, where n = 0, ..., N-M-1 for overlap with the prior block i – 1, and where n = M,...,N-1 for overlap with the following one i + 1, we must require that the sum of the windowed signal from both blocks add to the original signal. In terms of the window functions, this is equivalent to:

$$w_a^i[n] * w_s^i[n] + w_a^{i-1}[M + n] * w_s^{i-1}[M + n] = 1 \quad \text{for } n = 0,..., N - M - 1$$

Notice that this condition relates the right sides of the windows of one block with the left sides of the windows of the following block (see *Figure 7*). In some cases, we use this observation to allow ourselves to change window shapes on the fly by employing transition windows with left sides that match the prior block's windows and right sides that match the following block's windows.

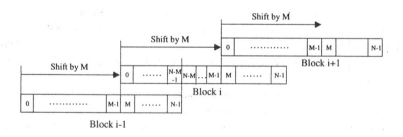

Figure 7. Overlap regions of the different blocks

If we choose to work with identical analysis and synthesis windows then we find that the perfect reconstruction conditions simplify to:

$$w^i[n]^2 + w^{i-1}[M+n]^2 = 1 \quad \text{for } n = 0,\ldots,N-M-1$$
$$w^i[n]^2 = 1 \quad \text{for } n = N-M,\ldots,M-1$$

We can immediately write down one simple window that easily satisfies the overlap-and-add perfect reconstruction conditions:

$$w[n] = \begin{cases} \sin\left[\dfrac{\pi}{2}\dfrac{n+\frac{1}{2}}{N-M}\right] & \text{for } n = 0,\ldots,N-M-1 \\[2ex] 1 & \text{for } n = N-M,\ldots,M-1 \\[2ex] \sin\left[\dfrac{\pi}{2}\dfrac{N-n-\frac{1}{2}}{N-M}\right] & \text{for } n = M,\ldots,N-1 \end{cases}$$

Notice that this window relies on the property of sine and cosine that $\sin(x)^2 + \cos(x)^2 = 1$ to achieve the perfect reconstruction condition in the overlap region.

The sine-based overlap-and-add window may not provide the resolution versus leakage trade-off needed for a particular application. Are there other windows that satisfy the overlap-and-add perfect reconstruction requirement? In fact, we can apply a normalization procedure by which any window function can be modified to satisfy the overlap-and-add conditions. Namely, we can take any initial window kernel $w'[n]$ of length $N - M + 1$, where N and M are even numbers, and create a length N overlap-and-add window $w[n]$ as follows:

$$w[n] = \begin{cases} \sqrt{\dfrac{\sum\limits_{p=0}^{n} w'[p]}{\sum\limits_{p=0}^{N-M} w'[p]}} & \text{for } n = 0,\ldots,N-M-1 \\[3ex] 1 & \text{for } n = N-M,\ldots,M-1 \\[3ex] \sqrt{\dfrac{\sum\limits_{p=n-M+1}^{N-M} w'[p]}{\sum\limits_{p=0}^{N-M} w'[p]}} & \text{for } n = M,\ldots,N-1 \end{cases}$$

Notice how this window satisfies the condition that $w[n]^2 + w[M+n]^2 = 1$ through the normalization procedure. If we start with an initial window kernal $w'[n]$ that has parameters controlling its shape, we end up with a corresponding normalized window for each parameter setting. We can then use the parameters to tune the normalized window so that it has appropriate frequency resolution and leakage properties.

The window normalization procedure described above can be carried out using a Kaiser-Bessel window as the kernel window with 50% overlap between adjacent blocks to create the so-called "Kaiser-Bessel derived" or KBD window used in the Dolby AC family of coders [Fielder et al. 96] and in MPEG AAC [Bosi et al. 97]. For example, *Figure 8* shows the window shape of the $\alpha = 4$ KBD window as compared with both the $\alpha = 4$ KB window and the sine window. Notice how the KBD window is shaped much more like the sine window than the corresponding KB window, however, the KBD window has a much broader top followed by faster drop-off than the sine window. *Figure 9* shows the frequency response corresponding to these three windows. Again we see that the $\alpha = 4$ KBD window is much more similar to the sine window than to the $\alpha = 4$ KB window. Notice also that the smooth edges of the KBD window leads to faster side lobe drop-off in the frequency response than the sine window but the narrower average width of the window leads to slightly worse frequency localization.

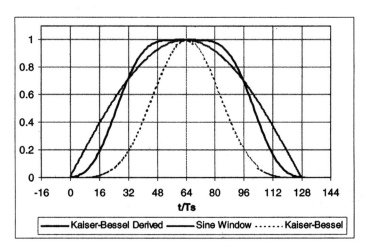

Figure 8. Time domain comparison of the α=4 KBD window with the sine window and the α=4 KB window

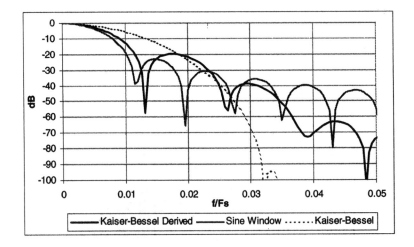

Figure 9. Frequency response of the α=4 KBD window compared with that of the sine window and the α =4 KB window

3.1 Window Considerations in Perceptual Audio Coding

Some of the major factors that come into play in the design of filter banks for audio coding is the ability to maximize the frequency separation of the filter bank and the ability to minimize the effects of audible blocking artifacts. As we saw in the previous sections, two window parameters are directly linked to these properties, namely the selected window length and shape. Given a certain block size for the input data to the filter bank, the selection of the window shape determines the degree of spectral separation of the filter bank. For example, the sine window ensures a better close-selectivity than the $\alpha = 4$ KBD window (see *Figure 9*), i.e. the sine window main lobe is narrower than the $\alpha = 4$ KBD window main lobe. On the other hand, the ultimate rejection, i.e. the amount of attenuation in the side lobes energy, of the sine window is worse than ultimate rejection of the $\alpha = 4$ KBD window (see *Figure 9*).

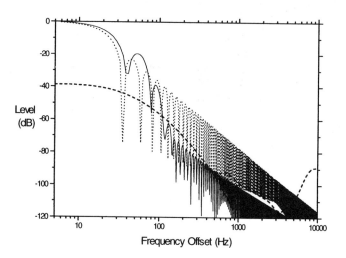

Figure 10. Comparison of the minimum masking template with α = 4 KBD and sine windows for Fs = 48 kHz (Solid line - KBD window; Dotted line - Sine Window; Dashed line - Minimum masking template) from [Bosi et al. 97]

Depending on the characteristics of the input audio signal either the sine or the α = 4 KBD window may provide better frequency resolution for the signal representation. If we consider a highly tonal signal with closely spaced picket-fence spectral structure (such as a harpsichord excerpt for example) then close selectivity plays a more important role than ultimate rejection in the frequency representation of the signal, given the superimposition of masking effects (see next chapter for a detailed discussion on masking effects) due to different parts of the signal spectrum. If instead the signal exhibits wide separation among its frequency components (such as a glockenspiel excerpt for example) higher ultimate rejection allows for better exploitation of the signal components masking.

In *Figure 10* a comparison of the frequency response of an N = 2048-point sine window, an α = 4 KBD window with a 50% overlapping region, and a particularly demanding masking template (see next chapter for details on masking curves) is shown. The sampling frequency utilized is 48 kHz. If windowing spreads a masker's energy to other frequencies above the masking curve, it will be impossible to see if the signal in that frequency region is being masked. Notice how the close selectivity of the sine window is better than the α = 4 KBD window, however the ultimate rejection of the sine window falls short of the requirements for the minimum masking threshold. The KBD window satisfies much better this requirement.

In summary, no single-shape window is optimal for all signals. Based on the signal characteristics one should dynamically select the window shape while satisfying the perfect reconstruction conditions on the window.

3.2   Window-Shape Switching

In general the trick is to recognize that the overlap-and-add perfect reconstruction condition is actual a requirement that involves the overlapping region, i.e. the right side of each window in conjunction with the left side of the subsequent one. When we use a single window type, this becomes a condition on the right and left sides of that window but nothing says that a single window type is necessary or that the windows utilized need to be symmetric. This allows us to change from a series of KBD windows to a series of sine windows or from a series of long windows into a series of shorter windows provided that we appropriately handle the overlap-and-add conditions for each overlap region.

We accomplish this by designing a pair of "transition windows" for which one side of each overlaps correctly with the previous window series' shape and length while the other side of each overlaps correctly with the following window series' shape and length. These asymmetrical windows are constructed for example by concatenating the left half KBD window with the right half sine window and vice versa. An example of a window shape sequence where the KBD window is alternated with the sine window is shown in *Figure 11*. In this figure, the amount of overlapping between adjacent blocks is 50% (i.e. M = N/2). Notice how, in order to satisfy the perfect reconstruction requirement, during the transition from the KBD window to the sine window and vice versa, asymmetrical hybrid transition windows are employed.

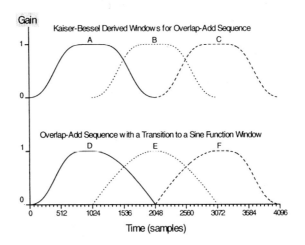

Figure 11. Window shape sequence from [Bosi et al. 97]

3.3   Block Switching

In order to adjust the frequency selectivity of the filter bank, we can also change window lengths during the coding of a signal without losing the perfect reconstruction property. Such ability can be useful when transient behavior is detected in the input signal. Although long smooth windows reduce leakage and provide high frequency resolution, they tend to blur time resolution leading to artifacts where quantization noise is spread to times prior to sharp attacks. These artifacts are known in the literature as pre-echo effects (see next chapter for a description of these temporal effects). To better handle transient conditions it is helpful to use very short analysis windows. During more steady state conditions we would like to keep the high frequency resolution found in long windows.   To satisfy both conditions, the coder can use long windows until a transient is detected. When a transient is approaching, the coder can use a "start" window to shift into short window operation until the transient is past. Once the transient is past, the coder can use a "stop" window to return to normal long window operation. For a long to short length transition during the course of a transient signal component, the start window will have a left side that overlaps with a long window and a right side that overlaps with a short window, while the stop window will be the reverse (see *Figure 12*).

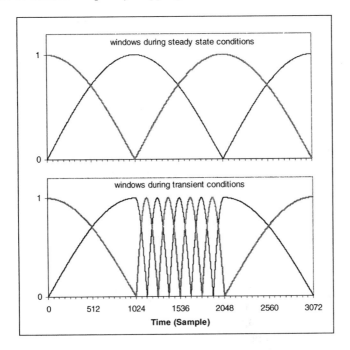

Figure 12. Window switching to better model transient conditions

One of the important window properties discussed early on in this chapter is the ability to reduce blocking artifacts. In order to reduce blocking effects from windowing, we want the transition down to zero at the edges or our windows to be as gradual as possible. This means that blocking effects are most reduced when we are set up for large overlap regions. If, for example, we set $M = N-M = N/2$, what is the implication for the system data rate? In general, each block of N samples that we encode starts M samples after the prior block. This means that we have to code and transmit/store N samples of transformed data for every M samples of new time samples fed to the coder. In other words, we are increasing our data rate in this approach by a factor of N/M prior to any coding gain from redundancy and irrelevancy removal. In the case of 50% overlap ($M = N/2$), we are doubling our data rate prior to any coding gain. This puts a high hurdle on any coding scheme! How much easier it would be if we could find some way of doing these types of transforms on blocks of data without sacrificing data rate! In fact, such a method has been found and is the subject of the next section.

4. THE MODIFIED DISCRETE COSINE TRANSFORM, MDCT

We saw in the previous sections that we can develop a good frequency representation of audio signals by taking finite length blocks of time-sampled data and transforming the data into a finite length of discrete frequency domain samples. We can then quantize those samples based on psychoacoustics principles, transmit/store the data, and recover a dequantized version of the frequency-domain samples. As discussed in the previous section, we can restore a good approximation of the original time-domain samples from the dequantized frequency samples through an overlap-and-add procedure. The main problem with implementing such a coding scheme is that the overlap-and-add procedure increases the data rate of the frequency-domain signal prior to any coding gains.

In coding applications, it is desirable that the analysis/synthesis system be designed so that the overall rate at the output of the analysis stage is equal to rate of the input signal. Systems that satisfy this condition are described as being critically sampled. When we transform the signal via the DFT, even a small amount of overlapping between adjacent blocks increases the data rate of the spectral representation of the signal, yet, in order to reduce blocking artifacts, we would like to apply maximum overlapping. With 50% overlap between adjoining blocks we end up doubling our data rate prior to quantization. The modified discrete cosine transform or MDCT is an alternative transform to the DFT utilized in state-of-the-art audio coders. One of the main advantages of the MDCT is that it allows for a 50% overlap between blocks without increasing the data rate.

The MDCT is an example of a class of transforms called Time Domain Aliasing Cancellation (TDAC) [Princen and Bradley 86, Princen, Johnson and Bradley 87]. Specifically, the MDCT is sometimes referred to as an oddly-stacked TDAC, OTDAC [Princen, Johnson and Bradley 87], as opposed to the evenly-stacked TDAC, ETDAC, which consists of alternate series of MDCT and modified discrete sine transforms, MDST [Princen and Bradley 86]. These transforms do not invert like the DFT to recover the original signal but rather invert to recover a signal that has some of the prior and following blocks' signal mixed into it. This mixing of subsequent blocks' data is called "time-domain aliasing" and it is analogous to the frequency-domain aliasing that occurs when under-sampling mixes data from frequencies outside of the block of frequencies from –Fs/2 to Fs/2 into that frequency block. The TDAC transforms, however, are designed so that the overlap-and-add procedure exactly cancels out the time-domain aliasing that occurs. Therefore, although they are not invertible as a stand-alone transforms, they still allow perfect reconstruction of an input signal.

Moreover, for a real-valued signal, only half of the N frequency-domain samples from an N-point TDAC are independent implying that the transform for audio signals only requires N/2 frequency samples from each data block for full signal recovery. This means that we can design a coder with 50% overlap (best for eliminating blocking effects) that does not increase the data rate, i.e. is critically-sampled – a solution to the problems with the DFT coder.

4.1 Matrix Derivation of Time Domain Aliasing Cancellation and Perfect Reconstruction Conditions

Let's see how transforms based on time domain aliasing cancellation can lead to perfect reconstruction filter banks without increasing the data rate. The matrix structure of a transform that converts N input samples into N/2 frequency domain samples and then back into N output time domain samples is shown in *Figure 13* (see also [Vetterli and Kovačević 95]). In this figure, the input samples on the right-hand-side are indexed in groups of N/2. The current and prior input groups (indices i and i-1) considered as a single block are windowed with a length N analysis window, $W_i\,^A{}_R$ and $W_i\,^A{}_L$ where the indices R and L indicate the right and left part of the analysis window for the current block i. They are then transformed into only N/2 frequency samples with the matrix kernels A_1 and A_2 and inverse transformed back into N time samples with the matrix kernels B_1 and B_2. Finally, they are windowed with a length N synthesis window, $W_i\,^S{}_R$ and $W_i\,^S{}_L$ where the indices R and L indicate the right and left part of the synthesis window for the current block i. The result of this matrix multiplication is then added to the result from prior analysis (block i – 1) and the transform process then continues for another pass with the index i incremented by 1. If we multiply out the matrices and add the result to that of all the other input blocks we find that the net result is that the input data is multiplied by the band diagonal matrix shown in *Figure 14*.

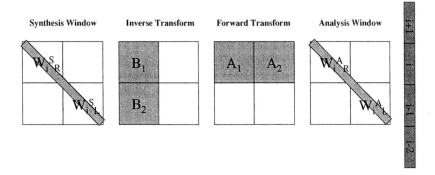

Figure 13. Matrix structure of the ith pass through the TDAC transform (gray areas show location of non-zero entries)

In order to recover the input signal after the transform process shown in *Figure 13* (after overlap-and-add) we must require that the matrix shown in *Figure 14* be equal to the identity matrix. We need therefore to impose the following matrix conditions:

$$W_{i\,R}^{S} B_1 A_2 W_{i\,L}^{A} = W_{i\,L}^{S} B_2 A_1 W_{i\,R}^{A} = \mathbf{0}$$
$$W_{i\,L}^{S} B_2 A_2 W_{i\,L}^{A} + W_{i-1R}^{S} B_1 A_1 W_{i-1R}^{A} = \mathbf{1}$$

for all blocks i

where $\mathbf{1}$ is the N/2 by N/2 identity matrix and $\mathbf{0}$ is the N/2 by N/2 zero matrix. These conditions constrain both our choice of window function and our choice of transform. The MDCT provides a particular solution to these equations.

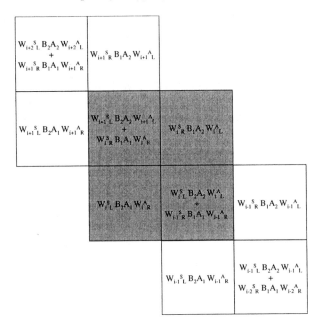

Figure 14. Matrix structure of the TDAC transforms after overlap-and-add

To better understand how such a solution can come about, consider a set of transform matrices that satisfy the following properties (see *Figure 16* below):

$$B_1 A_2 = B_2 A_1 = 0$$
$$B_1 A_1 = 1 + J$$
$$B_2 A_2 = 1 - J$$

where J is the N/2 by N/2 antidiagonal matrix (all ones on the other diagonal). The first property is an anti-aliasing condition that eliminates the blocks above and below the diagonal (top-right and lower-left gray blocks in *Figure 14*) and makes sure that the time aliasing cancellation condition is met for any window function. Applying the second property, i.e. the properties on the matrix product A_1B_1 and A_2B_2, allows us to satisfy the perfect reconstruction condition by requiring the following two window conditions:

$$W_{i\,L}^S W_{i\,L}^A + W_{i-1R}^S W_{i-1R}^A = 1$$
$$W_{i\,L}^S J W_{i\,L}^A = W_{i-1R}^S J W_{i-1R}^A$$

The first of the perfect reconstruction window conditions is the same perfect reconstruction condition we had for the DFT case. The second window condition is a new condition and adds further constraints to the window functions. This condition is linked to the cancellation of the time domain aliasing terms. Recognizing that $\mathbf{JDJ}$ time-reverses any diagonal matrix $\mathbf{D}$ and that $\mathbf{JJ=1}$, the second window condition can be met by requiring that the analysis and synthesis windows be time-reversed copies of each other:

$$W_{i\ L}^{A} = \mathbf{J}W_{i-1R}^{S}\mathbf{J}$$

$$W_{i\ L}^{S} = \mathbf{J}W_{i-1R}^{A}\mathbf{J}$$

We can rewrite these conditions in a more familiar form (i.e. without matrix notation) as:

$$w_{a}^{i}[n] * w_{s}^{i}[n] + w_{a}^{i-1}[N/2+n] * w_{s}^{i-1}[N/2+n] = 1$$

$$w_{a}^{i}[n] = w_{s}^{i-1}[N-1-n] \qquad \text{for n=0,...,N/2-1}$$

$$w_{s}^{i}[n] = w_{a}^{i-1}[N-1-n]$$

It is worth noting that the new time-reversal conditions linking the analysis and synthesis windows are similar to conditions we have earlier seen being required of sub-band coders (e.g., CQF) for perfect reconstruction.

Having seen how to select windows to achieve perfect reconstruction, let's look in detail at how the MDCT transform kernel satisfies its perfect reconstruction conditions. The MDCT forward transform takes a block of N time samples $x_i[n]$ and transforms them into N/2 frequency samples $X_i[k]$ according to:

$$X_{i}[k] = \sum_{n=0}^{N-1} w_{a}^{i}[n]x_{i}[n]\cos(\tfrac{2\pi}{N}(n+n_0)(k+\tfrac{1}{2})) \quad \text{for k = 0,..., N/2 - 1}$$

where

$$n_0 = (\tfrac{N}{2}+1)/2$$

is a phase term that ensures alias cancellation. The MDCT inverse transform then takes the N/2 frequency samples and transforms them back into N time samples $x'_i[n]$ according to:

$$x'_i[n] = w^i_s[n] \frac{4}{N} \sum_{k=0}^{N/2-1} X_i[k]\cos(\frac{2\pi}{N}(n + n_0)(k + \frac{1}{2})) \qquad \text{for } n = 0,\dots, N\text{-}1$$

In terms of the prior matrix notation, we therefore have that:

$$A_{1kn} = \cos\left(\frac{2\pi}{N}(n + N/2 + n_0)(k + \frac{1}{2})\right)$$

$$A_{2kn} = \cos\left(\frac{2\pi}{N}(n + n_0)(k + \frac{1}{2})\right)$$

$$B_1 = \frac{4}{N}A_1^T$$

$$B_2 = \frac{4}{N}A_2^T$$

It is a straightforward exercise to show that these matrices satisfy the necessary conditions that:

$$B_1A_2 = B_2A_1 = 0$$

$$B_1A_1 = 1 + J$$

$$B_2A_2 = 1 - J$$

when n_0 has the correct anti-aliasing value. This result is most easily derived by using the fact that:

$$\sum_{k=0}^{N/2-1} \cos\left(\frac{2\pi}{N}(k + \frac{1}{2})n\right)\cos\left(\frac{2\pi}{N}(k + \frac{1}{2})m\right) = \frac{N}{4} \sum_{p=-\infty}^{\infty} (-1)^p (\delta_{n,m+pN} + \delta_{n,-m+pN})$$

To better understand the behavior of the MDCT, we can look in detail at the rows of the matrices A_1 and A_2. For example, *Figure 15* shows the rows of the matrices A_1 and A_2 for N=16 (in which case they are 8 by 8 matrices). Notice how the rows of A_1 are symmetric around the center while the rows of A_2 are antisymmetric – the results for B_1A_1 and B_2A_2 are a direct result of these symmetries. Multiplying the data by A_1 will destroy the antisymmetric part of the data so that the closest to perfect reconstruction we could possibly recover from B_1A_1 is $1+J$. Likewise, multiplying by A_2 will eliminate the symmetric part of the data so we can't get any closer to 1 from B_2A_2 than $1-J$.

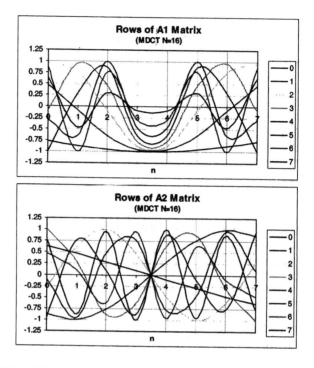

Figure 15. Rows of the MDCT transform matrices A_1 and A_2 for N=16

To summarize this section, the MDCT transform allows us to have 50% overlap between successive data blocks without increasing the overall data rate. Given a set of analysis and synthesis windows that satisfy the perfect reconstruction conditions we can transform N inputs from the ith and $(i-1)^{th}$ sets of N/2 inputs into N/2 frequency domain outputs according to:

$$X[k] = \sum_{n=0}^{N-1} x[n] w_a[n] \cos(\tfrac{2\pi}{N}(n + n_0)(k + \tfrac{1}{2})) \quad \text{for } k = 0,\ldots, N/2 - 1$$

and then return them into N time domain samples ready to be overlapped-and-added using:

$$x'[n] = w_s[n] \tfrac{4}{N} \sum_{k=0}^{N/2-1} X[k] \cos(\tfrac{2\pi}{N}(n + n_0)(k + \tfrac{1}{2})) \qquad \text{for } n = 0,\ldots, N-1$$

The portions of the analysis and synthesis windows that overlap between adjacent blocks should be time-reversals of each other. In addition, the windows should satisfy the following perfect reconstruction condition:

$$w_a^i[n] * w_s^i[n] + w_a^{i-1}[N/2 + n] * w_s^{i-1}[N/2 + n] = 1 \text{ for } n=0,...,N/2-1$$

Since this latter condition is the same one we faced for the overlap-and-add DFT, we can use the windows discussed there (e.g., sine window, KBD window) for the MDCT as well. We can also make use of the same tricks we saw earlier to change window shape in differing blocks if we keep the same window length and continue to overlap by 50%. However, the requirements for time domain aliasing across blocks require us to be very careful in designing windows to change time resolution (i.e. block size) or overlap region. In the next section we take a look at what is required to change block size after a transient is detected in an MDCT-based coder.

4.2   Changing Block Size with the TDAC Transforms

We saw in the previous section that the cosine functions in the OTDAC transforms have phases specifically chosen so that a single pass through the MDCT followed by an inverse MDCT, IMDCT, leads to the matrix structure shown in *Figure 16*. Similar results apply to the ETDAC transforms [Princen and Bradley 86]. The window conditions are such that overlap and add with the prior and subsequent windows leads to the matrix structure becoming just the identity matrix. Making the analysis and synthesis windows of overlapping blocks time reverses of each other is enough to cancel out the antidiagonal parts of the single pass matrix (the **J** parts) and the perfect reconstruction condition then makes sure the resulting diagonal matrix is equal to the identity matrix. The challenge in developing window functions that support changing block size is to ensure that it remains possible to cancel out the antidiagonal parts of the single pass matrix. A number of approaches have been proposed in the literature, see for example [Edler 89], [Sugiyama, Hazu and Iwadare 90], [Bosi and Davidson 92], and [Princen and Johnston 95]. In this section, we explore in detail two different methods [Edler 89, Bosi and Davidson 92] that have been used to create transition windows that allow a change of block size while still maintaining time-domain aliasing cancellation so perfect reconstruction is still achieved. These methods or variants of these methods are currently in use state-of-the-art coders such as MPEG Layer III, MPEG AAC and Dolby AC-3.

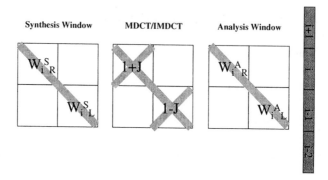

Figure 16. Matrix structure of the MDCT/IMDCT transform before overlap-and-add

Let's consider what's needed in changing from a series of long blocks to a series of shorter blocks. *Figure 17* shows the matrix structure on either side of the transition block. Overlap and add for the prior series of long blocks leads to the identity matrix along the diagonal other than the last $N_{long}/2$ by $N_{long}/2$ section which is equal to $W_{i-1}{}^S_R$ $(1+J)$ $W_{i-1}{}^A_R$ while overlap and add for the following series of short blocks leads to the identity matrix along the diagonal other than the first $N_{short}/2$ by $N_{short}/2$ section which is equal to $W_{i+1}{}^S_L$ $(1-J)$ $W_{i+1}{}^A_L$. Any transition block needs to be designed to match both sides of this matrix and cancel out time-domain aliasing.

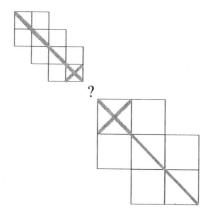

Figure 17. Matrix structure on either side of a transition from a series of long blocks into a series of shorter ones (shown for the case where long blocks are twice as long as short blocks)

One of the first solutions to the transition block was developed by Edler [Edler 89]. This solution was based on the observation that zero values in the analysis and synthesis windows can project out part of the unwanted matrix structure. In particular, although doing a long block MDCT for a transition window can easily cancel out aliasing in the prior long block by setting the left side windows to the normal long block left side windows, the antidiagonal part is too long to easily cancel out the antidiagonal part in the subsequent short block. In terms of *Figure 17*, the gray antidiagonal bar in the bottom right of the transition is too long for the gray antidiagonal bar at the top left of the transition. However, making part of the window in the long block equal to zero can shorten the antidiagonal content of the matrix and therefore make the problem easier.

Edler's solution is that you can create a transition block out of a long block MDCT by setting the window values to zero in the right side of the analysis and synthesis windows for the last $N_{long}/4 - N_{short}/4$ entries. Doing so will reduce the antidiagonal part that needs to cancel out the antidiagonal part of the short window from $N_{long}/2$ non-zero entries down to only $N_{short}/2$ non-zero entries. Now that the antidiagonal part is the right length, it can be cancelled out by aligning the subsequent short block with the center of the right side of the transition block. The perfect reconstruction conditions then become that the $N_{short}/2$ entries at the center of the right side of the transition block should be the same as the right side of a short block while the first $N_{long}/4 - N_{short}/4$ entries (which don't overlap with the short block should be equal to one (see *Figure 18*). Similarly, the transition block back from short blocks to long blocks has windows that are just the time-reverse of those for the transition from long to short. *Figure 19* shows an example sequence of windows from long blocks to short blocks and back again to long blocks using the Edler solution.

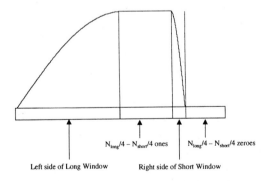

$N_{long}/4 - N_{short}/4$ ones $N_{long}/4 - N_{short}/4$ zeroes

Left side of Long Window   Right side of Short Window

Figure 18. Structure of the Edler transition window from long blocks to short blocks

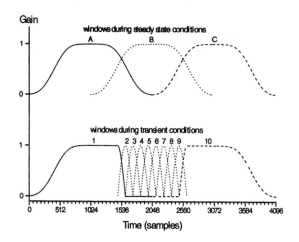

Figure 19. Transition between long and short blocks using Edler transition windows from
[Bosi et al. 97]

One last thing to note about the Edler transition block structure is its effect on data rate. In going from long blocks to short blocks the Edler transition block takes a length N_{long} MDCT which requires passing $N_{long}/2$ frequency samples corresponding to the $N_{long}/2$ new time samples needed for this block. However, each of the short blocks that overlaps the right side of the window (including the zero entries) also passes $N_{short}/2$ frequency samples corresponding to the same set of time samples resulting in an excess of $N_{long}/4$ extra frequency samples. This implies that going through an Edler transition from long blocks to short blocks actually increases the data rate and the critical sampling property is lost in that region. However, in going through an Edler transition from short blocks to long blocks reduces the data rate exactly enough to cancel out any net increase in going from long blocks to short blocks and then back again to long blocks. The data rate reduction in going from short to long blocks results from that fact that $3N_{long}/4$ new time samples are needed to perform the length N_{long} transition block MDCT while only $N_{long}/2$ new frequency samples are passed by the block – a reduction of $N_{long}/4$ frequency samples as required. The net result is that using Edler transition blocks to switch to short blocks only for the duration of transient behavior in the input signal (for which shorter blocks give better time resolution) and then going back to long blocks (with their better frequency resolution) keeps the coder critically sampled overall.

Applications of the Edler method are found in the MPEG Layer III and AAC coders.

A second solution to the transition block problem was provided by the Dolby AC-2A team [Bosi and Davidson 92]. In the AC-2A solution, originally developed for ETDAC transforms, the transition windows from long to short are left sides of long windows on the left and right sides of short windows on the right (see *Figure 20*). The transition windows from short to long are the time-reverses of the windows from long to short. So how does time-domain aliasing cancellation come about? It comes about by changing the kernel of the MDCT transform. Namely, the MDCT is now carried out with length ½ ($N_{long} + N_{short}$) with a phase term $n_0 = b/2 + ½$ where b is the length of the right side of the window, i.e. b = $N_{short}/2$ for a transition from long to short and b = $N_{long}/2$ for a transition from short to long. For this choice of phase, the MDCT followed by the IMDCT gives exactly the matrix structure needed to tie together the two parts of *Figure 17* (without any space between). As in the normal MDCT case, this fact can easily be verified by writing out the MDCT followed by the IMDCT as a matrix multiplication and then using the cosine orthogonality condition

$$\sum_{k=0}^{N/2-1}\cos\left(\tfrac{2\pi}{N}(k+\tfrac{1}{2})n\right)\cos\left(\tfrac{2\pi}{N}(k+\tfrac{1}{2})m\right)=\frac{N}{4}\sum_{p=-\infty}^{\infty}(-1)^{p}\,(\delta_{n,m+pN}+\delta_{n,-m+pN})$$

Figure 12 earlier in this chapter showed an example sequence of windows from long blocks to short blocks and back again to long blocks using the AC-2A solution.

One last thing to note about the AC-2A solution is that, like in the Edler solution, the data rate increases going from long blocks to short blocks but there is no net increase in overall data rate in following a subsequent return to long block operation. In going from long blocks to short blocks the AC-2A transition block sends ¼ ($N_{long} + N_{short}$) frequency samples although it only uses $N_{short}/2$ new time samples. The AC-2A transition block from short blocks to long blocks only sends ¼ ($N_{long} + N_{short}$) frequency samples while it actually needed $N_{long}/2$ new time samples for its MDCT. The net result is that using the AC-2A transition blocks to switch to short blocks only for the duration of transient behavior in the input signal also keeps the coder critically sampled overall. A variant of this method is utilized in AC-3 (see also *Figure 4* in Chapter 14). In AC-3 the transition and short blocks equal half of the long block and the overlap region for these blocks alternates between 0 and half of the long block length, so that b in the MDCT phase term n_0 equals either 0 or $N_{long}/2$.

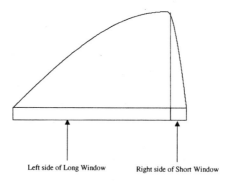

Left side of Long Window   Right side of Short Window

Figure 20. Structure of the AC-2A transition window from long blocks to short blocks

4.3   MDCT and PQMF Filter Banks

While historically the PQMF and the MDCT filter banks were developed independently, [Malvar 92] showed how these approaches can be unified. In fact, Malvar showed that the MDCT is a special case of the PQMF filterbank with filter length N = 2K where K is the number of frequency channels. (Note that this relationship does not apply to the ETDAC). In this section we first rewrite the N=2K PQMF filterbank in terms of an overlap-and-add block transform. We then compare it with the MDCT to better understand what it means when we say that the MDCT is a PQMF filterbank.

Recall that the PQMF filterbank was defined as a set of K paired analysis and synthesis filters $h_k[n]$ and $g_k[n]$, respectively, equal to

$$h_k[n] = h[n]\cos\left(\pi\left(\frac{k+\frac{1}{2}}{K}\right)\left(n - \frac{(N-1)}{2}\right) + \phi_k\right) \text{ for k=0,...,K-1}$$

$$g_k[n] = h_k[N-1-n]$$

for a particular phase term ϕ_k selected to ensure alias cancellation between adjacent frequency bands. The PQMF filterbank is then carried out by passing the input signal in parallel through each of the analysis filters $h_k[n]$; down-sampling the result by a factor of K to keep the total data rate constant; quantizing, transmitting/storing, and dequantizing the output $y_k[n]$; up-sampling the recovered $y_k[n]$ values by a factor of K; passing that through each corresponding synthesis filter $g_k[n]$; and, finally, adding the outputs of each of the K channels (see *Figure 1* in the prior chapter). Let's explore in detail what this process looks like for the specific case of N=2K.

If our input signal is x[n] then the action of filter $h_k[n]$ is the convolution of the input sequence and the filter impulse response. Down-sampling this result by a factor of K then just gives us the outputs of this process taken at points n that are multiples of K (i.e. n = iK for integer i). Therefore, we find that $y_k[i]$ is equal to:

$$y_k[i] = \sum_{m=0}^{N-1} x[iK - m]h_k[m]$$

$$= \sum_{n=0}^{N-1} x[(i-2)K + 1 + n]h_k[N-1-n]$$

where in the second line we made the change of variables to n = N-1-m and recognized that N = 2K.

Suppose we choose to group our input signal x[n] into blocks of length N where each successive block is the input signal slid forward by K = N/2 samples. We number our input samples so that the first non-zero value is defined as x[1] and we number the blocks such that the block with i=1 contains the first K non-zero samples. In other words, we define the ith block of the input signal $x_i[n]$ as the following:

$$x_i[n] = x[(i-2)K + 1 + n] \quad \text{for n=0,...,N-1}$$

Notice how block i=0 ends with the last zero value x[0] while block i=1 goes up the Kth non-zero value x[K].

With this block definition we can rewrite the outputs of the analysis stage $y_k[i]$ as the following:

$$y_k[i] = \sum_{n=0}^{N-1} x_i[n]h_k[N-1-n]$$

Substituting the specific form of $h_k[n]$ for the PQMF filterbank into this expression gives that:

$$y_k[i] = \sum_{n=0}^{N-1} h[N-1-n]x_i[n]\cos\left(\tfrac{2\pi}{N}\left(k+\tfrac{1}{2}\right)\left(n - \tfrac{(N-1)}{2}\right) - \phi_k\right) \quad \text{for k = 0 ,...,K-1}$$

This expression shows us that the analysis stage of the N = 2K PQMF can be written as a block transform on blocks of input samples where each successive block contains the next K = N/2 samples. After writing out an expression for the synthesis stage of the N = 2K PQMF, we return to this expression and compare it to the forward transform of the MDCT.

In the synthesis stage of the PQMF, we up-sample each of the analysis channels $y_k[i]$ by a factor of K, filter them with the corresponding synthesis filter $g_k[n]$, and sum the results. Up-sampling intersperses K-1 zeros between each i value of $y_k[i]$ in channel k. The synthesis filters in each k channel are only N = 2K samples long and so each new set of outputs from the filters will only feel the last two $y_k[i]$ values from the analysis stage and their interspersed zeros. For example, when $y_k[i]$ is the newest element fed to the kth synthesis filter (preceded by the prior interspersed K-1 zeros, $y_k[i-1]$, and followed by K-1 more zeros), the filter output will be $y_k[i]g_k[0]+ y_k[i-1]g_k[K]$. The next filter input will be the first zero interspersed after $y_k[i]$ and the filter output will be $y_k[i]g_k[1]+ y_k[i-1]g_k[K+1]$, etc. In other words, each new input value $y_k[i]$ from the analysis stage will lead after up-sampling and filtering to K new outputs from each channel where the nth output is equal to $y_k[i]g_k[n]+ y_k[i-1]g_k[K+n]$.

Adding up the channel outputs sums this over k and we find the following result for the ith block of K new outputs from the synthesis stage $\hat{x}_i[n]$:

$$\hat{x}_i[n] = \sum_{k=0}^{K-1} y_k[i]g_k[n] + y_k[i-1]g_k[K+n] \quad \text{for } n = 0,...,K-1$$

Notice that the first term in this sum applies the first K elements of the synthesis filters to the current $y_k[i]$ synthesis stage inputs while the second turn applies the second half of the synthesis filters to the last set of synthesis stage inputs $y_k[i-1]$.

We can get the same result by instead extending the length of $\hat{x}_i[n]$ to N elements and filling it using only the current synthesis inputs $y_k[i]$ if we then overlap and add the results of successive $\hat{x}_i[n]$ blocks. In specific, we instead define the extended $\hat{x}_i[n]$ as

$$\hat{x}_i[n] = \sum_{k=0}^{K-1} y_k[i]g_k[n] \quad \text{for } n = 0,...,N-1$$

and for each block of K synthesis inputs $y_k[i]$ we output the first half of the extended $\hat{x}_i[n]$ added to the second half of the prior extended $\hat{x}_{i-1}[n]$. Rewriting this expression with the specific form of the PQMF synthesis filters leads to the following result:

$$\hat{x}_i[n] = h[N-1-n]\sum_{k=0}^{K-1} y_k[i]\cos\left(\tfrac{2\pi}{N}\left(k+\tfrac{1}{2}\right)\left(n-\tfrac{(N-1)}{2}\right)-\phi_k\right) \text{ for } n = 0,\ldots,N\text{-}1$$

for the result of the synthesis stage prior to overlapping and adding. Notice again that this is a block transform from the block of K synthesis stage inputs $y_k[i]$ onto a block of N outputs $\hat{x}_i[n]$.

Having rewritten an arbitrary $N = 2K$ PQMF filterbank in terms of a pair of block transforms, we can compare the specific form of the MDCT transform to this and see how they are related. In particular, we would like to compare the pair of MDCT transform equations

$$X_i[k] = \sum_{n=0}^{N-1} x_i[n]w_a^i[n]\cos(\tfrac{2\pi}{N}(n+n_0)(k+\tfrac{1}{2})) \text{ for } k = 0,\ldots,K\text{-}1$$

$$x_i[n] = \tfrac{4}{N} w_s^i[n]\sum_{k=0}^{K-1} X_i[k]\cos(\tfrac{2\pi}{N}(n+n_0)(k+\tfrac{1}{2})) \text{ for } n = 0,\ldots,N\text{-}1$$

to the pair of $N = 2K$ PQMF transform equations

$$y_k[i] = \sum_{n=0}^{N-1} h[N-1-n]x_i[n]\cos\left(\tfrac{2\pi}{N}\left(k+\tfrac{1}{2}\right)\left(n-\tfrac{(N-1)}{2}\right)-\phi_k\right) \text{ for } n = 0,\ldots,K\text{-}1$$

$$\hat{x}_i[n] = h[N-1-n]\sum_{k=0}^{K-1} y_k[i]\cos\left(\tfrac{2\pi}{N}\left(k+\tfrac{1}{2}\right)\left(n-\tfrac{(N-1)}{2}\right)-\phi_k\right) \text{ for } k = 0,\ldots,N\text{-}1$$

We first note that the arguments of the cosines are identical if

$$\phi_k = -\tfrac{2\pi}{N}(k+\tfrac{1}{2})(n_0 + \tfrac{(N-1)}{2})$$

The MDCT required us to set $n_0 = N/4+1/2$ in order to ensure time aliasing – this implies that the MDCT has the same cosine as the $N = 2K$ PQMF with a phase term ϕ_k equal to

$$\phi_k = -\tfrac{3\pi}{2}(k+\tfrac{1}{2})$$

For the MDCT to be a PQMF filterbank we need to have the phase term ϕ_k satisfy

$\phi_k - \phi_{k-1} = \frac{\pi}{2}(2r+1)$

for integer r to ensure that aliasing terms are cancelled between adjacent frequency bands – clearly a condition met by the MDCT phase term. In fact, it turns out that the MDCT phase term cancels frequency aliasing between all pairs of frequency bands, not merely adjacent ones. We can therefore conclude that the MDCT does take the form of an N = 2K PQMF filterbank.

Next we can relate the MDCT window functions to the prototype filter h[n] for a PQMF filterbank. We should note that the outputs of the analysis stages are defined with slightly different normalizations and are related by

$y_k[i] = \frac{2}{\sqrt{N}} X_i[k]$

Noting this change of normalization allows us to see that the MDCT has the same form as a PQMF filterbank when the MDCT windows are related to the PQMF prototype filter h[n] through

$w_a^i[n] = \frac{\sqrt{N}}{2} h[N-1-n]$

$w_s^i[n] = \frac{\sqrt{N}}{2} h[N-1-n]$

Notice that these relationships require that the MDCT analysis and synthesis windows be equal to each other. Now, for a time-invariant filterbank, i.e. for a single window type, perfect reconstruction requires that the MDCT analysis and synthesis windows be time reverses of each other. The MDCT with a time-invariant filterbank can only satisfy both of the window conditions with symmetric window functions.

Finally, the MDCT requires the window condition to satisfy the perfect reconstruction condition that

$w_a^i[n]*w_s^i[n] + w_a^{i-1}[N/2+n]*w_s^{i-1}[N/2+n] = 1$ for $n = 0,...,N/2-1$

while a PQMF prototype filter is designed to satisfy the power complementarity condition:

$|H(f)|^2 + |H(-F_s/2K+f)|^2 = 2/F_s^2$ for $0 \le f \le F_s/4K$

and required to be effectively zero beyond $|f| = F_s/2K$. We need to ask ourselves how, if at all, these conditions are related.

In the case where the MDCT windows are symmetric, they correspond to a symmetric PQMF prototype filter h[n] that satisfies the following perfect reconstruction condition

$$h[n]^2 + h[n + N/2]^2 = 4/N \quad \text{for n=0,...,N/2-1}$$

It can be shown that this perfect reconstruction condition implies the following power complementarity condition in the frequency domain:

$$\frac{1}{N} \sum_{k=0}^{N-1} |H(f + kF_s/N)|^2 = 2/F_s^2$$

Now if an MDCT window is well enough localized in the frequency domain that it has negligible power beyond $|f| = F_s/2K$ then we find that this condition is equivalent to the PQMF power complementarity condition other than a trivial difference in gain of 1/N between the two cases. However, any window that satisfies the MDCT perfect reconstruction condition leads to perfect reconstruction even if it is not localized enough to satisfy the PQMF condition. For example, the sine window satisfies the MDCT conditions and leads to perfect reconstruction but is not localized enough to be a PQMF prototype filter. What we see is that the MDCT transform is a version of the N = 2K PQMF filterbank that produces exact perfect reconstruction while allowing a wider set of prototype filters than generally allowed in PQMF filter banks.

In this section, we have tied together the two storylines of this and the previous chapter. We first met the MDCT as a transform that solves a number of problems in transform coding by allowing for 50% overlap without increasing the data rate. We have now seen that the MDCT is also a sub-band coding filterbank that allows for perfect reconstruction with lower cost filters than the PQMF due to its lesser requirements on stopband attenuation. The ease of window design (e.g., just use a sine or KBD window) and the ability to adapt the filterbank resolution by simply altering a single parameter (the window length N) have made the MDCT the transform of choice for most of the newer coders.

4.4 Implementation of the MDCT via the FFT

We have seen how the MDCT solves the data rate problem inherent in DFT coders, however, to be truly useful for implementation it needs to have a fast transform method so that the runtime doesn't grow like N^2 for large block sizes. In fact, the FFT can be leveraged to create a fast version of the MDCT through a rewriting of the transform as follows:

$$X[k] = \sum_{n=0}^{N-1} x[n]w[n]\cos(\tfrac{2\pi}{N}(n + n_0)(k + \tfrac{1}{2}))$$

$$= \text{Re}\{\sum_{n=0}^{N-1} x[n]w[n]e^{-j\frac{2\pi}{N}(n+n_0)(k+\frac{1}{2})}\}$$

$$= \text{Re}\{e^{-j\frac{2\pi}{N}n_0(k+\frac{1}{2})}\sum_{n=0}^{N-1}[x[n]w[n]e^{-j\frac{2\pi n}{2N}}]e^{-j\frac{2\pi kn}{N}}\}$$

This rewriting tells us that we can implement the forward MDCT by carrying out the following steps:

1. "Pre-twiddle" the input samples by (complex) multiplying with the factor

$$e^{-j\frac{2\pi n}{2N}}$$

2. Perform an N-point FFT on the pre-twiddled data
3. "Post-twiddle" the transformed data for k values from 0 to N/2-1 by taking the real part of the transformed data times the factor

$$e^{-j\frac{2\pi}{N}n_0(k+\frac{1}{2})} .$$

Notice that rather than growing like N^2, the fast implementation runtime only grows like $N*\log_2(N)$ (from the FFT, the pre- and post-twiddle operations only grow like N).

Analogously, we can rewrite the inverse transformation to make use of the FFT as follows:

$$x'[n] = w[n]\tfrac{4}{N}\sum_{k=0}^{N/2-1} X[k]\cos(\tfrac{2\pi}{N}(n + n_0)(k + \tfrac{1}{2}))$$

$$= w[n]\tfrac{2}{N}\sum_{k=0}^{N-1} X[k]\cos(\tfrac{2\pi}{N}(n + n_0)(k + \tfrac{1}{2}))$$

$$= w[n]\text{Re}\{\tfrac{2}{N}\sum_{k=0}^{N-1} X[k]e^{j\frac{2\pi}{N}(n+n_0)(k+\frac{1}{2})}\}$$

$$= 2w[n]\text{Re}\{e^{j\frac{2\pi}{2N}(n+n_0)}\tfrac{1}{N}\sum_{k=0}^{N-1}[X[k]e^{j\frac{2\pi}{N}kn_0}]e^{j\frac{2\pi}{N}kn}\}$$

This rewriting tells us that we can implement the inverse MDCT by carrying out the following steps:

1. Pre-twiddle the frequency samples (note that we now go from k=0,...,N-1 – use $X[N-1-k] = -X[k]$ for $k \geq N/2$) by (complex) multiplying with the factor

$$e^{j\frac{2\pi}{N}kn_0}$$

2. Perform an N-point inverse FFT on the pre-twiddled data
3. Post-twiddle the inverse transformed data by taking the real part of the inverse transformed data times the factor

$$e^{j\frac{2\pi}{2N}(n+n_0)}$$

and then multiply by two times the synthesis window.

Again we can notice that use of the FFT algorithm allows us to reduce the number of operations to only being of order $N*\log_2(N)$.

Please be aware that what we've seen in this section is the simplest example of converting the MDCT implementation from order N^2 to order $N*\log_2(N)$ but not necessarily the fastest. A bit of work can get the implementation even faster. The interested reader can refer to [Malvar 92], [Duhamel, Mahieux and Petit 91], and [Bosi 99].

5. SUMMARY

In summary for this and the prior chapter, we have reviewed two approaches for parsing input signals into frequency components to be encoded: the sub-band filter bank approach where a PQMF is used to segment frequency components, and the transform approach where a modulated transform is used to segment frequency components. We have seen that the PQMF methods of creating large numbers of frequency channels rely on modulating a prototype low-pass filter into the appropriate frequency band location. The PQMF is utilized in the time to frequency mapping of MPEG-1 and 2 Layers I and II (see Chapters 11 and 12). We have also seen that blocking effects require transform methods to window the input signals with smooth windows and to overlap-and-add the output data to reconstruct the signal. Finally, we have learned about the MDCT which can easily be seen as either a PQMF sub-band method or a type of windowed transform showing that these two approaches are just different faces of a single underlying multi-channel approach to signal processing. Moreover, the MDCT achieves perfect reconstruction without adding to overall system data rate, and has a fast implementation allowing easy scaling up to large numbers of channels. The MDCT is utilized in the time to frequency mapping of MPEG-2 and 4 AAC (see Chapters 13 and 15) and AC-3 (see Chapter 14). A hybrid filter bank, which cascades the PQMF stage with an MDCT stage, is utilized in MPEG Layer III (see Chapter 9).

In general, in the design of time to frequency mapping for audio coding, the main goal is to maximize the ability to separate the frequency components of the signal while minimizing the audibility of blocking artifacts. Critical sampling, although not a strict requirement, is highly desirable. Perfect reconstruction or nearly perfect reconstruction filter banks simplify the design of the coding systems. Time delay and computational complexity are also important factors in the choice of filter banks. The filter bank is a central step in coding systems, setting the stage for the extraction of redundancies and irrelevancies of the signal. Before examining in detail the allocation of the bit pool in order to achieve this goal, we turn our attention in the next chapter to models of the human ear so that we can better understand what signal components and quantization noise will not be audible in a signal.

6.   REFERENCES

[Bosi 99]: M. Bosi, "Analysis/Synthesis system with efficient oddly stacked single-band filter using time-domain aliasing cancellation," Patent number 5,890,106, March 1999.

[Bosi and Davidson 92]: M. Bosi and G. A. Davidson, "High-Quality, Low-Rate Audio Transform Coding for Transmission and Multimedia Applications", presented at the 93rd AES Convention, J. Audio Eng. Soc. (Abstracts), vol. 40, P. 1041, Preprint 3365, December 1992.

[Bosi et al. 97]: M. Bosi, K. Brandenburg, S. Quackenbush, K. Akagiri, H. Fuchs, J. Herre, L. Fielder, M. Dietz, Y. Oikawa, G. Davidson, "ISO/IEC MPEG-2 Advanced Audio Coding", JAES, 51, 780 - 792, October 1997.

[Brigham 74]: E. Oran Brigham, *The Fast Fourier Transform*, Prentice Hall Englewood Cliffs, 1974.

[Cooley and Tukey 65]: J. W. Cooley and J. W. Tukey, "An Algorithm for Machine Calculation of Complex Fourier Series", Math. Computation, vol. 19, pp. 297-301, April 1965.

[Duhamel 86]: P. Duhamel, "Implementation of split radix FFT for complex, real and real-symmetrical data", IEEE Transactions on Acoustics, Speech, and Signal Processing, vol. ASSP-34, pp. 285 – 295 , April 1986.

[Duhamel, Mahieux and Petit 91]: P. Duhamel, Y. Mahieux, and J. P. Petit, " A fast algorithm for the implementation of filter banks based on time domain aliasing cancellation", IEEE Transactions on Acoustics Speech, and Signal Processing, vol. ASSP-34, pp. 285 – 295, April 1986.

[Edler 89]: B. Edler, "Coding of Audio Signals with Overlapping Transform and Adaptive Window Shape" (in German), Frequenz, Vol. 43, No. 9, pp. 252-256, September 1989.

[Fielder et al. 96]: L. D. Fielder, M. Bosi, G. A. Davidson, M. Davis, C. Todd, and S. Vernon" AC-2 and AC-3: Low Complexity Transform-Based Audio Coding," in N. Gielchrist and C. Grewin (ed.), *Collected Papers on Digital Audio Bit-Rate Reduction*, pp. 54-72, AES 1996.

[Harris 78]: F. J. Harris, "On the Use of Windows for Harmonic Analysis with the Discrete Fourier Transform", Proc. of the IEEE,Vol. 66, no. 1, pp. 51-84, January 1978.

[ISO/IEC 11172-3]: ISO/IEC 11172, Information Technology, "Coding of moving pictures and associated audio for digital storage media at up to about 1.5 Mbit/s, Part 3: Audio", 1993.

[ISO/IEC 13818-7]: ISO/IEC 13818-7, Information Technology, "Generic coding of moving pictures and associated audio, Part 7: Advanced Audio Coding", 1997.

[Malvar 90]: H. S. Malvar, "Lapped Transforms for efficient transform/sub-band coding," IEEE Transactions on Acoustics, Speech, and Signal Processing, vol. ASSP-38, pp. 969 – 978, June 1990.

[Malvar 92]: H. S. Malvar, *Signal Processing with Lapped Transforms*, Artech House, Norwood, MA, 1992.

[Oppenheim and Schafer 75]: A. V. Oppenheim and R. W. Schafer, *Digital Signal Processing*, Prentice Hall, Englewood Cliffs, 1975.

[Portnoff 80]: M. R. Portnoff, "Time-Frequency Representation of Digital Signals and Systems Based on Short-Time Fourier Analysis", IEEE Transactions on Acoustics, Speech, and Signal Processing, vol. ASSP-28, no. 1, pp. 55 – 69, February 1980.

[Princen and Bradley 86]: J. P. Princen and A. B. Bradley, "Analysis/Synthesis Filter Bank Design Based on Time Domain Aliasing Cancellation," IEEE

Transactions on Acoustics, Speech, and Signal Processing, vol. ASSP-34, no. 5, pp. 1153 – 1161, October 1986.

[Princen and Johnston 95]: J. Princen, J. D. Johnston, "Audio Coding with Signal Adaptive Filter banks," IEEE Proc. of ICASSP, pp. 3071 – 3074, 1995.

[Princen, Johnson and Bradley 87]: J. P. Princen, A. Johnson and A. B. Bradley, "Subband/Transform Coding Using Filter Bank Designs Based on Time Domain Aliasing Cancellation", Proc. of the ICASSP, pp. 2161-2164, 1987.

[Sinha, Johnston, Dorward and Quackenbush 98]: D. Sinha, J. D. Johnston, S. Dorward and S. R. Quackenbush, "The perceptual Audio Coder (PAC)", in *The Digital Signal Processing Handbook*, V. Madisetti and D. Williams (ed.), CRC Press, pp. 42.1-42.18, 1998.

[Sugiyama, Hazu and Iwadare 90]: A. Sugiyama, F. Hazu, M. Iwadare, "Adaptive Transform Coding with an Adaptive Block Size (ATCABS)", Proc. of the ICASSP, Albuquerque, pp. 1093 – 1096, 1990.

[Vetterli and Kovačević 95]: M. Vetterli and J. Kovačević, *Wavelets and Subband Coding*, Prentice Hall, Englewood Cliffs, 1995.

7.   EXERCISES

a) MDCT:

In this exercise you will implement a time-to-frequency mapping using the MDCT. You will verify that the mapping leads to perfect reconstruction and that the fast implementation is significantly faster than straightforward implementation. Your fast MDCT/IMDCT implementation will be useful for later exercises.

1.   Program functions to carry out the MDCT and IMDCT using the transform definitions.

2.   Use your MDCT/IMDCT functions to implement a 50% overlap analysis/synthesis system. Allow for arbitrary block sizes N and do your windowing using the sine window.

3.   Verify that your system leads to perfect reconstruction by testing it using N = 2048 length transforms and the following test signals:
$x[n] = \cos(2\pi n/44.1)$ [1 kHz tone sampled at 44.1 kHz]
$x[n] = \theta(n)$ [step function]

4.   Program new functions to carry out the MDCT and IMDCT using the FFT-based fast implementation. To do so, you will need a routine for implementing the FFT/IFFT. Source code for such routines is readily

available (e.g., see the Numerical Recipes book), but you will need to check that the conventions for sign (-j in the forward transform) and normalization factor (1/N for inverse transform) are consistent with our usage. Verify that your new routines are correct by using them in your analysis/synthesis system with the above test signals.

5.   Compare the execution time of your analysis/synthesis system when using the fast implementation versus using the straightforward implementation.

b) A Frequency Domain Audio Coder:

In this exercise you will convert the audio coder you developed in Chapter 2 into a frequency domain coder using the MDCT as a time-to-frequency mapping.

1.   Write an audio coder that reads in 16 bit PCM audio files, transforms sine-windowed blocks of N time samples into N/2 frequency components, quantizes those frequency components, packs and writes the quantized frequency components into coded files, reads your coded files, dequantizes and inverse transforms blocks of N/2 frequency components into N time samples, overlaps and adds the time samples with 50% overlap, and writes the decoded signal into a 16 bit PCM audio file you can play. Verify that your coder is bug-free by making sure that files coded using 16-bit midtread uniform quantization do not sound degraded when decoded.

2. Test your codec on some sound samples using N=256 and N=2048 while using 1) 4-bit midtread uniform quantization, 2) 4-bit midrise uniform quantization, 3) 8-bit midtread uniform quantization, and 4) 3 scale bits, 5 mantissa bits midtread floating point quantization. What compression ratios do you get? Describe the quantization noise you hear. How does the quantization noise differ from what you heard at the same quantization in the Chapter 2 coder?

Chapter 6

Introduction to Psychoacoustics

1. INTRODUCTION

In the introduction to this book, we saw that the last stage in the coding chain is the human ear. A good understanding of how the human ear works can be a powerful tool in the design of audio codecs. The general idea is that quantization noise can be placed in areas of the signal spectrum where it least affects the fidelity of the signal, so that the data rate can be reduced without introducing audible distortion.

In this chapter, we examine the main aspects of psychoacoustics (the science that studies the statistical relationships between acoustical stimuli and hearing sensations) that are useful in the design of perceptual audio coders. The main goal of this chapter is to introduce the basic principles and data behind the masking models currently utilized in state-of-the-art audio coders. First, units for sound pressure level measurements and the range of human hearing are introduced. The hearing threshold and masking phenomena are discussed and their main empirical properties presented. We then examine the underlying mechanism of the hearing process and how the ear acts as a spectrum analyzer, analyzing sound in specific frequency units called critical bands. This will provide us with the foundation for developing psychoacoustic models, which link empirical masking data with the sound hearing sensation.

2.   SOUND PRESSURE LEVELS

As we saw in Chapter 1, sound can be represented as a function of time. Sound reaches the human ear in the form of a pressure wave. It can be represented as the variation of the air pressure in time, $p(t)$, where the pressure is defined as force per unit area. The unit of pressure in the MKS system is the Pascal (Pa) where 1 Pa = 1 N/m^2. Relevant values of sound pressure for audio applications vary between 10^{-5} Pa, which is close to the limits of human hearing at the most sensitive frequencies, and 10^2 Pa, which corresponds to the threshold of pain.

To describe such a wide range of relevant sound pressures, we usually choose to work in logarithmic units and define the sound pressure level, SPL, in units of dB as

$$SPL = 10 \log_{10} (p/p_o)^2$$

where p_o = 20 μ Pa is roughly equal to the sound pressure at the hearing threshold for tone frequencies around 2 kHz [Zwicker and Fastl 90].

We often also describe sounds in terms of the sound intensity. The sound intensity, I, is the power per unit area in the sound wave and it is proportional to p^2. The SPL (in units of dB) can also be calculated in terms of sound intensity as:

$$SPL = 10 \log_{10} (I/I_o)$$

The intensity I is measured in MKS units in terms of W/m^2 (1W = 1N m/s) and the reference sound intensity $I_o = 10^{-12}$ W/m^2 corresponds to a wave with the reference pressure p_o.

3.   LOUDNESS

The hearing sensation that corresponds to sound levels is the loudness of the sound. The concept of loudness was first introduced by Barkhausen in the 1920s as a means to describe perceived sound intensities. The loudness level is defined as the level of a 1 kHz sound tone that is perceived as loud as the sound under examination for frontally incident plane fields. In general, the loudness of an audio signal depends on its duration and its temporal and spectral structure in addition to its intensity. The loudness unit is the phon, where the phon describes a curve of equal loudness as a function of frequency. It is interesting to note that the difference between values for the loudness measured in phones and values for the intensity measures in dB decreases at high levels (see *Figure 1*). For example a 1 kHz tone at 100 dB is perceived almost as loud as a 100 Hz tone at 100 dB, while 1 kHz tone at 40 dB is perceived as about 20 dB louder than a 100 Hz tone at 40 dB [Fletcher and Munson 33]. It should be noted that, depending on how the equal loudness contours are measured, there might be differences in the data. Some of these differences can be accounted for by considering an attenuation factor necessary to produce equal loudness from frontally incident plane fields versus diffused sound fields [Zwicker and Fastl 90].

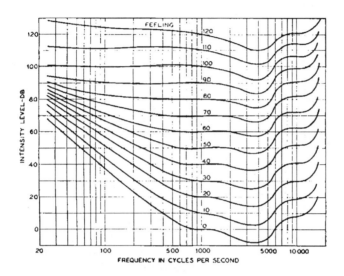

Figure 1. Loudness contours from [Fletcher and Munson 33].

4. HEARING RANGE

The human ear can cover a wide range of SPLs. *Figure 2* shows the hearing area of a typical human ear [Zwicker and Fastl 90]. The graph illustrates different SPL curves as function of frequency. The frequencies shown in the abscissa vary between 20 Hz and 20 kHz, which is generally considered the frequency range of audible sounds. It should be noted, however, that recent findings imply that particularly sensitive subjects can hear sounds at frequencies above 20 kHz.

The curve in the lower part of the graph represents the threshold in quiet, which is the level of audibility for pure tones in steady state conditions. The dotted line extending upwards from the threshold in quiet between 2 and 20 kHz represents the hearing loss commonly seen in subjects exposed to loud sounds in the mid-range frequency region. The threshold of pain is the dashed line at the top of the diagram. The area between the threshold in quiet and the threshold of pain represents the human hearing range.

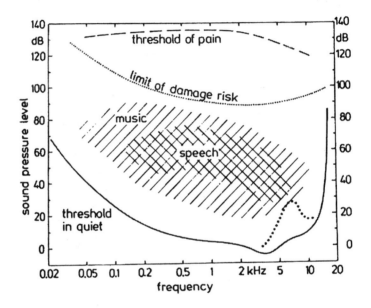

Figure 2. Hearing area from [Zwicker and Fastl 90]

Human speech typically falls into the frequency range comprised between 100 Hz and 8 kHz, and has SPLs ranging from about 30 dB up to around 70 dB, with typical conversation levels at values at about 50-60 dB. Music typically has a wider range in both frequency and SPLs than speech. For example, the A_0 note in the piano is tuned at 27.5 Hz (C_0 is tuned at about 16 Hz) while the highest note of the piccolo is at a frequency of about 8.4 kHz. Moreover, harmonics of musical instruments such as the violin and cymbals can reach frequencies above 15 kHz. The dynamic range for music typically varies between 20 dB and 95 dB. Around 100 dB is the onset of risk for hearing damage. At about 120 dB is the threshold of pain.

5. HEARING THRESHOLD

The hearing threshold, or threshold in quiet, represents the lowest sound level that can be heard at a given frequency. Even in extremely quiet conditions, the human ear cannot detect sounds at SPLs below the threshold in quiet. This curve is extremely important for audio coding since frequency components in a signal that fall below this level are irrelevant to our perception of sound and therefore they do not need to be transmitted. In addition, as long as the quantization noise in frequency components that are transmitted falls below this level, it will not be detectable by the human hearing process.

The threshold in quiet is also important in describing how loud we perceive sounds to be. In particular, the equal loudness contours display a shape that is nearly parallel to the threshold in quiet for low loudness levels (20 phones or below) suggesting that it is the difference between a sound and the threshold in quiet that determines the loudness for soft sounds. For loud sounds, the SPL itself plays a more important role in the determination of loudness. According to [Zwicker and Fast 90] the threshold in quiet corresponds to the equal loudness contour described by phon = 3.

The threshold in quiet can be measured by recording the sound pressure level of the lowest sound level that elicits a listener's response that the sound is audible. The frequency dependence can be tracked by giving the test subject a switch which changes between continuously incrementing and continuously decrementing the sound pressure level of a test tone whose frequency is slowly sweeping from low to high values and vice-versa. The test subject is instructed to switch to decrementing the sound pressure level when the sound is definitely audible and to switch to incrementing the pressure level when the sound is definitely inaudible. Typically, the results produce zigzag curves such as that in *Figure 3* from [Zwicker and Fastl 90] with a range of roughly 6 dB between the point where the sound is definitely audible and where it is definitely inaudible. The average of the two curves marking the top and bottom of the zigzags is used as the assessment of the threshold in quiet. According to [Zwicker and Fastl 90], the reproducibility of the threshold in quiet for a single subject is within ±3 dB. In addition, the frequency dependence of this curve has been recorded in a similar manner for many subjects with normal hearing.

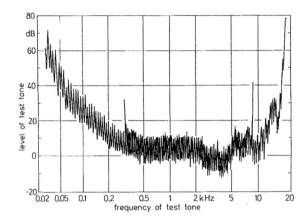

Figure 3. Sample results from an experimental assessment of the threshold in quiet from
[Zwicker and Fastl 90]

The frequency dependence of the hearing threshold has been fairly well established. The threshold is relatively high at low frequencies. It is at an SPL of around 40 dB at 50 Hz and almost drops to 0 dB by 500 Hz. It remains almost constant near 0 dB between 500 and 2 kHz. It can then drop below zero between 2 kHz and 5 kHz for listeners with good hearing. For frequency above 5 kHz, there are peaks and valleys that vary from subject to subject but the threshold is generally rising. Typically, the threshold then increases quite rapidly above 16 kHz. While for frequencies below 2 kHz the threshold seems to be largely independent of age, above 2 kHz it is shifted to a value almost 30 dB higher at 10 kHz for 60-year old subjects than for 20 year old subjects. *Figure 4* shows a comparison plot of the threshold in quiet for test subjects of various ages [Zwicker and Fastl 90].

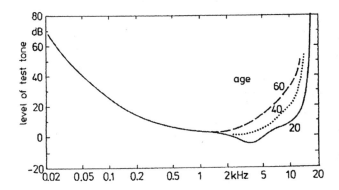

Figure 4. Threshold in quiet for listeners of different ages from [Zwicker and Fastl 90]

As shown in [Terhardt 79], one can obtain a good approximation of the threshold in quiet by utilizing the following frequency dependent function:

$$A(f)/dB = 3.64(f/kHz)^{-0.8} - 6.5e^{-0.6(f/kHz-3.3)^2} + 10^{-3}(f/kHz)^4$$

where the threshold in quiet is modeled by taking into consideration the transfer function of the outer and middle ear and the effect of the neural suppression of internal noise in the inner ear (see also Section 9 later in this chapter). A graph of the frequency dependence of this function can be seen in *Figure 5*. Notice how it reasonably mimics the behavior of the experimentally derived curves shown in the prior figures.

One should be aware that, in order to be able to compare a signal with the threshold in quiet, it is important to know the exact playback level of the audio signal. In general, the playback level is not known a priori in the design of a perceptual audio coder. A common assumption is to consider the playback level as such that the smallest possible signal represented in the audio coding system under design will be presented close to 0 dB. This is equivalent to aligning the fairly flat bottom of the threshold in quiet, corresponding to frequencies of roughly 500 Hz to 2 kHz, with the energy level represented by the least significant bit of the spectral signal amplitude in the system under design.

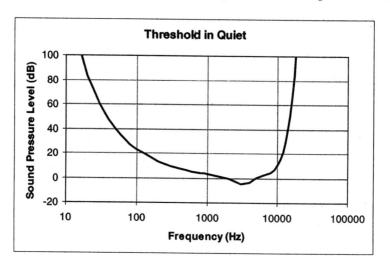

Figure 5. Approximate formula for the threshold in quiet

6.   THE MASKING PHENOMENON

Masking of soft sounds by louder ones is part of our everyday experience. For example, if we are engaged in a conversation while walking on the street, we typically cease conversation while a loud truck passes since we are not be able to hear speech over the truck noise. This can be seen as an example of masking: when the louder masking sound (the truck) occurs at the same time as the maskee sound (the conversation), it is no longer possible to hear the normally audible maskee. This phenomenon is called simultaneous or frequency masking. Another example of frequency masking occurs when in a performance one loud instrument (masker) masks a softer one (maskee) that is producing sounds close in frequency. In general simultaneous masking phenomena can be explained by the fact that a masker creates an excitation in the cochlea's basilar membrane (see also next sections) that prevents the detection of a weaker sound exciting the basilar membrane in the same area.

Masking can also take place when the masker and the maskee sounds are not presented simultaneously. It this case we refer to this phenomenon as temporal masking. For example, in speech a loud vowel preceding a plosive consonant tends to mask the consonant. Temporal masking is the dominant effect for sounds that present transients, while frequency masking is dominant in steady state conditions. For example, in coding sharp instrument attacks like those of castanets, glockenspiel, temporal masking plays a more important role than frequency masking.

6.1   Frequency Masking

Figure 6 illustrates frequency masking. In this figure, we see a loud signal masking two other signals at nearby frequencies. In addition to the curve showing the threshold in quiet, the figure shows a curve marked "masking threshold"[2] that represents the audibility threshold for signals in the presence of the masking signal. Other signals or frequency components that are below this curve will not be heard when the masker is present. In the example shown in *Figure 6*, the two other signals fall below the masking threshold, so they are not heard even though they are both well above the threshold in quiet. Just like with the threshold in quiet, we can exploit the masking thresholds in coding to identify signal components that do not need to be transmitted and to determine how much inaudible quantization noise is allowed for signal components that are transmitted.

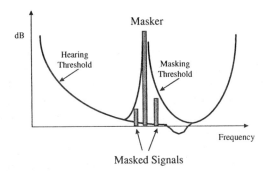

Figure 6. Example of frequency masking

6.2   Temporal Masking

In addition to simultaneous masking, masking phenomena can extend in time outside the period when the masker is present. Masking can occur prior to and after the presence of the masker. Accordingly, two types of temporal

---

[2] We shall refer to "masking thresholds" or "masking curves" to indicate the elevation of the hearing threshold due to the presence of one or more masker sounds. We define the "masked threshold" or "masked curve" as the combination of the hearing threshold and the masking threshold.

masking are generally encountered: pre-masking and post-masking. Pre-masking takes place before the onset of the masker; post-masking takes place after the masker is removed. Pre-masking is somewhat an unexpected phenomenon since it takes place before the masker is switched on. In general, temporal masking can be explained if we consider the fact that the auditory system requires a certain integration time to build the perception of sound and by the fact that louder sounds require longer integration intervals than softer ones.

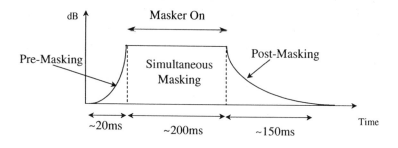

Figure 7. Example of temporal masking

In *Figure 7*, an example of temporal masking is shown [Zwicker and Fastl 90]. A 200 ms masker masks a short tone burst with very small duration relative to the masker. In the figure pre-masking lasts about 20 ms, but it is most effective only in the few milliseconds preceding the onset of the masker. There is no conclusive experimental data that link the duration of pre-masking effects with the duration of the masker. Although pre-masking is a less dramatic effect than post or simultaneous masking, it is nevertheless an important issue in the design of perceptual audio codecs since it is related to the audibility of "pre-noise" or "pre-echo" effects caused by encoding blocks of input samples. Pre-noise or pre-echo distortion occurs when the energy of the coded signal is spread in time prior to the onset of the attack. This effect is taken into consideration in the design of several perceptual audio coding systems both in terms of psychoacoustics models and analysis/synthesis signal adaptive filter design.

Figure 8 from [Bosi and Davidson 92] shows an example of a castanet signal (*Figure 8* (a)) in which encoding with a fixed block length led to a spread of energy in the 5 ms prior the onset of the transient (*Figure 8* (b)). This effect is perceived as a distortion sometimes described as a "double attack" and it is known in literature as pre-echo. Although some pre-masking effects can last on the order of tens of milliseconds, pre-masking is most effective only a few milliseconds. It should also be noted that pre-masking is less effective with trained subjects. In order to correct for pre-echo distortion, adaptive filter banks (see Chapter 5) are often adopted in perceptual audio coding. *Figure 8* (c) shows the reduction in pre-echo distortion that resulted from using an adaptive filter bank to adjust the block length in the presence of the transient signal.

Post-masking is a better understood phenomenon. It reflects the gradual decrease of the masking level after the masker is switched off. Post-masking is a stronger effect than pre-masking and has a much longer duration. In *Figure 7* post-masking lasts about 150 ms. Post-masking depends on the masker level, duration, and relative frequency of masker and probe.

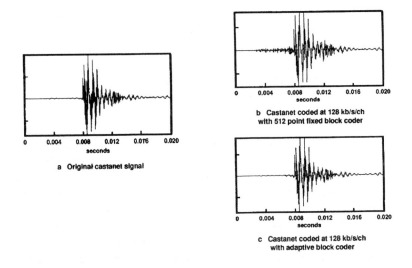

Figure 8. Example of pre-echo effects in a transient signal coded with a fixed (b) versus adaptive (c) resolution filter bank; the original signal is shown in (a). In (b) the amount of energy spread in time prior to the onset of the signal is perceived as pre-echo distortion and it is not temporally masked by the signal from [Bosi and Davidson 92]

An important question in the design of perceptual audio coders is how to account for masking effects. Masking curves are typically measured only for very simple maskers and maskees (either pure tones or narrow-band noise). In perceptual audio coding the assumption is that masking effects derived from simple maskers can be extended to a complex signal. Masking thresholds are computed by: a) identifying masking signals in the frequency domain representation of the data, b) developing frequency and temporal masking curves based on the characteristics of each identified masker, and c) combining the individual masking curves with each other and with the threshold in quiet to create an overall threshold representing audibility for the signal. This overall audibility threshold or masked threshold is then used to identify inaudible signal components and to determine the number of bits needed to quantize audible signal components.

7.   MEASURING MASKING CURVES

Masking curve data are collected by performing experiments on subjects that record what are the limits of audibility for a test signal (or probe) in the presence of a masking signal. The masking threshold varies dramatically depending on the nature and the characteristics of the masker and of the probe. Typically, for frequency masking measurements, the probe and the masker can be a sinusoidal tone or narrow band noise of extended duration. For temporal masking measurements, a short burst or sound impulse is used as a probe and the masker is of limited duration.

One way to measure a masking curve is to use a variant of the tracking method described for measuring the threshold in quiet. In this case, however, a masking signal will be played as the subject tries to identify the audibility limits for test signals. *Figure 9* shows an example of the masking curve that results from such an experiment. In this example, the masking signal is a pure tone a 1 kHz with an SPL of 60 dB. The lower zigzag line is the threshold in quiet for this test subject measured in the absence of the masking signal. The upper zigzag line is the audibility threshold when the masking signal is playing. Notice how masking in this case is strongest at frequencies near the masker's frequency and how it drops off quickly as the test signal moves away from the masker frequency in either direction – these features tend to be quite general results. Notice also that the highest masking level is roughly 15 dB below the masker level and that the drop-off rate is much quicker moving to low frequencies than moving to higher ones – these features tend to be very dependent on the specifics of both the masking signal and the test signal. In the following sub-sections, we summarize some of the main features of frequency masking curves as determined by similar experiments on test subjects.

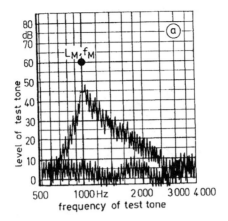

Figure 9. Sample results from experimental determination of a masking curve from [Zwicker and Fastl 90]

7.1   Narrow-Band Noise Masking Tones

In the case of narrow-band noise masking tones, the masker is noise with a bandwidth equal to or smaller than a critical band (see the definition of critical bandwidth in the next sections). *Figure 10* shows measured masking thresholds for tones masked by narrow-band noise centered at 250 Hz, 1 kHz, and 4 kHz [Zwicker and Fastl 90]. The noise bandwidths are 100 Hz, 160 Hz, and 700 Hz respectively. The slopes of the noise above and below the center frequency are very steep, dropping more than 200 dB per octave. The level of the masker is 60 dB, computed based on the noise intensity density and bandwidth. The horizontal dashed line shows the noise level in the figure. The solid lines in the figure show the levels of the pure tone probe in order to be just audible. The dashed curve at the bottom represents the threshold in quiet.

The masking threshold curves present different characteristics depending on the frequency of the masker. While the frequency dependence of the threshold masked by the 1 kHz and the 4 kHz narrow-band noise are similar, the 250 Hz threshold appear to be much broader. In general, masking thresholds are broader for low frequency maskers (when graphed, as is customary, using a logarithmic frequency scale). The masking thresholds reach a maximum near the masker center frequency. Their slopes can be very steep ascending from low frequencies (over 100 dB per octave), and present a somewhat gentler decrease after reaching the maximum. This steep rise creates the need for very good frequency resolution in the analysis of the audio signals, otherwise errors will be made in the evaluation of masking effects.

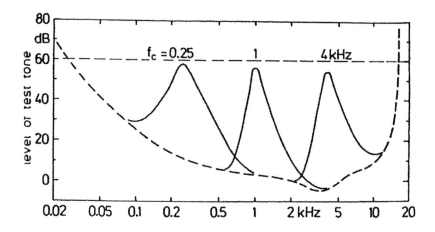

Figure 10. Masking thresholds for 60 dB narrow-band noise masking tones from [Zwicker and Fastl 90]

The difference in level between a signal component and the masking threshold at a certain frequency is sometimes referred to as the signal to mask ratio, SMR. Higher SMR levels indicate less masking. The minimum SMR between an masker and the masking curve it generates is a very important parameter in the design of audio coders. The minimum SMR values for a given masker tend to increase as the masker frequency increases. For example, in *Figure 10* we have a minimum SMR value of 2 dB for a noise masker with 250 Hz center frequency, 3 dB for the 1 kHz masker, and 5 dB for the 4 kHz masker.

In *Figure 11*, the masking threshold for narrow-band noise centered at 1 kHz is shown for different masker SPLs. The minimum SMR stays constant at around 3 dB for all levels. At frequencies lower than the masker, each of the measured masking curves has a very steep slope that seems to be independent of the masker SPL. In contrast, the slope in the masking curve towards higher frequencies shows noticeable sensitivity to the level of the masking signal. Notice that the slope appears to get shallower as the masking level is increased. In general, the frequency dependence of the masking curves is level sensitive, i.e. non-linear. The dips in *Figure 11* are caused by non-linear effects in the hearing system driven by the high level of the noise masker and the probe.

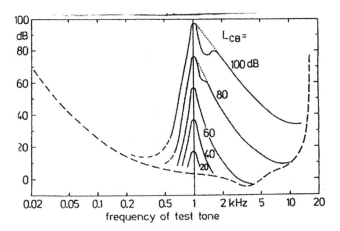

Figure 11. Masking thresholds for a 1 kHz narrow-band noise masker at different levels masking tones from [Zwicker and Fastl 90]

7.2 Tones Masking Tones

Although much of the early work on masking phenomena was based on measurements of pure tones masking pure tones, such masking experiments present greater difficulties than noise masking experiments due to the phenomenon of beating. In such experiments, the subjects also sometimes perceive additional tones besides the masker and probe. The most dominant effect, the beating effect, is localized in the neighborhood of the masker frequency and it depends on the masker level. *Figure 12* shows the results for a 1 kHz masker at different levels. In this particular experiment [Zwicker and Fastl 90] the probe was set 90 degrees out of phase with the masker when it reached the frequency of 1 kHz (equal to the masker frequency) to avoid beating in that area. It is interesting to notice that at low masking levels, there is a greater spreading of the masking curves towards lower frequencies than higher frequencies. The situation is reversed at high masking levels, where there is a greater spreading towards high frequencies than lower frequencies.

In general, the minimum masker SMRs are larger in experiments on tones masking tones than in experiments of noise masking tones. For example, we can see that the 90 dB masking curve in *Figure 12* peaks at roughly 75 dB implying a minimum SMR of roughly 15 dB. These types of results have been reproduced many times and the implication seems to be that noise is a better masker than tones. This phenomenon is referred to in the literature as the "asymmetry of masking" [Hellmann 72 and Hall 97].

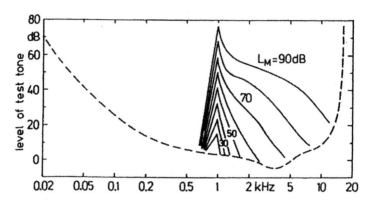

Figure 12. Masking thresholds for a 1 kHz tone masker at different levels masking tones from [Zwicker and Fastl 90]

7.3 Narrow-Band Noise or Tones Masking Narrow-Band Noise

Masking models exploited in perceptual audio coding rely upon the assumption that quantization noise can be masked by the signal. Often the codecs' quantization noise is spectrally complex rather than tonal. In this context, therefore, a suitable masking model might be better derived from experimental data collected in the case of narrow-band noise probes masked by narrow-band noise or tonal maskers. Unfortunately, there is very little data in the literature that address this issue. In the case of narrow-band noise probes masked by narrow-band noise maskers, phase relationships between the masker and the probe largely affect the results. According to [Hall 98] and based on [Miller 47], measurements for wide-band noise lead to minimum SMRs of about 26 dB.

In the case of tones masking narrow-band noise, early work by Zwicker and later others [Schroeder, Atal and Hall 79 and Fielder 87], suggest that the minimum SMR levels are between 20 and 30 dB. In general, it appears that when the masker is tonal, the minimum SMR levels are higher than when the masker is noise-like.

8. CRITICAL BANDWIDTHS

In measuring frequency masking curves, it was discovered that there is a narrow frequency range around the masker frequency where the masking threshold is flat rather than dropping off. For example, *Figure 13* shows the masking threshold for narrow-band noise at 2 kHz centered between two tonal maskers at 50 dB SPL as a function of the frequency separation of the two maskers. Notice how the masking threshold is flat at about 33 dB until the maskers are about 150 Hz away from the test tone (i.e. about 300 Hz away from each other) at which point it drops-off rapidly.

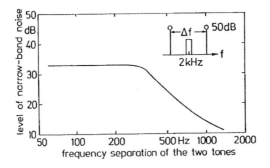

Figure 13. Threshold of a narrow-band noise centered between two sinusoidal maskers at a level of 50 dB as a function of the frequency separation between the two sinusoids from [Zwicker and Fastl 90]

Figure 14 shows analogous results for the case where the maskers are narrow-band noise and the test signal is tonal. Notice how the masking threshold is again flat until the maskers are about 150 Hz away from the maskee. Notice also that the level of masking at low frequency separations from these noise maskers is at roughly 46 dB (versus only roughly 33 dB when tonal maskers are employed), consistently with our earlier findings that noise-like maskers provide greater masking than tonal maskers. The main point, however, is that there is a so-called "critical bandwidth" around a masker that exhibits a constant level of masking regardless of the type of masker. The concept of critical bandwidth was first introduced by Harvey Fletcher in 1940 [Fletcher 40]. Fletcher's measurements and assumption led him to model the auditory system as an array of band-pass filters with continuously overlapping pass-bands of bandwidths equal to critical bandwidths. Experiments have shown that the critical bandwidth depends on the frequency of the masker. However, the exact form of the relationship between critical bandwidth and masker frequency is somewhat subject to controversy since differing results have been obtained using different types of measurements.

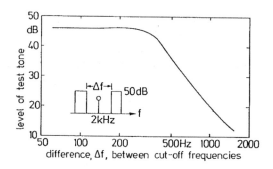

Figure 14. Threshold of a sinusoid centered between two narrow-band noise maskers at a level of 50 dB as a function of the frequency separation between the cut-off frequencies of the noise maskers from [Zwicker and Fastl 90]

Since the early work by Fletcher, different methods for measuring critical bandwidths have been developed and the resulting empirical data seem to differ substantially for frequencies below 500 Hz. In the pioneering work of Fletcher and later work by Zwicker [Zwicker 61], the critical bandwidth was estimated to be constant at about 100 Hz up to masker frequencies of 500 Hz, and to be roughly equal to 1/5 of the frequency of the masker for higher frequencies. An analytical expression that smoothly describes the variation of critical bandwidth Δf as a function of the masker center frequency f_c is given by [Zwicker and Fastl 90]:

$$\Delta f / Hz = 25 + 75\left[1 + 1.4\left(f_c / kHz\right)^2\right]^{0.69}$$

This formula for critical bandwidths is widely accepted as the standard description of them.

8.1   Equivalent Rectangular Bandwidth

A number of articles including Greenwood [Greenwood 61], Scharf [Scharf 70], Patterson [Patterson 76], Moore and Glasberg [Moore and Glasberg 83] disagree in their estimation of the critical bandwidths with that of the standard formula, especially below 500 Hz. In particular, Moore and Glasberg measure a quantity they define called the "equivalent rectangular bandwidth", ERB, which should be equivalent to the critical bandwidth. Their experiments were designed to provide an estimate of the auditory filter shapes by detecting the threshold of a sinusoidal signal masked by notched noise as a function of the width of the notch. The ERB as defined by Moore and Glasberg is about 11% greater than the -3 dB bandwidth of the auditory filter under consideration. The ERB, as a function of the center frequency f_c of the noise masker, is well fit by the function [Moore 96]:

$$ERB/Hz = 24.7 \ (4.37 \ f_c/kHz + 1)$$

The ERB function seems to provide values closer to the critical bandwidth measurements of Greenwood [Greenwood 61] than of Flecther or Zwicker at low frequencies. *Figure 15* compares the standard critical bandwidth formula with Moore's ERB formula and with other experimental measurements of critical bandwidth. Notice that the critical bandwidths predicted by the ERB formula are much narrower at frequencies below 500 Hz than implied by the standard critical bandwidth formula. Since the critical bandwidth represents the width of high-level masking from a signal, narrower critical bandwidth estimates put stronger requirements on a coder's frequency resolution.

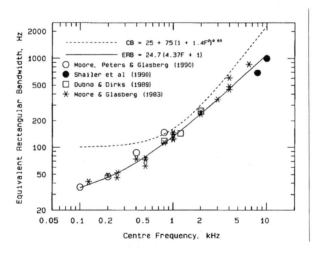

Figure 15. Critical bandwidth function and the ERB function plotted versus different experimental data for critical bandwidth from [Moore 96]

In summary, we have found that we can measure frequency masking curves for various masking and test signals. In all cases, we find that the masking curve levels are highest at frequencies near the masker frequency and drop off rapidly as the test signal frequency moves more than a critical bandwidth away from the masker frequency. We have seen that the shape of the masking curves depend on the frequency of the masker and its level. We have also seen that the masking curves depend strongly on whether or not the masker is tonal or noise-like, where we have seen that much greater masking is created by noise-like maskers. We now turn to describe how hearing works to help us interpret the empirical data we have just seen and create models that link them together.

9.   HOW HEARING WORKS

A schematic diagram of the human ear is shown in *Figure 16*. The outer, middle, and inner ear regions are shown. The main role of the outer ear is to collect sound and funnel it down the ear canal to the middle ear via the eardrum. The middle ear translates the pressure wave impinging on the eardrum into fluid motions in the inner ear's cochlea. The cochlea then translates its fluid motions into electrical signals entering the auditory nerve.

We can distinguish two distinct regions in the auditory system where audio stimuli are processed:
1. The peripheral region where the stimuli are pre-processed but retain their original character
2. The sensory cells which create the auditory sensation by using neural processing.

The peripheral region consists of the proximity zone of the listener where reflections and shadowing take place through the outer ear and ear canal to the middle ear. The sensory processing takes place in the inner ear.

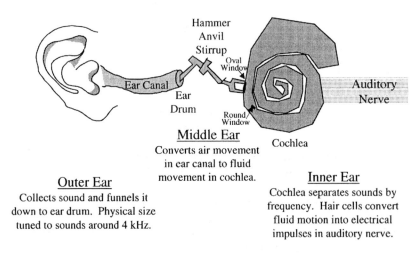

Figure 16. Outer, middle, and inner ear diagram.

9.1 Outer Ear

A sound field is normally approximated by a plane wave as it approaches the listener. The presence of the head and shoulders then distorts this sound field prior to entering the ear. They cause shadowing and reflections in the wave at frequencies above roughly 1500 Hz. This frequency corresponds to a wavelength of about 22 cm, which is considered a typical head diameter. The outer ear and ear canal also influence the sound pressure level at the eardrum. The outer ear's main function is to collect and channel the sound down to the eardrum but some filtering effects take place that can serve as an aid for sound localization. The ear canal acts like an open pipe of length roughly equal to 2 cm, which has a primary resonant mode at 4 kHz (see *Figure 17*). One can argue that the ear canal is "tuned" to frequency near its resonant mode. This observation is confirmed by the measurements of the threshold in quiet, which shows a minimum, i.e. maximum sensitivity, in that frequency region.

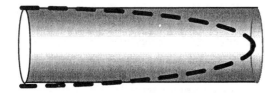

Figure 17. Outer ear model as an open pipe of length of about 2 cm

9.2 Middle Ear

The middle ear converts air movement in the ear canal into fluid movement in the cochlea. The hammer, anvil, and stirrup combination acts as lever and fulcrum to convert large, low-force displacements of air particles against the eardrum into small, high-force fluid motions in the cochlea. To avoid loss of energy transmission due to impedance mismatch between air and fluid, the middle ear mechanically matches impedances through the relative areas of the eardrum and stirrup footplate, and with the leverage ratio between the hammer and anvil arm. This mechanical transformer provides its best match in the impedances of air and cochlear fluid at frequencies of about 1 kHz. The stirrup footplate and a ring-shaped membrane at the base of the stirrup called the oval window provide the means by which the sound waves are transmitted into the inner ear. The frequency response of the filtering caused by the outer and middle ear can be described by the following function [Thiede et al. 00]:

$$A'(f)/dB = 0.6 * 3.64(f/kHz)^{-0.8} - 6.5e^{-0.6(f/kHz-3.3)^2} + 10^{-3}(f/kHz)$$

9.3   Inner Ear

The main organ in the inner ear is the cochlea. The cochlea is a long, thin tube wrapped around itself two and a half times into a spiral shape. Inside the cochlea there are three fluid-filled channels called "scalae" (see *Figure 18* for a cross sectional view): the scala vestibuli, the scala media, and the scala tympani. The scala vestibuli is in direct contact with the middle ear through the oval window. The scala media is separated from the scala vestibuli by a very thin membrane called the Reissner membrane. The scala tympani is separated from the scala media by the basilar membrane. From the functional point of view, we can view the scala media and the scala vestibuli as a single hydro-mechanical medium. The important functional effects involve the fluid motions across the basilar membrane. The basilar membrane is about 32 mm long and is relatively wide near the oval window while it becomes only one third as wide at the apex of the cochlea where the scala tympani is in direct fluid contact with the scala vestibuli through the helicotrema. The basilar membrane supports the organ of Corti (see *Figure 18*), which contains the sensory cells that transform fluid motions into electrical impulses for the auditory nerve.

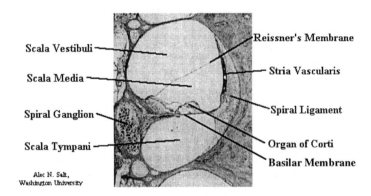

Figure 18. Cross section of the cochlea showing the scalae and organ of Corti. (Courtesy of Professor Alec N. Salt of Washington Unversity. Used with Permission.)

Figure 19 shows a functional diagram of the (unwrapped) cochlea [Pierce 83]. Fluid is displaced in the scala media/scala vestibuli by movements in the oval window driven by the middle ear. This fluid displacement is equalized by movement of the basilar membrane or, for low frequencies, by fluid flow into the scala tympani through the helicotrema. Finally, the scala tympani fluid flow is equalized by offsetting movements of the round window, which is localized at the base of the scala tympani. The delay between the presentation of the signal at the oval window and the response of the basilar membrane increases with distance from the oval window. Such delay varies between less than 1 ms for high frequencies to above 5 ms for low frequencies.

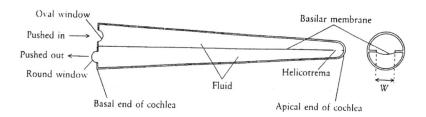

Figure 19. Functional diagram of the cochlea from [Pierce 83]

Georg von Békésy [von Békésy 60] experimentally studied fluid motions in the inner ear and proved a truly remarkable result previously proposed by von Helmholtz: the cochlea acts as a spectral analyzer. Sounds of a particular frequency lead to basilar membrane displacements with a small amplitude displacement at the oval window, increasing to peak displacements at a frequency-dependent point on the basilar membrane, and then dying out quickly in the direction of the helicotrema. *Figure 20* shows the displacement envelope that results from the motion of the basilar membrane in response to a 200 Hz frequency tone.

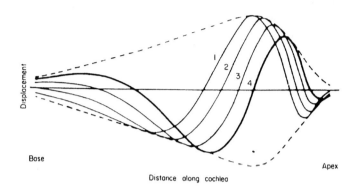

Figure 20. Traveling wave amplitude of the basilar membrane displacement relative to a 200 Hz frequency tone; the solid lines indicate the pattern at different instants in time; the dotted line indicates the displacement envelope from [von Békésy 60]

The experiments by von Békésy showed that low frequency signals induce oscillations that reach maximum displacement at the apex of the basilar membrane near the helicotrema while high frequency signals induce oscillations that reach maximum displacement at the base of the basilar membrane near the oval window. *Figure 21* shows the relative displacement envelopes of the basilar membrane for several different frequencies (50, 200, 800, 1600 Hz tones). *Figure 22* shows the locations of the displacement envelope peaks for differing frequencies along the basilar membrane from [Fletcher 40]. In this sense, it is often said that the cochlea performs a transformation that maps sound wave frequencies onto specific basilar membrane locations or a "frequency-space" transformation. The spectral mapping behavior of the cochlea is the basis for our understanding of the frequency dependence of critical bandwidths, which are believed to represent equal distances along the basilar membrane.

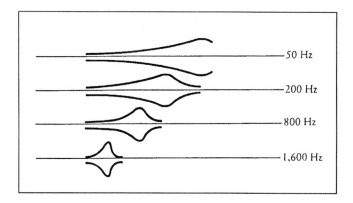

Figure 21. Plots of the relative amplitude of the basilar membrane response as a function of the basilar membrane location for different frequency tones; the left side of the plot is in proximity of the oval window, the right side of the plot is in proximity of the helicotrema from [Pierce 83]

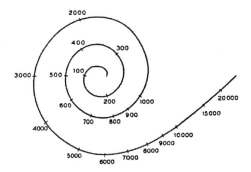

Figure 22. Frequency sensitivity along the basilar membrane from [Fletcher 40]. Copyright 1940 by the American Physical Society

On the basilar membrane, the organ of Corti transforms the mechanical oscillations of the basilar membrane into electrical signals that can be processed by the nervous system. The organ of Corti contains specialized cells called "hair cells" that translate fluid motions into firing of nerve cells in the auditory nerve. In the organ of Corti two types of sensory cells are contained: the inner and outer hair cells. Each hair cell contains a hair-like bundle of cilia that move when the basilar membrane oscillates. When the cilia move, ions are released into the hair cell. This release leads to neurotransmitters being sent to the attached auditory nerve cells. These nerve cells then send electrical impulses to the brain, which lead to the hearing sensation. The inner ear is connected to the brain by more than

30,000 auditory nerve fibers. The characteristic frequency of a fiber is determined by the part of the basilar membrane where it innervates a hair cell. Since the nerve fibers tend to maintain their spatial relation with one another, this results in a systematic arrangement of frequency responses according to location in the basilar membrane in all centers of the brain.

At high intensity levels, the basilar movement is sufficient to stimulate multiple nerve fibers while much fewer nerve fibers are stimulated at lower intensity levels. It appears that our hearing process is able to handle a wide dynamic range via non-linear effects (i.e., dynamic compression) in the inner ear. Structural differences between the inner and the outer hair cells indicate different functions for the two types of sensory cells. The inner hair cells play the dominant role for high-level sounds (the outer hair cells being mostly saturated for these levels). The outer hair cells play the dominant role at low levels, heavily interacting with the inner hair cells. In this case, the outer hair cells act as a non-linear amplifier to the inner hair cells with an active feedback loop and symmetrical saturation curves, allowing for the perception of very soft sounds.

It should be noted that in the inner ear a certain level of neural suppression of internal noise takes place. The effects of this noise suppression can be modeled by the following filtering of the signal [Thiede et al. 00]:

$$\text{Internal Noise} / dB = 0.4 * 3.64(f / kHz)^{-0.8}$$

Summing this expression with that of the transfer function for the outer and middle ear, A'(f), one can derive the analytical expression A(f) that fits the experimental data for the threshold in quiet.

Finally, it is worth mentioning that at low frequencies, the nerve fibers respond according to the instantaneous phase of the motion of the basilar membrane while at frequencies above 3500 Hz there is no phase synchronization. Comparing intensity, phase, and latency in each ear, we are provided physical clues as to a sound source's location.

10.   SUMMARY

In this chapter we have learned that the human ear can only hear sound louder than a frequency dependent threshold. We have seen that we can hear very little below 20 Hz and above 20 kHz. We extensively discussed the phenomenon of masking. Masking is one of the most important psychoacoustics effects used in the design of perceptual audio coders since it identifies signal components that are irrelevant to human perception. Masking depends on the spectral composition of both the masker and maskee, on their temporal characteristics and intensity, and it can occur before and after the masking signal is present (temporal masking) and simultaneously with the masker. The experiments we have reviewed show that frequency masking is most pronounced at the frequency of the masker with rapid drop off as the frequency departs from there and that the ear has a frequency dependent limit to its frequency resolution in that masking is flat within a "critical band" of a masker. We discussed how the auditory system can be described as a set of overlapping band-pass filters with bandwidths equal to critical bandwidths. Examining how the hearing process works, we found that air oscillations at the eardrum are converted into oscillations of the basilar membrane, where different parts of the basilar membrane are excited depending on the frequency content of the signal, and then into auditory sensation sent to the brain. In the next chapter, we will show how to put these observations to use in audio coding.

11. REFERENCES

[Bosi and Davidson 92]: M. Bosi and G. A. Davidson, "High-Quality, Low-Rate Audio Transform Coding for Transmission and Multimedia Applications", Presented at the 93rd AES Convention, J. Audio Eng. Soc. (Abstracts), vol. 40, P. 1041, Preprint 3365, December 1992.

[Fielder 87]: Louis D. Fielder, "Evaluation of the Audible Distortion and Noise Produced by Digital Audio Converters", J. Audio Eng. Soc., Vol. 35, no. 7/8, pp. 517-535, July/August 1987.

[Fletcher 40]: H. Fletcher, "Auditory Patterns", Rev. Mod. Phys., Vol. 12, pp.47-55, January 1940.

[Fletcher and Munson 33]: H. Fletcher and W. A. Munson, "Loudness, Its Definition, Measurement and Calculation ", J. Acoust. Soc. Am., Vol. 5, pp. 82-108, October 1933.

[Greenwood 61]: D. Greenwood, "Critical Bandwidth and the Frequency Coordinates of the Basilar Membrane", J. Acoust. Soc. Am., Vol. 33 no. 10, pp. 1344-1356, October 1961.

[Hall 97]: J. L. Hall, "Asymmetry of Masking Revisited: Generalization of Masker and Probe Bandwidth", J. Acoust. Soc. Am., Vol. 101 no. 2, pp. 1023-1033, February 1997.

[Hall 98]: J. L. Hall, "Auditory Psychophysics for Coding Applications", in *The Digital Signal Processing Handbook*, V. Madisetti and D. Williams, CRC Press, pp. 39.1-39.25, 1998.

[Hellman 72]: R. Hellman, "Asymmetry of Masking Between Noise and Tone", Percep. Psychphys., Vol. 11, pp. 241-246, 1972.

[Miller 47]: G. A. Miller, "Sensitivity to Changes in the Intensity of White Noise and its Relation to Masking and Loudness", J. Acoust. Soc. Am., Vol. 19 no. 4, pp. 609-619, July 1947.

[Moore 96]: B. C. J. Moore, "Masking in the Human Auditory System", in N. Gilchrist and C. Gerwin (ed.), *Collected Papers on Digital Audio Bit-Rate Reduction*, pp. 9-19, AES 1996.

[Moore and Glasberg 83]: B. C. J. Moore and B. R. Glasberg, "Suggested Formulae for Calculating Auditory-Filter Bandwidths and Excitation Patterns ", J. Acoust. Soc. Am., Vol. 74 no. 3, pp. 750-753, September 1983.

[Patterson 76]: R. D. Patterson, "Auditory Filter Shapes Derived with Noise Stimuli", J. Acoust. Soc. Am., Vol. 59 no. 3, pp. 640-650, March 1976.

[Pierce 83]: J. Pierce, *The Science of Musical Sound*, W. H. Freeman, 1983.

[Scharf 70]: B. Scharf, "Critical Bands", in *Foundation of Modern Auditory Theory*, New York Academic, 1970.

[Terhardt 79]: E. Terhardt, "Calculating Virtual Pitch", Hearing Res., Vol. 1, pp. 155-182, 1979.

[Thiede et al. 00]: T. Thiede, W. Treurniet, R. Bitto, C. Schmidmer, T. Sporer, J. Beerends, C. Colomes, M. Keyhl, G. Stoll, K. Brandenburg and B. Feiten, "PEAQ-The ITU Standard for Objective Measurement of Perceived Audio Quality", J. Audio Eng. Soc., Vol. 48, no. 1/2, pp. 3-29, January/February 2000.

[von Békésy 60]: G. von Békésy, *Experiments in Hearing*, McGraw-Hill, 1960.

[Zwicker 61]: E. Zwicker, "Subdivision of the Audible Frequency Range into Critical Bands (Frequenzgruppen)," J. Acoust. Soc. of Am., Vol. 33, p. 248, February 1961.

[Zwicker and Fastl 90]: E. Zwicker and H. Fastl, *Psychoacoustics: Facts and Models*, Springer-Verlag, Berlin Heidelberg 1990.

12. EXERCISES

Masking Curves Framework:
In this exercise you will develop the framework for computing the masking curve for a test signal. We will return to this test signal in the next chapters to complete the masking curve calculation and utilize these results to guide the bit allocation for this signal.

1. Use an FFT to map a 1 kHz sine wave with amplitude equal to 1.0 into the frequency domain. Use a sample rate of 48 kHz and a block length of N = 2048. Do your windowing using a sine window. How wide is the peak? What is the sum of the spectral density $|X[k]|^2$ over the peak? Try dividing this sum by $N^2/8$, how does the result relate to the amplitude of the input sine wave? (Check that you're right by changing the amplitude to ½ and summing over the peak again.) If we define this signal as having an SPL of 96 dB, how can you estimate the SPL of other peaks you see in a test signal analyzed with the same FFT?

2. Use the same FFT to analyze the following signal:

$$x[n] = A_0 \cos(2\pi 440n / F_s) + A_1 \cos(2\pi 554n / F_s)$$
$$+ A_2 \cos(2\pi 660n / F_s) + A_3 \cos(2\pi 880n / F_s)$$
$$+ A_4 \cos(2\pi 4400n / F_s) + A_5 \cos(2\pi 8800n / F_s)$$

where $A_0 = 0.6$, $A_1 = 0.55$, $A_2 = 0.55$, $A_3 = 0.15$, $A_4 = 0.1$, $A_5 = 0.05$, and F_S is the sample rate of 48 kHz. Using the FFT results, identify the peaks in the signal and estimate their SPLs and frequencies. How do these results compare with what you know the answer to be based on the signal definition?

3. Apply the threshold in quiet to this spectrum. Create a graph comparing the test signal's frequency spectrum (measured in dB) with the threshold in quiet.

Chapter 7

Psychoacoustic Models for Audio Coding

1. INTRODUCTION

In the prior chapter we learned about the limits to human hearing. We learned about the threshold in quiet or hearing threshold below which sounds are inaudible. The hearing threshold is very important to coder design because it represents frequency-dependent levels below which quantization noise levels will be inaudible. The implication in the coded representation of the signal is that certain frequency components can be quantized with a relatively small number of bits without introducing audible distortion.

We learned about the phenomenon of masking where loud sounds can cause other normally audible sounds to become inaudible. Frequency masking effects temporarily raise the hearing threshold in certain areas of the spectrum near the masker, allowing for larger levels of quantization noise localized in these portions of the spectrum to be inaudible.

Finally, we learned that the ear acts as a spectrum analyzer mapping frequencies into critical bandwidths, which correspond to physical locations along the basilar membrane. This suggests that some frequency dependant aspects of human hearing may be more naturally represented in terms of physical distance along the basilar membrane rather than in terms of frequency.

In this chapter we present a heuristic model of simultaneous masking based on our limited ability to distinguish small changes in the basilar membrane excitation. Such a model is characterized by the "shape" of a sound excitation pattern, defined as the activity or excitation produced by that sound in the basilar membrane, and by the minimum amount of detectable change in this excitation pattern. These parameters correspond to the shape of the masking curves relative to a sound masker and the minimum SMR we discussed in Chapter 6. Moreover, this model suggests that masking curves are represented more naturally in terms of distances along the basilar membrane rather than in terms of frequency. We define a critical-band rate known as the Bark scale to map frequency values onto values in the Bark scale and then represent masking curves on that scale. We then introduce the main masking curves shapes or "spreading functions" commonly used in audio coding and discuss how they are used to create an overall masking threshold to guide bit allocation in an audio coder.

2. EXCITATION PATTERNS AND MASKING MODELS

In this section we consider a heuristic model to explain frequency masking. Consider a signal that creates a certain excitation pattern in the basilar membrane. Since our sound intensity detection mostly operates on a logarithmic scale of sensation, we will assume that:

1) We "feel" the excitation pattern in dB units and
2) We cannot detect changes in the pattern that are smaller than a certain threshold value, ΔL_{min}, measured in dB.

We define the mapping z(f) from frequency to space to identify the location z along the basilar membrane that has the largest excitation from a signal of frequency f. The change in dB of the excitation pattern at basilar membrane location z resulting from the addition of a second, uncorrelated test signal will be equal to:

$$\Delta L(z) = 10 \log_{10}\left(A(z)^2 + B(z)^2\right) - 10 \log_{10} A(z)^2 = 10 \log_{10}\left(\frac{A(z)^2 + B(z)^2}{A(z)^2}\right)$$

$$\approx \frac{10}{\ln(10)} \frac{B(z)^2}{A(z)^2}$$

where A(z), B(z) are excitation amplitudes at location z of the original signal and the test signal, respectively.

A test tone will become unmasked when the peak of its excitation pattern causes ΔL to exceed the threshold value ΔL_{min}. We would expect the peak of a signal's excitation pattern to be proportional to the signal intensity, so that at the z corresponding to the peak excitation of the test signal we should have that

$$\frac{B(z(f))^2}{A(z(f))^2} = \frac{I_B}{I_A F(z(f))}$$

where I_A, I_B are the intensities of the original signal A and the test signal B, respectively, $F(z)$ is a function describing the shape of the original signal's excitation pattern along the basilar membrane, and $z(f)$ represents the location along the basilar membrane of the peak excitation from a signal at frequency f. The function $F(z)$ is normalized to have a peak value equal to 1 at the z corresponding to the peak of the original signal's excitation pattern. At the point where the test signal just becomes unmasked we have that

$$\Delta L_{min} = \frac{10}{\ln(10)} \frac{I_B}{I_A F(z(f))}$$

or equivalently that

$$I_B = I_A \left(\frac{\ln(10)}{10} \Delta L_{min} \right) F(z(f))$$

In units of SPL this can be written as

$$SPL_B = SPL_A + 10 \log_{10} \left(\frac{\ln(10)}{10} \Delta L_{min} \right) + 10 \log_{10} \left(F(z(f)) \right)$$

where the fact the $F(z)$ is normalized to peak at one implies that the last term will have a peak value of zero. Test signals at levels below SPL_B will be masked by the masker A. In other words, the above equation shows that the masking curve relative to the masker A can be derived at each frequency location from the SPL of the masker A by:

a) Down-shifting it by a constant that depends on ΔL_{min} evaluated for the masker A and

b) Adding a frequency dependent function that describes the spreading of the masker's excitation energy along the basilar membrane.

The down-shift described by the second term of the equation represents the minimum SMR of the masker. We saw in the last chapter that it depends both on the characteristics of the masker, namely whether it is noise-like or tone-like, and its frequency. The last term in the equation is usually referred to as the masker "spreading function" and it is determined based on experimental masking curves.

We now turn to characterizing the mapping from frequency f onto basilar membrane distance z and see how the representation of masking curves is greatly simplified when shown in terms of this scale rather than frequency.

Then we present models commonly used to describe the spreading function and minimum SMR in creating a masking curve from a single masking component. Finally we address the issue of how to combine the masking curves from multiple maskers.

3. THE BARK SCALE

The critical bandwidth formula introduced in the last chapter gives us a method for mapping frequency onto a linear distance measure along the basilar membrane. Assuming that each critical bandwidth corresponds to a fixed distance along the basilar membrane, we can define the unit of length in our basilar distance measure z(f) to be one critical bandwidth. This unit is known as the "Bark" in honor of Barkhausen, an early researcher in the field.

The critical bandwidth formula represents the quantity df/dz at each frequency point f, which just tells us that it represents the change in frequency per unit length along the basilar membrane. We can invert and integrate this formula as a function of f to create a distance mapping z(f). We call this mapping function z(f) the "critical band rate". We can approximate the critical band rate z(f) using the following expression [Zwicker and Fastl 90]:

$$z / \text{Bark} = 13 \arctan(0.76 \, f \, / 1 \, \text{kHz}) + 3.5 \arctan((f \, / 7.5 \, \text{kHz})^2)$$

Table 1 shows the frequency ranges corresponding to each unit of basilar distance up to an upper frequency of 15,500 Hz, which is near the upper limit of human hearing. The frequency range corresponding to each unit of basilar distance is called a "critical band" and the Bark scale measure z represents the critical band number as a function of critical band lower frequency f_l. If we assume that the basilar membrane is about 25 critical bands long, then clinical measurements showing that the membrane is actually about 32 mm long imply that each critical band represents roughly 1.3 mm in basilar membrane distance.

Table 1. Critical bands and corresponding lower frequency f_l, upper frequency f_u, center frequency f_c and critical bandwidth, Δf from [Zwicker and Fastl 90]

z (Bark)	f_l (Hz)	f_u (Hz)	f_c (Hz)	Δf (Hz)	z (Bark)	f_l (Hz)	f_u (Hz)	f_c (Hz)	Δf (Hz)
0	0	100	50	100	13	2000	2320	2150	320
1	100	200	150	100	14	2320	2700	2500	380
2	200	300	250	100	15	2700	3150	2900	450
3	300	400	350	100	16	3150	3700	3400	550
4	400	510	450	110	17	3700	4400	4000	700

z (Bark)	f_l (Hz)	f_u (Hz)	f_c (Hz)	Δf (Hz)	z (Bark)	f_l (Hz)	f_u (Hz)	f_c (Hz)	Δf (Hz)
5	510	630	570	120	18	4400	5300	4800	900
6	630	770	700	140	19	5300	6400	5800	1100
7	770	920	840	150	20	6400	7700	7000	1300
8	920	1080	1000	160	21	7700	9500	8500	1800
9	1080	1270	1170	190	22	9500	12000	10500	2500
10	1270	1480	1370	210	23	12000	15500	13500	3500
11	1480	1720	1600	240	24	15500			
12	1720	2000	1850	280					

4.   MODELS FOR THE SPREADING OF MASKING

Given the transformation between frequency and the Bark scale, we can see how masking looks when transformed to frequency units that are linearly related to basilar membrane distances. Not surprisingly, the masking curve shapes are much simpler to describe when shown in the Bark scale. For example, *Figure 1* shows the excitation patterns that arise from narrow-band noise maskers at various frequencies. Excitation patterns are derived from experimental masking curves by shifting them up to the SPL of the masker and then graphing them on the Bark scale. The slopes towards low frequencies are fairly independent of the masker center frequency at roughly 27 dB per bark. The upper slopes are steeper for frequencies below 200 Hz, but remain constant above that frequency. Compare the similarity of shape across all these curves with how different the curves looked in *Figure 10* of Chapter 6 using normal frequency units. The transformation to the Bark scale suggests that much of the shape change in masking curves with masker frequency is an artifact of our measurement units – if we define the frequency dependence of our masking curve in the Bark scale then the shape is fairly independent of masker frequency.

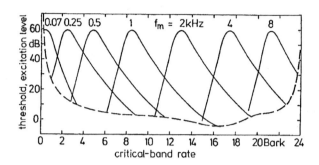

Figure 1. Excitation patterns for narrow-band noise signals centered at different frequencies and at a level of 60 dB from [Zwicker and Fastl 90]

Although we can reasonably assume that the excitation pattern is independent of frequency when described in terms of the Bark scale, we cannot necessarily make a similar assumption for the level dependence. For example, *Figure 2* shows the excitation patterns from 1 kHz narrow-band noise at various masker levels. Notice how the shape changes from symmetric patterns at low levels to very asymmetric ones at higher levels. For levels below 40 dB the slopes are symmetrical dropping at about 27 dB per bark while at higher levels the slope towards higher frequencies ranges from about –5 dB per bark for a noise masker at 100 dB to –27 dB per bark for a noise masker at less than 40 dB.

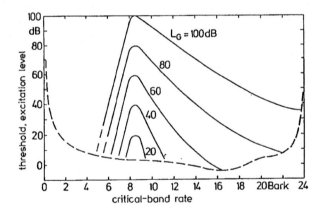

Figure 2. Excitation patterns for narrow-band noise signals centered at 1 kHz and at different levels from [Zwicker and Fastl 90]

As a first approximation, a representation of the spreading function that can be utilized to create excitation patterns is given by a triangular function. We can write this spreading function in terms of the Bark scale difference between the maskee and masker frequency dz = $z(f_{maskee}) - z(f_{masker})$ as follows:

$$10 \log_{10}\left(F(dz, L_M)\right) = \left(-27 + 0.37\, \text{MAX}\{L_M - 40, 0\}\, \theta(dz)\right) |dz|$$

where L_M is the masker's SPL and $\theta(dz)$ is the step function equal to zero for negative values of dz and equal to one for positive values of dz. Notice that dz assumes positive values when the masker is located at a lower frequency than the maskee and negative values when the masker is located at a higher frequency than the maskee. In *Figure 3*, this spreading function is shown for different levels of the masker L_M.

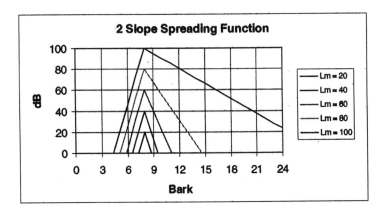

Figure 3. Spreading function described by the two slopes derived from narrow-band noise masking data for different levels of the masker

There are a number of other spreading functions found in the literature. For example Schroeder [Schroeder, Atal and Hall 79], suggested the use of the following analytical function for the spreading function:

$$10 \log_{10} F(dz) = 15.81 + 7.5\,(dz + 0.474) - 17.5\,(1 + (dz + 0.474)^2)^{1/2}$$

This spreading function was used in some of the earliest works on perceptual coding applied to speech signals. A similar spreading function was later adopted in ISO/IEC MPEG Psychoacoustic Model 2. *Figure 4* shows a plot of the Schroeder spreading function. It should be noted that this spreading function is independent of the masker level. Ignoring the dependence of the spreading function on the masker level allows for the computation of the overall masking curve as a simple convolution operation between F(z) and the signal intensity spectrum rather than a multiplication of (potentially) different spreading functions with the different masking components of the signal expressed in dB units and then an addition of the different components spread intensities. The advantage of the Schroeder approach is that the result of the convolution incorporates an intensity summation of all maskers' contributions, so that there is no need to perform an additional sum to obtain the final excitation pattern (see also next sections).

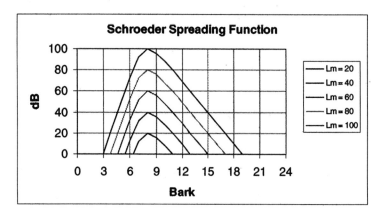

Figure 4. Schroeder spreading function

A modification of the Schroeder spreading function was later introduced that takes the masker level into consideration as follows:

$$10 \log_{10}\left(F(dz, L_M, f)\right) = (15.81 - I(L_M, f)) + 7.5(dz + 0.474)$$
$$- (17.5 - I(L_M, f))(1 + (dz + 0.474)^2)^{\frac{1}{2}}$$

where the level adjustment function $I(L_M, f)$ is defined as

$$I(L_M, f) = \min\{5 \ 10^{(Lm-96)/10} \ df/dz, 2\}.$$

which also has a slight frequency dependence due to the variation of critical bandwidth df/dz with frequency. As shown in *Figure 5*, the effect of the modification is to include a dependence of the spreading function on the masker level. In this case, and consistently with the experimental data, increasing the level of the masker translates in a decrease of the slopes of the spreading function, i.e. increased masking.

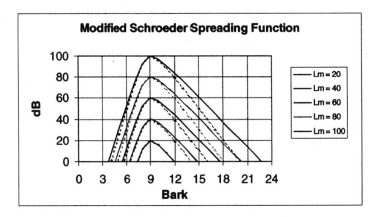

Figure 5. Modified Schroeder spreading function (solid lines) compared with the original level-independent Schroeder spreading function (dashed lines)

The ISO/IEC MPEG Psychoacoustic Model 2 spreading function (see *Figure 6*), which is derived from the Schroeder spreading function, is given by

$$10 \log_{10}\big(F(dz)\big) = 15.8111389 + 7.5*(1.05*dz + 0.474) - 17.5*$$
$$\sqrt{1.0 + (1.05*dz + 0.474)^2} + 8*\text{MIN}(0,(1.05*dz - 0.5)^2 - 2*(1.05*dz - 0.5))$$

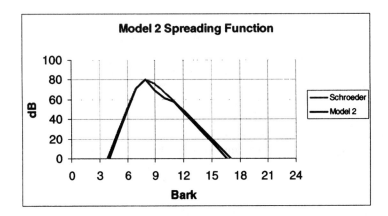

Figure 6. Basic MPEG Psychoacoustic Model 2 spreading function compared with the Schroeder spreading function for a masker with an SPL of 80 dB

Another example of a spreading function is given by the function adopted in ISO/IEC MPEG Psychoacoustic Model 1:

$$10\log_{10}(F(dz, L_M)) = \begin{cases} -17\,dz + 0.15\,L_M\,(dz-1)\,\theta(dz-1) & \text{for } dz \geq 0 \\ -(6+0.4\,L_M)|dz| - (11+0.4\,L_M)\,(|dz|-1)\,\theta(|dz|-1) & \text{for } dz < 0 \end{cases}$$

Figure 7 shows the ISO/IEC MPEG Psychoacoustic Model 1 spreading function. Notice how this spreading function starts symmetric at low levels while has a great deal more masking at higher frequencies than lower frequencies when the masker level is high. The two-piece linear spreading function for upper and lower frequencies in Model 1 is meant to mimic the masking data for tones masking tones (see also *Figure 12* in Chapter 6).

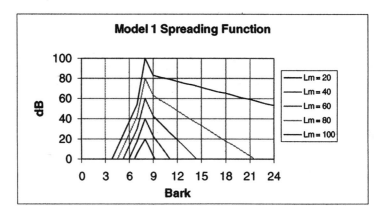

Figure 7. Two-piece linear spreading function for upper and lower frequencies adopted in ISO/IEC MPEG Psychoacoustic Model 1

In *Figure 8* a comparison of the three approaches described above for the spreading function relative to an 80 dB masker is shown. Model 1 spreading function allows for a larger amount of upward spreading of masking than the triangular function or the Schroeder function. From *Figure 8*, it is clear that, of the three spreading functions introduced, the triangular offers the most conservative approach to determine irrelevancies in the signal.

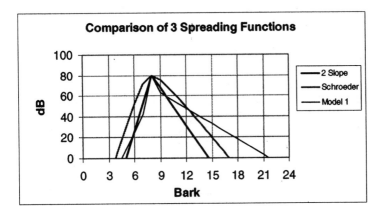

Figure 8. Comparison of three different approaches for the spreading function relative to an 80 dB masker

The spreading functions shown in *Figure 8* do not depend upon the masker center frequency. While this is correct as a first approximation, there appears to be some frequency dependence according to some experimental data. For example, the masking slope towards higher frequencies is less shallow for masker frequencies below 200 Hz than for frequencies above that value. A frequency dependence that reflects this behavior is built in the following triangular approximation to the spreading function introduced by [Terhardt 79]:

$$10 \log_{10}\left(F(dz, L_M)\right) = \left(-24 + \left(0.2\,L_M + 230\,Hz/f\right)\theta(dz)\right) dz$$

In *Figure 9* the Terhardt spreading function for different masker center frequencies is shown. Notice that the spreading functions are superimposed in the bark scale to facilitate the comparison. The Terhardt spreading functions is adopted in objective perceptual measurement models see for example [Thiede et al. 00] (see also Chapter 10).

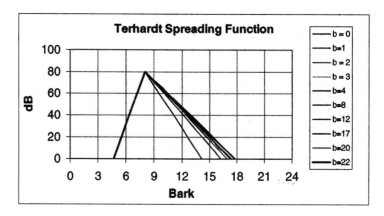

Figure 9. Terhardt spreading function for different masker center frequencies superimposed at 8 Bark

5.   MASKING CURVES

We have seen empirically that the peak of the masking curve is shifted down from the masker SPL by an amount that depends on the type of masker. Our understanding in terms of excitation patterns relates this downward shift to the minimum changes in excitation pattern ΔL_{min} that we can detect. By looking at experimental values for the difference between the masker level and the maximum threshold value, one can get a feel for how ΔL_{min} varies with different types of signals. For example, Zwicker suggested a value of 1 dB for ΔL_{min}. Adopting this value implies that the peak of the masking curve should be about 6 dB below the SPL level of the masker. This value reflects experimental data for narrow-band noise masking tones. Moore suggested a value of ΔL_{min} equal to 0.1 dB as more appropriate, corresponding to a difference of about 16 dB between the masker and the maskee levels. This value is close to experimental data for tone maskers.

A number of factors should be considered in modeling the offset between the peak of the masking curve and the masker SPL. First, from experimental data, our ability to detect changes in excitation level is reduced at low frequencies, i.e. the difference in peak level between masker and masking curve increases with increasing frequency. Second, given the asymmetry of masking, depending on whether the masker is noise-like or tone-like this difference is bigger or smaller, noise being a "better masker". Finally, when the masker is a complex sound, information can be combined from several parts of the excitation pattern to improve the detection of the maskee.

In *Figure 10* a simple masking curve derived from the triangular spreading function is shown. First the masker level L_M is evaluated. The masker level L_M is then convolved with the triangular spreading function to reflect the spreading of excitation energy along the basilar membrane. Finally a down-shift by Δ is applied to predict the masking threshold relative to the masker under examination.

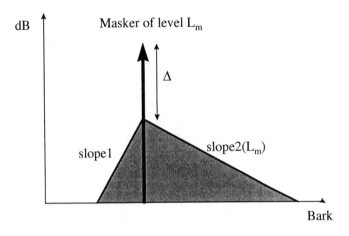

Figure 10. Example of a predicted masking threshold for the masker of level Lm

For example, in [Jayant, Johnston and Safranek 93] the difference Δ in the case of a tone-like signal masking a noise-like signal and in the case of a noise-like signal masking a tone-like signal are given respectively by:

$$\Delta_{\text{tone masking noise}} = 14.5 + z \ dB$$

$$\Delta_{\text{noise masking tone}} = C \ dB$$

where C varies between 3 and 6 dB depending upon the experimental data. Notice, however, that according to the data presented in *Figure 10* of Chapter 6 there appears to be some degree of frequency dependence of the Δ even in the case of noise masking tones. In general, depending on the nature of the masker, Δ assumes different values. It is a good exercise to test the design of the coder varying Δ to optimize the coder output.

Typically, audio sounds may contain several tone-like and noise-like components. Once the different components are identified at a certain time interval, the individual masking patterns are derived and the relative masking thresholds are computed by shifting masking patterns down by the appropriate amount Δ. We discuss next how these thresholds are combined to create a global masked threshold.

6.   "ADDITION" OF MASKING

In general, when a complex sound is presented to the ear, we bear the concurrent effects of more than one single masker. As a first approximation, we can identify the individual masking components of the signal on a critical band scale and create their respective masking curves as if they acted independently from each other. The issue then becomes how should we combine these masking curves together at particular frequency locations.

A natural way to expect these masking curves to combine is to assume that their intensities simply sum up. In this case, two masking curves of equal intensity would combine to give a combined effect 3 dB higher than either curve. Another plausible addition rule is to assume that the highest masking curve dominates at each frequency location. In this case, two equal masking curves would just lead to a combined effect equal to the maximum value of the two curves at each frequency location. Either of these cases can be described according to the summation formula

$$I_N = \left(\sum_{n=0}^{N-1} I_n^\alpha \right)^{\frac{1}{\alpha}} \qquad 1 \le \alpha \le \infty$$

where I_N represents the intensity of the masking curve that results from the combination of N individual masking curves with intensities I_n for n = 0, ..., N-1, and α is a parameter that defines the way the curves add. In this equation, setting $\alpha = 1$ corresponds to intensity addition while taking the limit as $\alpha \to \infty$ corresponds to using the highest masking curve. Setting α to values between 1 and ∞ gives results intermediate to these two cases. One could also choose to set α lower than 1, in which case the combined effect of two equal maskers is greater than the sum of their intensities, however one would need to be careful because this sum rule becomes ill-defined as α approaches zero. How should we set α in a coder? Of course, the way to decide is to turn to the experimental literature for the effects of maskers addition.

Some studies in the literature, see for example [Lufti 83], suggest that "addition" of masking for maskers of comparable intensities is best described using values of $\alpha \approx 0.33$, implying that two equal masking curves have a combined effect equal to a single masking curve with an intensity 8 times either curve. Such a result follows rules closer to the addition of specific loudnesses rather than the addition of intensities and it is quite surprising. Further review of the literature, shows numerous experimental results where an increased amount of masking with respect to intensity summation of the single maskers was detected, see for example [Green 67]. In these experiments, the combined effect produced 6 to 14 dB more masking than simple intensity addition of the two equal-level maskers would predict. This empirical result was observed for both narrow-band noise and sinusoidal maskers, independently of the maskers absolute level, and for cases of up to four maskers [Lufti 83]. It should be noted that there is not complete agreement in the literature on this subject but the fact that maskers over-add in at least some cases does seem to be a real empirical effect.

Figure 11 shows the implication of using Lufti's value of α ≈ 0.33 for adding two masking curves of various intensities. The curve shows how many dB higher the combined masking curve is above the greater of the two masking added curves for various relative intensities. Notice that two equal curves combine equivalently to a single curve about 9 dB higher (rather than 3 dB higher as expected from intensity addition). In fact, adding a second masker about 20 dB below a masker is roughly equivalent to doubling the intensity of the first masker. It should be noted however, that most of the experiments directly test the area around 0 dB difference in masking curves. The rest of the curve is based on extrapolating these results using α = 0.33 for adding maskers with higher-level differences.

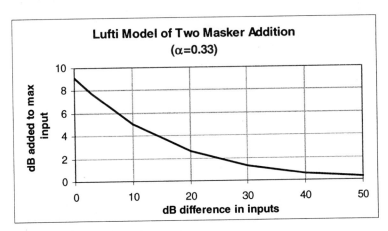

Figure 11. Lufti's model for the addition of two masking curves

Another issue to be aware of is the implication of adopting the summation formula for the addition of multiple (i.e., more than two) masking curves. Lufti [Lufti 83] cites experiments using 4 maskers that are better fit using the summation with a value of $\alpha = 0.33$ than by intensity addition. *Figure 12* shows how the results extend to arbitrary numbers of maskers. In *Figure 12* the summation formula for $\alpha \approx 0.33$ is compared with intensity addition for combining the effects of various numbers of equal intensity masking curves. Notice that 10 maskers combine to be equivalent to a masking curve 30 dB higher than each individual curve. Simpleminded use of this addition formula would suggest the nonsensical result that a large number of maskers could combine to mask themselves – an effect in which if you make enough noise and you can't hear it any more. The answer to this apparent paradox has to do with the integrative effects of critical bands (discussed below) wherein we learn that nearby frequencies are integrated in their impacts on masking by the ear. In other words, although you can subject the ear to an arbitrary number of tone sources, the ear will not deal with all of the tones as independent maskers when their separation in frequency is less than a critical bandwidth.

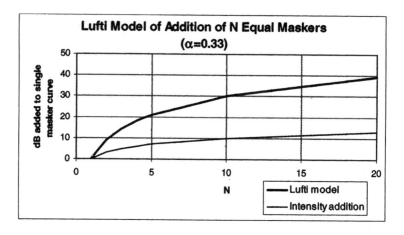

Figure 12. Implications of Lufti's model for the addition of large numbers of maskers

A number of psychoacoustics models in the literature, for example ISO/IEC MPEG Psychoacoustic Models 1 and 2, sum the intensities of the different masking components, i.e., a value for α equal to one is adopted. In some other cases, for example Dolby AC-2 and AC-2A (see also Chapter 14), the maximum value of the different masking components is retained, i.e. a value of $\alpha \to \infty$ is adopted. In [Baumgarte 95] the non-linear summation model is applied to ISO/IEC MPEG Psychoacoustics Model 2 with $\alpha = 0.33$.

A non-linear model for the addition of masking is also applied for sound quality measurement systems, see for example [Beerends and Stemerdink 92] and [Thiede et al. 00]. In [Thiede et al. 00] the value for α is set to 0.4.

It should be noted that the non-linear addition of masking models described above doesn't take into consideration cases where the maskers happen to be very close in frequency. In these instances, other phenomena like beating may cause masking to behave in a very different fashion and, in particular, to unmask some of the regions that are considered below their separate masking curves. In the latter case, current models may be erroneous in the description of the perceived signal.

Typically, audio sounds may contain several tone-like and noise like components. Once the thresholds are combined to create a global masked threshold, the threshold of hearing is also taken in consideration to derive the masked threshold for the signal during that time interval. Often in perceptual audio coding, the maximum value between the global masked threshold and the threshold of hearing is retained (see for example ISO/IEC MPEG Psychoacoustic Model 2 and AC-3, Chapters 11 and 14) as the masked threshold for the signal at that time interval. Portions of the signal below the masked threshold are considered irrelevant to the signal representation.

7. MODELING THE EFFECTS OF NON-SIMULTANEOUS (TEMPORAL) MASKING

In addition to simultaneous masking, perceptual models exploit also the effects of non-simultaneous or temporal masking. Modeling of frequency masking effects was described in detail in the previous sections. Modeling of temporal masking takes into consideration a time-sliding window. According to the experimental data described in Chapter 6, Section 5.2, a weighting function of time, which assigns a larger weight to events that occur near the center of the window as opposed to events that occur near its edges, is employed. It is in general assumed that this temporal smoothing occurs after the auditory filtering, i.e., it is applied to the signal spectrum, resulting in a smoothed version in time of the output signal. Examples of temporal windows can be found in [Plack and Moore 90, Moore 96]. Depending on the time resolution of the analysis filters used in the frequency representation of the signal, it may be possible to apply both backward and forward masking or forward masking only, where the time resolution needed to apply backward masking is very high, typically of the order of milliseconds. For example, in [Thiede et al. 00] a raised cosine FIR and a first order IIR low-pass filters are utilized to model backward and forward masking respectively. The time constant for the FIR filter was set at eight ms to mimic a backward masking of about two ms. The time constant τ for the IIR was set as follows:

$$\tau(f_c) = \tau_{min} + \frac{100\,Hz}{f_c}(\tau_{100} - \tau_{min})$$

where f_c is the center frequency of the auditory filter corresponding to the masker, $\tau_{min} = 4$ ms and $\tau_{100} = 50$ ms.

In addition to temporal smoothing, in perceptual audio coding, a measure of the temporal characteristics of the input signal is evaluated and utilized to adapt the analysis/synthesis properties of the overall system as described in Chapter 6, Section 5.2.

8.   PERCEPTUAL ENTROPY

Once the masked threshold is computed, the masked level values can be used to appropriately allocate quantization noise. It is assumed that the coding noise within a critical band will not be audible as long as the SNR resulting from an R-bit quantization of the signal in that critical band is higher than the SMR. Johnston [Johnston 88] introduced the concept of perceptual entropy to define the average minimum number of bits per frequency sample needed to code a signal without introducing any perceptual difference with respect to the original signal. Given a signal intensity, I, and the relative intensity of the masked threshold, I_T, at each frequency line f_i, the perceptual entropy (PE) for the signal at a determined time interval can be expressed as:

$$PE = \frac{1}{N}\sum_{i=0}^{N-1}\max\left\{0, \log_2\left(\sqrt{\frac{I(f_i)}{I_T(f_i)}}\right)\right\} \approx \frac{1}{N}\sum_{i=0}^{N-1}\log_2\left(1+\sqrt{\frac{I(f_i)}{I_T(f_i)}}\right)$$

where N is the number of the frequency lines in the signal representation. As shown in the above expression, the PE represents the logarithm of a geometric mean of the threshold-weighted energy across the frequency block. The perceptual entropy measure gives the lower bound estimate for the perceptual coding of audio signals based on the signal time-frequency analysis and the computed masked threshold.

9.   MASKED THRESHOLD AND ALLOCATION OF THE BIT POOL

In *Figure 13*, a simple example for different quantization SNR for a signal partially masked by a stronger one to its right is shown. The goal of bit allocation is to make sure that bits are allocated so that the SNR is greater than the SMR across the spectrum. The difference between the SMR and the SNR is referred to as the noise to mask ratio, NMR, and gives an indication of the rate of distortion with respect to the computed masked threshold (see *Figure 13*). Given a certain bit budget, when there are extra bits, they are allocated across the spectrum to create a coding margin. When there are not enough bits, bits are allocated to minimize the overall (positive) deviation between SMR and SNR or NMR.

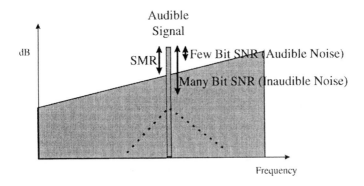

Figure 13. Example of different SNR values allocated to a signal component versus masked threshold

In *Figure 14* an example of a masked threshold and the associated bit allocation that introduces only "inaudible" quantization noise is shown for a mono signal with sinusoidal components at 440, 554, 660, 880, 4400, 8800 Hz sampled at 48 kHz. The signal analysis is performed by applying an FFT with a 2048-point sine window. The SPL curve shows the strength of the original signal. The spreading function adopted for the computation of the masking threshold (indicated by the line "mask" in *Figure 14*) is the two-slope function shown in *Figure 3*. The value selected for Δ is 15 dB. The masked threshold is computed as the maximum of the individual components masking thresholds and the threshold of hearing, corresponding to applying the summation rule with $\alpha=\infty$. The curve labeled "bits" shows the minimum number of bits allocated to the signal in order to keep the quantization noise below the masked threshold. In the next chapter, we describe how to optimally allocate bits given instead a fixed data rate constraint for generic audio signals.

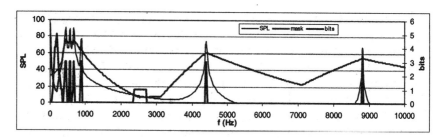

Figure 14. Example of masked threshold and bit allocation for a signal with sinusoidal components at 440, 554, 660, 880, 4400, 8800 Hz

10.   SUMMARY

In this chapter we have presented the main aspects of masking models used in the design of audio coders. While there is not complete agreement in literature on all aspects of auditory modeling, psychoacoustics tools represent a powerful aid in the coding of audio signals. By improving the masking models used by a coder, one can improve the quality of the codec at a given data rate. In this chapter we have seen that masking models are much simpler when defined using the Bark scale of frequency rather than in terms of frequency itself. We have discussed how to add up the effects of multiple maskers in a signal (although there is not complete agreement on this in the literature) to combine with the hearing threshold to get an overall masked threshold that can be used to guide quantization and bit allocation in an audio coder. In the next chapter we discuss how to use the masked thresholds developed here to optimally distribute the bit pool available.

11.   REFERENCES

[Baumgarte 95]: F. Baumgarte, C. Feredikis, H. Fuchs, "A Non-Linear Psychoacoustic Model Applied to the ISO MPEG Layer III Coder", Preprint 4087, October 1995.

[Beerends and Stemerdink 92]: J. G. Beerends and J. A. Stemerdink, "A Perceptual Audio Quality Measure Based on a Psychoacoustic Sound Representation", J. Audio Eng. Soc., Vol. 40, no. 12, pp. 963-978, December 1992.

[Green 67]: D. A. Green, "Additivity of Masking ", J. Acoust. Soc. Am., Vol. 41, no. 6, pp. 1517-1525, 1967.

[Jayant, Johnston and Safranek 93]: N. Jayant, J. Johnston and R. Safranek, "Signal Compression Based on Method of Human Perception", Proc. of IEEE, Vol. 81, no. 10, pp. 1385-1422, October 1993.

[Johnston 88]: J. D. Johnston, "Estimation of Perceptual Entropy Using Noise Masking Criteria", in Proc. of ICASSP, pp. 2524-2527, May 1988.

[Lufti 83]: R. A. Lufti, "Additivity of Simultaneous Masking ", J. Acoust. Soc. Am., Vol. 73 no. 1, pp. 162-267, January 1983.

[Lufti 85]: R. A. Lufti, "A Power Law Transformation Predicting Masking by Sounds with Complex Spectra", J. Acoust. Soc. Am., Vol. 77 no. 6, pp. 2128-2136, June 1985.

[Moore 96]: B. C. J. Moore, "Masking in the Human Auditory System", in *Collected Papers on Digital Audio Bit-Rate Reduction*, N. Gilchrist and C. Gerwin (ed.) pp. 9-19, AES 1996.

[Plack and Moore 90]: C. J. Plack and B. C. J. Moore, "Temporal Window Shape as a Function of Frequency and Level", J. Acoust. Soc. Am., Vol. 87 no. 5, pp. 2178-2187, May 1990.

[Schroeder, Atal and Hall 79]: M. R. Schroeder, B. S. Atal and J. L. Hall, "Optimizing Digital Speech Coders by Exploiting Masking Properties of the Human Ear", J. Acoust. Soc. Am., Vol. 66 no. 6, pp. 1647-1652, December 1979.

[Terhardt 79]: E. Terhardt, "Calculating Virtual Pitch", Hearing Res., Vol. 1, pp. 155-182, 1979.

[Thiede et al. 00]: T. Thiede, W. Treurniet, R. Bitto, C. Schmidmer, T. Sporer, J. Beerends, C. Colomes, M. Keyhl, G. Stoll, K. Brandenburg and B. Feiten, "PEAQ-The ITU Standard for Objective Measurement of Perceived Audio Quality", J. Audio Eng. Soc., Vol. 48, no. 1/2, pp. 3-29, January/February 2000.

[Zwicker and Fastl 90]: E. Zwicker and H. Fastl, "Psychoacoustics: Facts and Models", Springer-Verlag, Berlin Heidelberg 1990.

12.   EXERCISES

Masking Curves:
In this exercise you will develop masking curve for a test signal. We return to the test signal utilized in the exercise of Chapter 6:

$$x[n] = A_0 \cos(2\pi 440n / F_s) + A_1 \cos(2\pi 554n / F_s)$$
$$+ A_2 \cos(2\pi 660n / F_s) + A_3 \cos(2\pi 880n / F_s)$$
$$+ A_4 \cos(2\pi 4400n / F_s) + A_5 \cos(2\pi 8800n / F_s)$$

where $A_0 = 0.6$, $A_1 = 0.55$, $A_2 = 0.55$, $A_3 = 0.15$, $A_4 = 0.1$, $A_5 = 0.05$, and F_S is the sample rate of 48 kHz.

1. Using the FFT results and the identified peaks in the signal and their SPLs and frequencies, define a masking model and specify its parameters (spreading function and down-shift). Recall that masking models are simpler to define in the Bark scale.
2. Use the masking model to define masking curves for the different components of the test signal.
3. Create a masked threshold that combines the effects of all masking components and threshold in quiet and create a graph comparing the test signal's frequency spectrum (measured in dB) with the masked curve.
4. Are any of your signal peaks going to be masked? How about the rest of the frequency spectrum? What is the signal to mask ratio (SMR) of any of the unmasked signal peaks? Assuming 6 dB per bit, how many bits of resolution are needed for each unmasked signal peak for the resulting quantization noise to be inaudible, i.e., below the masked curve?

Chapter 8

Bit Allocation Strategies

1.   INTRODUCTION

The most common approach in perceptual coding of audio signals is to subdivide the input signal into frequency components and to encode each component separately. In Chapters 4 and 5 we discussed different time to frequency mapping techniques and how these techniques can represent the input signal in the frequency domain and allow for redundancy removal. Time domain based coding algorithms such as ADPCM can achieve similar results in terms of redundancy removal (see also Chapter 3). In this framework, typically the audio signal is treated as a single, wide-band signal and prediction and inverse filtering are adopted to describe it. In this context, the main difference between time-domain and frequency domain coding algorithms is the degree of redundancy removal and signal decorrelation.

One of the main advantages of frequency domain coding systems over time domain coding systems, however, is their ability to code each component separately with appropriate accuracy depending on its spectral strength. Bits can be allocated adaptively through the spectrum, where the bands that contain high-energy components are encoded with a large number of bits and bands that contain no components or components with very small energy may not be encoded at all. In this approach, quantization noise can be separately controlled in each band and the overall reconstruction noise spectrum is shaped in frequency. In addition, based on the power spectrum density of the signal, excitation patterns can be computed as described in Chapter 7 for each component from empirical masking data. By appropriately allocating bits through the spectrum and taking into consideration the masking patterns generated by each component, quantization noise can be shaped to be inaudible. We saw in the previous chapter how we can allocate bits based on the SMR of the signal under exam. The aim is to obtain a coded signal that is perceptually identical to the original signal. In this case, the quality of the coded signal is kept constant while distributing the bit pool through the signal spectrum. Maintaining a constant quality implies that the overall data rate of the system may vary. There are applications where a fixed data rate may be required. In this case, one would like to maximize the quality at the system specific data rate. The issue of maximizing the quality or equivalently minimizing the block distortion at a given rate and the resulting bit allocation strategies are the main topics we discuss in this chapter.

First, we introduce transmission data rates and an algorithm for optimal bit allocation that satisfies the data rate constraint. We compare the results of optimal bit allocation with uniform allocation. We define a measure of the potential gain introduced by the optimal bit allocation, the spectral flatness measure, which depends on the characteristics of the signal and is linked to the resolution of the signal representation, (see also [Jayant and Noll 84]). Finally, we apply psychoacoustics principles to the optimal bit allocation approach to spectrally confine the quantization noise in regions where it can't be detected. A new perceptual measure of the potential gain is introduced, the perceptual flatness measure, (see also [Bosi 99]). This approach based on perceptual models discussed in Chapter 57 allows for not only the removal of redundancies, but also the extraction of irrelevancies from the signal representation at a given data rate.

2. CODING DATA RATES

In the introduction to this book, we described how the main goal in coding audio signals is to maximize the perceived quality of the encoded sound while minimizing the data rate necessary to reproduce it. The coder data rate is probably one of the most important parameters in the design of the overall system. It is related to the overall system bandwidth and/or storage capacity. The operational data rate of a coder, I, depends on the rate F_s at which the time domain input is sampled the average number of bits per sample R, and the number of audio channels n. It is typically measured in bits per seconds and equals

$$I = n F_s R$$

For example, the CD format has a sampling rate of 44.1 kHz and uses 16 bits per sample for stereo sound. The CD data rate, therefore, is equal to I = 2*44,100*16 = 1.411 Mb/s. Based on these numbers, we can easily see how one hour of music encoded in the CD format requires 635 MB of storage.

In perceptual audio coders, the signal is typically represented in the frequency domain. Let's assume that we have K frequency sub-bands and that each frequency sub-band k is encoded with R_k bits where k = 0,...,K-1. The overall system data rate is given by the sum over all sub-bands of the rates needed to encode each individual sub-band. Assuming each sub-band output is sampled at a rate F_{sk}, we have

$$I = n \sum_{k=0}^{K-1} F_{sk} R_k$$

If we are critically sampled, for example by down-sampling by a factor of K when we divide the signal into the K sub-bands, then each sub-band has a sampling rate equal to

$$F_{sk} = F_s/K$$

The resulting system data rate I is then equal to

$$I = n\ F_s <R_k>$$

where $<R_k>$ represents the average number of bits used to encode a frequency sample:

$$\left\langle R_k \right\rangle = \frac{1}{K} \sum_{k=0}^{K-1} R_k$$

Typical data rates for high quality, state-of-the-art perceptual audio coders vary between 64 kb/s per channel up to 128 kb/s per channel. Let's consider, for example, a monophonic signal sampled with a sampling rate of 48 kHz, and mapped to the frequency domain using an MDCT with window length of N = 1024 and a system bit rate of 128 kb/s per channel. In this case, the average number of bits per frequency sample is given by

$$<R_k> = 128/48 = 2.666667 \text{ bits}$$

Given the data rate constraint of 128 kb/s per channel, the number of bits available for each new block of data is

$$\sum_{k=0}^{N/2-1} R_k = \langle R_k \rangle * 512 = 1365 \quad \text{bits/block per channel}$$

Given the constraint of the data rate, the aim is to allocate the available bit pool based on the spectral strength of the audio signal and its masking properties. In a perceptual audio coder one can maximize the quality versus the data rate by appropriately assigning the bit pool available through the signal spectrum. In the next sections of this chapter, we discuss methods for bit allocation to achieve these goals.

3.   A SIMPLE ALLOCATION OF THE BIT POOL

Once we know the number of bits available for each block of frequency samples, we need to decide how to allocate them. We start with a very simple scheme and then discuss ways to improve it. We first consider bit allocation to be binary (either a frequency component gets bits or it doesn't) and then move on to the more complicated case of variable number of bits per frequency sample.

In Chapter 3 we saw how by switching to the frequency domain representation of audio signals one can reduce redundancy for tonal signals; let's think through how we can carry out this redundancy reduction in a simple coder by appropriately allocating the bit pool available. Our time-frequency transformation is handing the bit allocation routine blocks of input data parsed into frequency sub-bands. If the signal is highly tonal we would expect most of the signal content to be located in only a few of the frequency sub-bands. We would like to exploit this fact to reduce the number of bits needed to quantize and pass along the signal.

We can reduce redundancy in our coder by only allocating bits to the sub-bands that contain useful data and not bothering to pass any bits for other sub-bands. Suppose we had 32 sub-bands and this block of data only had signal in five of the sub-bands – we would only need to pass five of the 32 sub-band samples on to the decoder. However, we would also need to tell the decoder which five sub-bands were the ones for which we are passing data or it wouldn't know what to do with the five samples! In other words, we need to also pass data telling the decoder how we allocated the bits. One way to do this is to allocate a single bit to each sub-band to tell the decoder whether or not a sample is being passed for that sub-band.

The next issue we face is how to decide which sub-bands contain useful data. For instance, we can set up a threshold and only pass data for sub-bands whose signal amplitude exceeds this threshold. We could throw away sub-band samples whose amplitude level is below the quantization noise level determined by the number of bits we are using for each coded sample. An issue with a fixed threshold, however, is that the number of sub-band samples that exceed the threshold and hence the bitrate of the coder differs from block to block. Although a variable data rate might be acceptable for some operations, it can be a problem for transmission or decode-on-playback applications. In general, given a predetermined bitrate, we could set the threshold to match the bitrate. For example, the sub-band amplitudes can be sorted from highest to lowest values and bits be distributed to the highest amplitude sub-bands until the bit pool for the block is exhausted. Note that in any such calculation, the bits needed to tell the decoder which sub-bands have data need to be taken out of the available bit pool first.

Removing redundancy in this manner decreases the system data rate only if the reduction in the data sample representation is higher than the bits allocated for the side information. For example, if our 32 sub-band coder utilizes 16 bits per sample then it would require 32 bits of side information per block to define the bit allocation and we would need to be able to expect to throw away more than 2 sub-bands of data per block on average to be better off encoding in this manner. This is a common theme in bit allocation routines – the more control there is on the encoder side of the bit allocation the more it costs in side information to be sent to the decoder.

4. OPTIMAL BIT ALLOCATION

Having described a simple binary bit-allocation scheme, let's discuss whether or not we can make it even better off by allocating variable numbers of bits to the sub-band samples for which we pass data. The question that we should ask ourselves is: "Can we increase coding gain by redistributing bits throughout the spectrum of the signal?" In general, there is a potential increase in coding gain if the signal spectrum is colored, i.e., certain spectral components are stronger than others. In this case, an increase in coding gain can be achieved by appropriately redistributing the bit pool throughout the spectrum. Given that the statistics of audio signals describe them as quasi-stationary, the assumption that audio signal spectra are colored is in general justified.

To better understand this issue, we should remind ourselves how much quantization error we can expect to have from a given number of bits. For the time being we ignore psychoacoustic masking effects and use the total block quantization error as our measure of signal distortion. Varying the number of bits can achieve coding gain relative to keeping the bits fixed if we can find a set of R_k, i.e. an appropriate bit allocation that reduces the average block squared error

$$\left\langle q^2 \right\rangle_{block} = \frac{1}{K} \sum_{k=0}^{K-1} < \varepsilon_k^2 >$$

relative to a fixed allocation of R bits to each sample, where ε_k is the quantization error for spectral sample k and $<\varepsilon_k^2>$ is the expected power of this quantization error.

Let's first look at the case of uniform quantization. From Chapter 2, we recall that the expected error power for a sample that is uniformly quantized with R bits is roughly equal to

$$< \varepsilon^2 >= \frac{1}{3*2^{2R}}$$

where our amplitude units have been chosen so that the maximum non-overload input x_{max} equals one. Unfortunately, the fact that 2^{-2R} is convex means that we cannot increase the system gain by shifting some bits from one sample to another since:

$$\frac{1}{2^{2(R+\delta)}} + \frac{1}{2^{2(R-\delta)}} \geq 2\frac{1}{2^{2R}}$$

for any shift δ in bits between samples. The net result for uniform quantization is that we reduce distortion by using the same number of bits for each spectral sample that we pass through the coder. Basically, we minimize the block error by keeping a constant error level across all samples for which we pass any data at all.

We can now look at the case of floating point quantization. In floating point quantization, the effect of the scale factor is to scale the quantizer maximum non-overload factor x_{max} to the order of the signal so that the expected error power in terms of the number of mantissa bits R_k is now roughly equal to:

$$< \varepsilon_k^2 >= \frac{x_k^2}{3*2^{2R_k}}$$

The average block squared error now becomes:

$$\left\langle q^2 \right\rangle_{block} = \frac{1}{K} \sum_{k=0}^{K-1} < \varepsilon_k^2 > = \frac{1}{K} \sum_{k=0}^{K-1} \frac{x_k^2}{3 * 2^{2R_k}}$$

where each term is now weighted by the signal power of the sub-band.

Again, we can increase the coding gain with dynamic bit allocation if we can find a set of R_k that decreases the average block squared error. In order to simplify this computation, one should remember that:

$$x_k^2 = 2^{\log_2(x_k^2)}$$

so we can rewrite the average block squared error as:

$$\left\langle q^2 \right\rangle_{block} = \frac{1}{K} \sum_{k=0}^{K-1} \frac{1}{3 * 2^{2R_k - \log_2(x_k^2)}}$$

We saw in the uniform quantization case that this is minimized when the exponent in the denominator is equal for all terms. This implies that we should allocate our mantissa bits R_k so that:

$$2R_k - \log_2(x_k^2) = C$$

or equivalently:

$$R_k = \frac{1}{2}\left(C + \log_2(x_k^2)\right)$$

for some constant C. The constant C is set based on the number of bits available to allocate to the mantissas in the block.

The above equation implies that we need to allocate more bits where the signal has higher amplitude. The reason for this is that the quantizer's x_{max} is large for such samples and so we need more mantissa bits to get down to the same error power as that from lower amplitude samples.

If we knew how many spectral samples were being passed and we didn't have to worry about capping the number of bits passed to any sample, we could relate C to the size of the bit pool and the signal spectrum. Suppose K_p of the K spectral samples are being passed to the decoder, the others being allocated zero mantissa bits. Suppose also that the bit pool for mantissas, i.e. total bit pool for the data block minus the bits needed for scale factors and for bit allocation information, is equal to P. If we averaged our allocation equation over all passed samples, we would find that

$$C = 2\left(\frac{P}{K_p}\right) - \frac{1}{K_p}\sum_{\text{passed } k}\log_2(x_k^2) = 2\left(\frac{P}{K_p}\right) - \log_2\left(\left|\prod_{\text{passed } k}x_k^2\right|^{1/K_p}\right)$$

Substituting this into our allocation equation and solving for R_k then gives us the following optimal bit allocation result:

$$R_k^{opt} = \left(\frac{P}{K_p}\right) + \frac{1}{2}\log_2\left(x_k^2\right) - \frac{1}{2}\log_2\left(\left(\prod_{\text{passed } k}x_k^2\right)^{1/K_p}\right)$$

$$= \left(\frac{P}{K_p}\right) + \frac{1}{2}\log_2\left(x_k^2\right) - \left(\frac{1}{K_p}\sum_{\text{passed } k}\frac{1}{2}\log_2(x_k^2)\right)$$

$$\equiv \left(\frac{P}{K_p}\right) + \frac{1}{2}\log_2\left(x_k^2\right) - <\frac{1}{2}\log_2\left(x_k^2\right)>_{\text{passed } k}$$

for all k bands with non-zero bit allocations.

The bit allocation equation tells us that each non-zero sample is allocated a number of bits that differs from the average number of mantissa bits available for non-zero samples, P/K_p, by an amount that depends on the ratio of the sample squared amplitude to the geometric mean of the non-zero sample squared amplitudes. The geometric mean of the power spectral densities reflects the contribution of the total block, not just the spectral sample under consideration, to the bit allocation for a particular sample.

4.1   A Mathematical Approach

A different method adopted to derive similar results for the optimal bit allocation is based on the solution of a set of equations that minimize the average block error power with the data rate constraint by means of Lagrange multipliers (see for example [Jayant and Noll 84]). While this method is mathematically rigorous, it gives less intuitive insights on practical implementation issues. We describe now this method for completeness. This optimization problem can be framed as follows:

$$\min_{\{R_k\}}\left\{\frac{1}{K}\sum_{k=0}^{K-1}\left(\frac{x_k^2}{3\cdot 2^{2R_k}}\right)\right\}\quad\text{such that}\quad\frac{1}{K}\sum_{k=0}^{K-1}R_k = R$$

This problem is a problem of constrained minimization and can be solved applying the following steps[3]:
1. Solve using a Lagrange multiplier λ to enforce the average data rate constraint.
2. Take the derivatives with respect to each R_k and with respect to λ.
3. Solve the resulting equation for R_k and then enforce the average data rate constraint.

We define the Lagrangian $L(\{R_k\}, \lambda)$

$$L[\{R_k\}, \lambda] = \left\{ \frac{1}{K} \sum_{k=0}^{K-1} \left(\frac{x_k^2}{3 \cdot 2^{2R_k}} \right) + \frac{\lambda}{K} \left(\sum_{k=0}^{K-1} R_k - KR \right) \right\}$$

By taking the derivatives of the Lagrangian with respect to each R_k and with respect to λ and setting these derivatives to zero we obtain the following equations:

$$\frac{\partial}{\partial R_m} \left\{ \frac{1}{K} \sum_{k=0}^{K-1} \left(\frac{x_k^2}{3 \cdot 2^{2R_k}} \right) + \frac{\lambda}{K} \left(\sum_{k=0}^{K-1} R_k - KR \right) \right\} = 0 \qquad m = 0,1,\ldots,K-1$$

$$\frac{\partial}{\partial \lambda} \left\{ \frac{1}{K} \sum_{k=0}^{K-1} \left(\frac{x_k^2}{3 \cdot 2^{2R_k}} \right) + \frac{\lambda}{K} \left(\sum_{k=0}^{K-1} R_k - KR \right) \right\} = 0$$

The above equations solve our minimization problem by finding the appropriate set of $\{R_k\}$ that satisfies these conditions. By taking the derivatives of the Lagrangian with respect to each R_m we have:

$$\frac{x_k^2}{3} \frac{\partial}{\partial R_k} 2^{-2R_k} + \lambda = -2\ln 2 \; \frac{x_k^2}{3} 2^{-2R_k} + \lambda = 0$$

from which we derive:

$$R_k = C + \frac{1}{2} \log_2 x_k^2$$

where

---

[3] We are assuming all $R_k >= 0$, so we don't need to apply Kuhn-Tucker multipliers, and ignoring the requirement that all R_k are integer so we can take derivatives. In applying the results we have to round the "optimal" R_k to the nearest integer and force any negative R_k up to zero (recovering the bits from other R_k, possibly by re-optimizing)

$$C = \frac{1}{2}\log_2 \frac{2\ln 2}{3\lambda}$$

The number of bits per frequency sample depends on the squared amplitude of that sample plus a constant throughout the block under exam. By taking the derivatives of the Lagrangian with respect to λ we have:

$$\sum_{k=0}^{K-1} R_k = KR$$

and by substituting into it the expression for R_k we obtain:

$$\sum_{k=0}^{K-1} \left(C + \log_2 x_k^2 \right) = KR$$

Using this result, we can solve for C to find:

$$C = R - \frac{1}{2}\log_2 \left(\prod_{l=0}^{K-1} x_l^2 \right)^{\frac{1}{K}}$$

As shown above, C depends on the average bits per sample, R, minus the $\log_2$ of the block geometric power spectral density. Finally substituting the expression for C into the expression for R_k we again obtain the optimal bit allocation:

$$R_k^{opt} = R - \frac{1}{2}\log_2 \left(\prod_{l=0}^{K-1} x_l^2 \right)^{\frac{1}{K}} + \frac{1}{2}\log_2 x_k^2 = R + \frac{1}{2}\log_2 x_k^2 - < \frac{1}{2}\log_2 x_k^2 >$$

4.2   Coding Gain and Spectral Flatness Measure

It is apparent that, for each block of samples, a bit allocation that varies based on the spectral energy distribution of the signal introduces an improvement with respect to uniform bit allocation when the geometric mean of the signal power spectral density is much smaller than the average of the signal power spectral density. If the signal presents a flat spectrum then the geometric mean of the signal power spectral density is equal to the average of the signal power spectral density. In this case, the optimal bit allocation R_k coincides with the uniform bit allocation R.

It is instructive to estimate the coding gains from using optimal bit allocation with that of uniform quantization. The quantization error from optimal bit allocation can be obtained by substituting the optimum R_k back into the expression of the average block squared error and we find that

$$\left\langle q^2 \right\rangle_{block}^{opt} = \frac{\left(\prod_{k=0}^{K-1} x_k^2 \right)^{\frac{1}{K}}}{3 \cdot 2^{2R}}$$

In contrast, the error from using uniform quantization is equal to

$$\left\langle q^2 \right\rangle_{block}^{uniform} = \frac{\frac{1}{K} \sum_{k=0}^{K-1} x_k^2}{3 \cdot 2^{2R}}$$

Optimal bit allocation performs better than uniform quantization when the ratio of these errors is less than one. In other words, the squared error for optimal bit allocation is decreased when the geometric mean of the signal power spectral density is less than its average through the block.

The ratio of the geometric mean of the signal power spectral density to the average of the signal power spectral density is a measure of the spectral flatness of the signal, sfm [Jayant and Noll 84]:

$$sfm = \frac{\left(\prod_{k=0}^{K-1} x_k^2 \right)^{\frac{1}{K}}}{\frac{1}{K} \sum_{k=0}^{K-1} x_k^2}$$

Notice that the sfm varies between 0 and 1, where the sfm assumes the value 1 when the spectrum is flat. It is worth noticing also that sfm depends not only on the spectral energy distribution of the signal but also on the resolution of the filter bank in terms of the total number of frequency channels K. If K is much bigger than 2, then, for a given signal, the sfm decreases by increasing the number of frequency channels K. Values for the sfm much smaller than 1, typical for audio signals, imply high coding gains from optimal bit allocation. Values of the sfm near 1, very flat spectra, imply low coding gains so the informational cost makes optimal bit allocation worse than uniform quantization.

4.3   Block Floating-Point Quantization

The bit optimal allocation equation above assumes that we are allocating bits independently to each spectral sample. This is typically the case for a small number of frequency bands, i.e. for typical sub-band coders. For large number of frequency bands, such as in transform coders, we normally group spectral samples into sub-bands containing multiple spectral samples and block floating point quantize the sub-band. We need to keep in mind that the x_k^2 terms in the bit allocation equation is inversely proportional to the quantizer spacing for that sample. The corresponding term for a block floating point quantized spectral sample is the peak value of x_k^2 for that sub-band. In the case of B sub-bands indexed by b with N_b spectral samples in sub-band b and with maximum value of x_k^2 for that sub-band denoted as $x_{\max b}^2$, the bit allocation equation for the spectral lines in sub-band b becomes:

$$
R_b^{opt} = \left(\frac{P}{K_p}\right) + \tfrac{1}{2}\log_2\left(x_{\max_b}^2\right) - \tfrac{1}{2}\log_2\left(\left(\prod_{\text{passed b}}(x_{\max_b}^2)^{N_b}\right)^{1/K_p}\right)
$$

$$
= \left(\frac{P}{K_p}\right) + \tfrac{1}{2}\log_2\left(x_{\max_b}^2\right) - \frac{1}{K_p}\sum_{\text{passed b}} N_b\,\tfrac{1}{2}\log_2\left(x_{\max_b}^2\right)
$$

Notice that this version of the equation also applies to sub-band coding where N_b usually equals 1 for each sub-band.

As an important note on optimal bit allocation, we do have to worry about how we pass bit allocation information to the decoder and about making sure that our bit allocation is feasible, i.e., non-negative. As opposed to the binary allocation described earlier, optimal bit allocation needs to pass information not only on whether bands are passed, but also how many bits are passed per band. If we allow for a large number of different bit allocations for a particular sub-band, more bits are needed to describe which allocation was chosen. In order to keep the bit allocation information to be transmitted to the decoder to a minimum, some predefined values can be incorporated in the decoder routines. For example, in MPEG Layer II (see also Chapter 11), depending on the sampling rate and data rate of the system and the known distribution of audio signals, a set of tables pre-defines the maximum number of bits that can be allocated to certain bands. In this fashion, the bit allocation information to be transmitted to the decoder is kept to a minimum.

We should also note that there is no difference between passing zero or one mantissa bits for a midtread quantizer (you need at least two mantissa bits to get a non-zero step) so you should not allow a midtread quantizer to ever be assigned only one bit.

A given number of bits used to describe the allocation limits the number of bits that can be assigned to any sub-band. When we apply our bit allocation equation, we likely find outcomes where some sub-bands are assigned more bits than we allow and while others have fewer than 2 bits assigned. In fact, depending on the data rate constraints, even negative numbers of bits can come out of the formula if a signal is particularly demanding or its spectrum is nearly flat. A natural way to fix this problem is to simultaneously raise a lower threshold while lowering an upper threshold, the maximum bit allocation being assigned for sub-band b when $\frac{1}{2}$ $\log_2(x_{max}{}^2{}_b)$ is above the upper threshold and no bits being assigned to sub-band b when $\frac{1}{2} \log_2(x_{max}{}^2{}_b)$ is below the lower one. The thresholds are set so that the residual mantissa bit pool can be allocated using the optimal bit allocation formula to all sub-bands whose $\frac{1}{2} \log_2(x_{max}{}^2{}_b)$ falls between the thresholds without leading to any allocations over the maximum bit allocation or below two bits. When doing so, it is important to keep in mind that an allocation of R_b bits per sample for a sub-band actually reduces the bit pool by $N_b R_b$ bits since there are N_b spectral samples in the sub-band.

Another way to fix the bit allocation problem is to do a "water-filling" allocation. The water-filling algorithm is an iterative approach wherein we allocate bits based on each sub-band's $\frac{1}{2} \log_2(x_{max}{}^2{}_b)$ relative to a threshold level. We start out by sorting the sub-bands based on $\frac{1}{2} \log_2(x_{max}{}^2{}_b)$, giving each sub-band a starting allocation of zero bits, and setting the threshold to the highest value of $\frac{1}{2} \log_2(x_{max}{}^2{}_b)$. At every iteration we lower the threshold by one and then we allocate one more bit to each sub-band for which $\frac{1}{2}$ $\log_2(x_{max}{}^2{}_b)$ is at or above the current threshold (but we stop giving additional bits to any sub-band that has hit the maximum bit allocation value). We stop the process when we run out of bits. In the water-filling case, when we run out of bits we may still have some sub-bands with just one bit each – we need to take lone bits away and either pair them up with other lone bits or throw them onto samples with more bits. Again, we need to keep in mind that an allocation of R_b bits per sample for a sub-band actually reduces the bit pool by $N_b * R_b$ bits. The choice between these and other methods is going to depend on the trade-offs you face on optimality versus complexity. The water-filling method is quite often used and seems to be a good compromise between accuracy and speed.

5.   TIME-DOMAIN DISTORTION

In the previous section, we showed that the block distortion (measured by the average block quantization error) of the frequency domain coefficients can be reduced by optimally allocating bits if the spectrum is not flat. Since ultimately the encoded signal will be presented to the listener in the time-domain, a natural question to ask is: "How does the block distortion in the frequency domain relate to the block distortion in the time domain?". Remarkably, for commonly used time-to-frequency mapping techniques, the time-domain distortion is equal to the frequency-domain distortion [Jayant and Noll 84] as we now show.

Suppose we start with a set of time domain samples x[n] for n = 0,...,N-1. We consider transforms of these samples to and from the frequency domain with a linear transform of the form:

$$y[k] = \sum_{n=0}^{N-1} A_{kn} x[n]$$

$$x[n] = \sum_{k=0}^{N-1} B_{nk} y[k]$$

where the inverse transform is such that $B_{nk} = A_{kn}^*$ (and where * represents the complex conjugate). We call such a transform a "unitary transform" since matrices that satisfy this condition (i.e., that their inverse is the complex conjugate of their transpose) are called "unitary matrices" and it turns out the DFT is such a transform. We can see this by writing the DFT in its symmetric form (in which we include a factor of $1/\sqrt{N}$ in the definition of the forward transform) for which $A_{kn} = e^{-j2\pi kn/N}/\sqrt{N}$ and $B_{nk} = e^{j2\pi kn/N}/\sqrt{N}$.

We now see how quantization error in the frequency domain samples translates back into quantization error in the time domain samples when we inverse transform. Suppose that quantization/dequantization changes the frequency domain samples from y[k] to y'[k] due to (possibly complex) quantization error ε_k. When we inverse transform back to the time domain the quantization error in y'[k] lead to output samples x'[n] containing quantization error ε_n where:

$$\varepsilon_n = x'[n] - x[n] = \left(\sum_{k=0}^{N-1} B_{nk} y'[k] \right) - x[n]$$

$$= \left(\sum_{k=0}^{N-1} B_{nk} y[k] + \sum_{k=0}^{N-1} B_{nk} \varepsilon_k \right) - x[n]$$

$$= \sum_{k=0}^{N-1} B_{nk} \varepsilon_k$$

Noting that ε_n is real-valued for real-valued input signals $x[n]$ and the quantization error is independent from sample to sample (so that we can assume $<\varepsilon_k \varepsilon_{k'}^*>$ is zero if $k \neq k'$), we can write the average block distortion in the time domain as:

$$\left\langle q^2 \right\rangle_{block}^{time\ domain} = \frac{1}{N} \sum_{n=0}^{N-1} <\varepsilon_n^2> = \frac{1}{N} \sum_{n=0}^{N-1} <\varepsilon_n \varepsilon_n^*>$$

$$= \frac{1}{N} \sum_{n=0}^{N-1} < \left(\sum_{k=0}^{N-1} B_{nk} \varepsilon_k \right) \left(\sum_{k'=0}^{N-1} B_{nk'}^* \varepsilon_{k'}^* \right) >$$

$$= \frac{1}{N} \sum_{n=0}^{N-1} \sum_{k=0}^{N-1} \sum_{k'=0}^{N-1} B_{nk} A_{k'n} <\varepsilon_k \varepsilon_{k'}^*>$$

$$= \frac{1}{N} \sum_{n=0}^{N-1} \sum_{k=0}^{N-1} B_{nk} A_{kn} <|\varepsilon_k|^2>$$

$$= \frac{1}{N} \sum_{k=0}^{N-1} <|\varepsilon_k|^2> \left(\sum_{n=0}^{N-1} A_{kn} B_{nk} \right)$$

$$= \frac{1}{N} \sum_{k=0}^{N-1} <|\varepsilon_k|^2>$$

$$= \left\langle q^2 \right\rangle_{block}^{freq\ domain}$$

where the transition to the second-to-last line is due to the fact that A_{kn} and B_{nk} are inverses of each other so that the quantity in parentheses is equal to one. Note that, for complex transforms, we need to worry about the quantization error in both the real and imaginary parts. However, since the quantization errors in the real and imaginary parts are independent of each other, the quantity $<|\varepsilon_k|^2>$ is just the sum of the expected quantization error power in the two parts.

This result tells us that the total block distortion in the time domain is equal to the block distortion in the frequency domain. Why then do we do our quantization in the frequency domain? Recall the main result from optimal bit allocation that the block distortion for a given number of bits is proportional to the spectral flatness measure. The reason we go to the frequency domain and do our quantization is that we expect most audio signals to be highly tonal. By highly tonal we mean that audio signals spectra have strong peaks. A very "peaky" signal has a low spectral flatness measure and therefore produces lower block distortion for a given number of bits per block. For example, consider a pure sinusoid. In the time domain the signal is spread out across the block while in the frequency domain its content is collapsed into two strong peaks (one at positive frequencies and one at negative frequencies). Clearly, the frequency domain representation is much less flat than the time domain representation.

We conclude that we go to the frequency domain because we expect the signal representation to be less flat than the time domain representation. Our calculations for optimal bit allocation tell us that we can reduce distortion *in the time domain output signal* by doing our quantization in a representation that is less flat than the time domain representation. This conclusion is the technical manner in which we "reduce redundancy" by changing signal representation.

As a final note, we mention the fact that the MDCT of a single block of samples is not a unitary transform due to time-domain aliasing effects. However, when we include the overlap-and-add to view the MDCT as a matrix transform on the overall input signal (see Chapter 5), it is a unitary transform. Therefore the conclusions above also apply to the MDCT with the caveat that, although the overall time domain distortion equals the frequency domain distortion when compared over the whole signal, there is not an exact balance on a block by block basis. Again, the fact that the MDCT is a frequency domain spectral representation implies that it is also peaky for highly tonal signals and as such it can be used to remove redundancy in the signal.

6.   OPTIMAL BIT ALLOCATION AND PERCEPTUAL MODELS

In perceptual audio coding, the goal is not just to remove redundancy from the source, but it is also to identify the irrelevant parts of the signal spectrum and extract them. This translates into not just trying to minimize the average error power per block, but also trying to confine the resulting quantization noise below the masking curves generated by the signal under examination. We no longer care just how large the error is but rather how large it is compared with the masking level at that frequency. We can keep the quantization noise imperceptible if we can keep all of the quantization noise localized below the masking curves. If, because of the data rate constraint, we don't have enough bits for imperceptible quantization, we want to keep the perceived noise at a minimum.

We can keep the perceived noise at a minimum by allocating bits to minimize the following measure of perceptible distortion:

$$\left\langle q^2 \right\rangle_{block}^{percept} = \frac{1}{K} \sum_{k=0}^{K-1} \frac{\varepsilon_k^2}{M_k^2}$$

where M_k is the amplitude equivalent to the masking level evaluated at frequency index k. Notice that this measure of distortion gives a lot of weight to quantization noise that is large compared to the masking level while very little weight to noise below the masking level.

Allocating bits to minimize this measure of distortion is almost identical to the problem we just studied other than the fact that now, when we substitute in our formula for the quantization noise from floating point quantization, the spectral sample amplitude x_k is always divided by the corresponding masking amplitude M_k. This means that we can make use of all of our prior results for optimal bit allocation if we make this substitution. The resulting perceptual bit allocation result is:

$$R_b^{opt} = \left(\frac{P}{K_p} \right) + \frac{1}{2} \log_2 \left(\frac{x_{max_b}^2}{M_b^2} \right) - \frac{1}{2} \log_2 \left(\left| \prod_{passed\ b} \left(\frac{x_{max_b}^2}{M_b^2} \right)^{N_b} \right|^{1/K_p} \right)$$

for all b with non-zero bit allocations (i.e., passed samples) where M_b is the amplitude corresponding to the masking level assumed to apply in sub-band b. Normally, our psychoacoustic model provides us with information on the signal-to-mask ratio for each sub-band. We can rewrite this equation in terms of each sub-band's SMR as

$$R_b^{opt} = \left(\frac{P}{K_p} \right) + \frac{\ln(10)}{20\ln(2)} \left(SMR_b - \frac{1}{K_p} \sum_{passed\ b} N_b SMR_b \right)$$

where SMR_b represents the SMR that applies to sub-band b.

Perceptual bit allocation proceeds very much analogously to optimal bit allocation. The main difference is that the masking thresholds and corresponding SMRs for the block need to be calculated prior to deciding how to allocate the bit pool. Given the SMRs, the bits are allocated exactly as in the bit allocation described in the previous section.

The effectiveness of carrying out perceptual bit allocation is measured by the perceptual spectral flatness measure, psfm, which can be described by [Bosi 99]:

$$\text{psfm} = \frac{\left(\prod_{k=0}^{K-1} \frac{x_k^2}{M_k^2}\right)^{\frac{1}{K}}}{\frac{1}{K}\sum_{k=0}^{K-1} \frac{x_k^2}{M_k^2}}$$

The psfm is analogous to the sfm in that it ranges between zero and 1 with low numbers implying the potential for high coding gain. Notice that the psfm depends on the spectral energy distribution of the signal weighted by the masking energy distribution.

7.   SUMMARY

In this chapter, we have brought together many of the themes discussed in prior chapters and shown how they fit together to reduce the bit rate of the system under consideration. We have seen how a transformation to the frequency domain can reduce bit rate for a highly tonal signal. We have then seen how the use of floating point quantization allows us to extract greater coding gain through optimal bit allocation. Finally, we have seen how perceptual measures of masking can be used to better allocate quantization noise and squeeze out more coding gain by removing irrelevant bits. Similarly to the procedures described in this chapter, many of the standard audio coding systems make use of bit allocation strategies based on the ratio of the signal versus its masking strength with a fixed data rate constraint (see for example the description of MPEG Layer I and II and Dolby AC-3 in later chapters). It should be mentioned, however, that the MPEG Layer III approach differs somewhat in that a locally variable data rate approach is adopted in order to accommodate particularly demanding audio signals (see also Chapter 11). Before examining the specific implementation of several state-of-the-art coders, we illustrate in the next chapter how all of the building blocks described in the previous chapters fit together into a coding system.

8. REFERENCES

[Bosi 99]: M. Bosi, "Filter Banks in perceptual Audio Coding", in Proc. of the AES 17th Intl. Conference, pp. 125-136, September 1999.

[Jayant and Noll 84]: N. Jayant, P. Noll, "Digital Coding of Waveforms: Principles and Applications to Speech and Video", Prentice-Hall, Englewood Cliffs, 1984.

9. EXERCISES

Bit Allocation:
In this exercise, you will compare bit allocation methods for the test signal studied in the exercises of Chapters 6 and 7. The goal is to gain an appreciation of how different bit allocations perform.

1. Define 25 sub-bands by mapping the N/2 frequency lines of a length N = 2048 MDCT onto the 25 critical bands. (Remember that $f_k = F_s \, k/N$ for k=0...N/2-1)

2. Consider the data rates I = 256 kb/s per channel and I = 128 kb/s per channel for the coded spectral lines of a length N=2048 MDCT.

    a) How many bits per line are available for coding the spectral data?

    b) If 4 bits/sub-band are used for a sub-band scale factor, how many bits per line remain for coding mantissas?

3. Write a function to allocate bits to a set of K sub-bands dividing up the N/2 frequency lines of a length N MDCT block so as to minimize the average block error. The lines in each sub-band will share a single scale factor represented with R_s bits and will all use the same number of mantissa bits. Also create a variant of this function to perform the allocation to minimize the average block perceptual error.

4. For the input signal used in Chapters 6 and 7:

$$x[n] = A_0 \cos(2\pi 440n / F_s) + A_1 \cos(2\pi 554n / F_s)$$
$$+ A_2 \cos(2\pi 660n / F_s) + A_3 \cos(2\pi 880n / F_s)$$
$$+ A_4 \cos(2\pi 4400n / F_s) + A_5 \cos(2\pi 8800n / F_s)$$

where $A_0 = 0.6$, $A_1 = 0.55$, $A_2 = 0.55$, $A_3 = 0.15$, $A_4 = 0.1$, $A_5 = 0.05$, and F_S is the sample rate of 48 kHz, and for both data rates above, quantize and inverse quantize each frequency output of an N = 2048 MDCT using "block" floating point, where each frequency sub-block has only one scale factor and the frequency sub-bands are the 25 sub-blocks defined in 1) above. Use 4 bits per scale factor and:

a) Uniformly distribute the remaining bits for the mantissas.
b) Optimally distribute the remaining bits for the mantissas based on signal amplitude.
c) Distribute the bits by hand to get the best sound you can.
d) Use the signal-to-masking level for each critical band calculated in Chapter 7 to optimally distribute the remaining bits for the mantissas.

Listen to the results of each bit allocation scheme above and comment on their relative performance. (Note: the maximum amplitude of this signal is 2.0. This implies that you should set x_{max} in your quantizer equal to 2.0 or, if your x_{max} is hard-coded to 1.0, you should divide the signal by 2.0 prior to quantizing it.)

Chapter 9

Building a Perceptual Audio Coder

1. INTRODUCTION

In this chapter we discuss how the coder building blocks described in the prior chapters can be fit together into a working perceptual audio coder. Particular attention is given to how to create masking curves for use in bit allocation. We also discuss issues in setting up standardized bitstream formats so that coded data can be decoded using decoders provided from a variety of vendors.

2. OVERVIEW OF THE CODER BUILDING BLOCKS

Figure 1 shows the basic building blocks of a perceptual audio encoder. Typically, the input data is an audio PCM input signal (rather than the original analogue input). This signal has its content mapped into the frequency domain using some type of filter bank, for example PQMF or MDCT. The frequency domain data is then quantized and packed into a bitstream. The quantization is carried out using a bit allocation that is designed to maximize the overall signal to noise ratio (SNR) minus the signal to mask ratio (SMR) of each block of data. The psychoacoustic model stage analyzes the input signal, determines the masking level at each frequency component, and computes the SMR values. The bit allocation routine allocates a limited number of mantissa bits to the frequency-domain data based on the signal components strength and the their relative SMR values. The encoded bitstream includes both the coded audio data, i.e., mantissas, scale factors, and bit allocation. In addition any control parameters including, for example, block length, type of windowing, etc. needed to tell the decoder how to decode the data is included in the coded bitstream. Synchronization word, sampling rate, data rate, etc. are typically contained in the data header and passed to the decoder at certain time intervals. Finally, error correction codes, time-synchronization stamps, and other auxiliary or ancillary data can also be multiplexed in the data stream. The result is an encoded bitstream that can be stored or directly transmitted to the decoder.

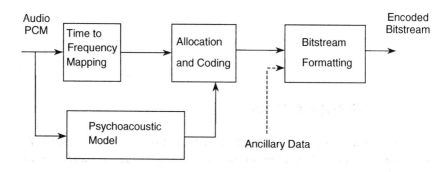

Figure 1. Basic building blocks for a perceptual audio encoder

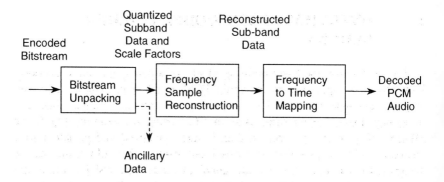

Figure 2. Basic building blocks for a perceptual audio decoder

The basic building blocks of a perceptual audio decoder are shown in *Figure 2*. First, the encoded bitstream is unpacked into its constituent parts, i.e., audio data, control parameters, and ancillary data. The bit allocation information is used to dequantize the audio data and recover as best as possible the frequency-domain representation of the original audio data. The reconstructed frequency-domain data contain quantization noise but, if the psychoacoustic model has correctly done its job, that noise is inaudible or as close to inaudible as possible given the data rate constraint. The frequency-domain data is returned to the time-domain using the appropriate filter bank, for example a synthesis bank of PQMF or an IMDCT, and finally converted into an audio PCM output data stream. It should be noted that the psychoacoustic model computation and relative bit allocation is shown only in the encoder side of the audio coding system. While for most state-of-the-art audio coding schemes this is the case, there are instances, like for example AC-3 (see also Chapter 14), in which the bit allocation routine is computed both in the encoder and, at least a sub-set of the routine, the decoder. In this approach the allocation side information to be transmitted to the decoder is minimized at the expense, however, of an increased layer of complexity for the decoder.

We've already discussed alternatives for time-to-frequency mapping tools, how to allocate bits given masking curves, and how to quantize the data. What we still need to explore in a bit more detail is how to use the psychoacoustic properties of hearing to create the masking curves and how to design a bitstream format. We'll first turn to the issue of computing a masking curve.

3.   COMPUTING MASKING CURVES

We've already discussed how masking models can be used to reduce the precision in the representation of frequency-domain data without introducing perceptual differences between the coded and the original signal. Time-domain masking is typically exploited in defining the time resolution of the coder, i.e., to control the system input block-size so that quantization errors are confined in time regions where they do not create audible artifacts (pre-echo). We also discussed measurements of the hearing threshold and developed models of frequency-domain masking – what is there still left to talk about? The main issues we still need to address revolve around bringing together the information contained in the computed masking curves relative to the input signal and the frequency representation of the signal in the coder's main-path time-to-frequency mapping stage.

We've seen that frequency-domain masking drops off very sharply in frequency, especially towards lower frequencies. This rapid drop off means that we potentially can introduce large errors in the masking levels at particular signal components if we don't know the frequency locations of both the masker and the maskee with reasonably high accuracy. In contrast, the time-to-frequency mapping used by the coder may not have adequate frequency resolution for this purpose. Moreover, the frequency-domain representation of the signal may have significant aliasing that, although it may disappear in the synthesis stage, could potentially lead to errors in estimating the masking curves.

Typically, perceptual audio coders perform a high-resolution DFT (using the FFT algorithm) with blocks of input data solely for use in the psychoacoustic model. The results of this high frequency resolution DFT are then employed to determine the masking curve for each block of coded data. An immediate issue that arises in this approach is making sure that the DFT data is time-synchronized with the data block being quantized. If it isn't, the DFT may show too much (or too little) frequency content from outside of the time region of interest. This issue is usually addressed by selecting a large enough data block input to the DFT and by centering it on the data block being quantized. Note also that, as usual, we don't want the DFT to be corrupted by edge effects so we need to window the data block prior to performing the DFT. Any of the windows we discussed in Chapter 5 can be used for this purpose, with the Hanning window a common choice (see for example the description of ISO/IEC MPEG Psychoacoustic Models 1 and 2 in ISO/IEC 11172-3 and in Chapter 11).

Having performed a DFT with adequate frequency resolution, we can use our frequency-domain masking models to determine the masking level at each DFT frequency line. The most straightforward approach for doing this is to loop over all signal frequency content represented on a bark scale, compute the masking curve from each signal component, and appropriately sum up the curves. Recall from Chapter 7 that the masking curve from a single component is created by convolving that component with an appropriate spreading function (i.e., by applying the spreading function shape to the component level at its frequency location) and then lowering the resulting curve level by a shift Δ that depends on the tonality of the masker component and its frequency position. The masking from different signal components is then added in the appropriate manner and combined with the hearing threshold, where usually the largest individual curve is used or the intensities are added.

Applying a straightforward implementation of the masking models takes order N^2 operations to carry out where N is the number of DFT frequency lines (presumably large). Two different solutions to the runtime problem are typically used: 1) limit the number of maskers, and 2) create the masking curves using convolutions rather than a loop over maskers.

The first solution to the runtime problem, i.e., to limit the number of maskers by developing curves only for the main maskers, is based on the idea that most of the masking is performed by a few strong components, which, if identified, are the only components that need to have masking curves created. One way to carry this out is to look for local maxima in the frequency spectrum and, if they are tonal, i.e., the spectrum drops off fast enough near them, to use the largest of them as tonal maskers. The remaining components can then be lumped together into groups, for example by critical bands or, at high frequencies where critical bands are quite wide, by 1/3 of a critical band, to use as noise-like maskers. In this manner, the number of components that need to have masking curves created and summed is limited to a number that can be computed in reasonable runtime (see also ISO/IEC MPEG Psychoacoustic Model 1 description in Chapter 11).

The second solution to the runtime problem is to create the overall masking curve as a convolution over the entire spectrum (see also [Schroeder, Atal and Hall 79]) rather than summing separately over all frequency lines. For example, suppose that the level shift Δ is independent of the type of masker, i.e., it does not depend on whether the masker is tonal or noise-like or on its frequency location, and that the spreading function shape is independent of the masker level. In this case, the masking curve from each component could be created by convolving the full spectrum with an appropriate spreading function and then shifting the result down by a constant Δ. The benefit of this approach is that the convolution theorem can be used to convert this frequency-domain convolution (naively requiring order N^2 operations) into a faster procedure in the time domain. Changing to and from the time domain requires order $N*\log_2(N)$ operations while implementing the convolution as a time-domain multiplication requires order N operations – leading to a total operation count of order $N + 2N*\log_2(N) = N*(1 + 2\log_2(N)) \approx 2N* \log_2(N)$. This can be a big reduction from order N^2 when N is large! Of course, the problem with this approach is that, as we saw in Chapter 7, the masking curves are very dependent on whether or not the masker is tonal. One solution to this problem is to ignore the difference and compromise by using a single shift Δ regardless of the masker's tonality.

A clever solution to this problem is adopted in ISO/IEC MPEG Psychoacoustic Model 2 (see for example 11172-3 or Chapter 11). For each block of data, Model 2 computes a tonality measure that is then convolved with the spreading function to create a frequency-dependent "spread" tonality measure. This spread tonality measure determines how tonal the dominant maskers are at each frequency location. Notice that also this second convolution can be carried out as a time-domain multiplication for order $2N*\log_2(N)$ operations. The shift Δ then depends on the spread tonality measure at each frequency location. In this manner, portions of the signal spectrum that are mostly masked by tonal components have their relative excitation patterns shifted downward by a Δ appropriate for tonal masking. Vice-versa, portions of the signal spectrum mostly masked by noise-like components have their relative excitation patterns shifted downward by a Δ appropriate for noise masking (see Chapter 11 for further details).

Having created the masking curves at each frequency line of the psychoacoustic DFT stage, we are now faced with the challenge of mapping them back into signal-to-mask ratios (SMRs) to use for the frequency bands in the coder's main path time-to-frequency mapping. In a sub-band coder, for example PQMF, the frequency bands are typically the pass bands of each of the K modulated prototype filters. In transform coders typically a single scale factor is used for multiple frequency lines, so that the frequency bands are the frequency ranges spanned by the sets of lines sharing a single scale factor. We typically find that the coder's frequency bands are wide compared to the ear's critical bands at low frequencies, where critical bands are narrow, and narrow compared to critical bands at high frequencies, where critical bands are wide. Since masking effects tend to be constant within a critical band, one way to do the mapping is to choose

  a)   the average masking level in the critical band containing the coder's frequency band when the coder's band is narrow compared with the ear's critical bands

  b)   the lowest masking level in the coder's frequency band when the coder's band is wide compared with the ear's critical bands, so that the masking level represents the most sensitive critical band in that coder band.

In the second case, the coder's frequency resolution is considered to be sub-optimal since its frequency bands span more than one critical band. It this case, additional bits may need to be allocated to the coder's bands with bandwidths larger than critical bandwidths in order to compensate for the coder's lack of frequency resolution.

Once the masking level is set for the coder's frequency band, we then set the SMR for that frequency band based on the amplitude of the largest spectral line in the band, or, if our scale factor is at the maximum value, so that the quantizer cannot adjust its spacing to any smaller value, based on the amplitude of a line whose amplitude corresponded to the maximum scale factor.

3.1 Absolute Sound Pressure Levels

Another issue that needs to be addressed is how absolute sound pressure levels (SPLs) can be defined based on the computed signal intensity in order to align the hearing threshold with the signal's spectrum and for intensity-dependent masking models. Although masking depends mostly on the relative intensities of masker and maskee, the hearing threshold is defined in terms of absolute SPL. In addition, the shape of the spreading functions is modeled as depending on the absolute pressure level of the sound. Unfortunately, the absolute SPL of the signal depends on the gain settings used on playback – higher volume settings lead to higher SPLs reaching the listener's ears- which are not known a priori.

Since we can't be assured exactly what gain settings are used on playback, we are forced to make an assumption about the target playback gain for the signal. The assumption usually made is that the input PCM data has been recorded and quantized so that the quantization error falls near the bottom of the hearing threshold at normal playback levels. In particular, we usually define a sinusoid with amplitude equal to ½ the PCM quantizer spacing as having an SPL equal to 0 dB. Recall that the hearing threshold has its minimum value at about -4 dB for young listeners, so this definition implies that some listeners would be hearing a bit of quantization noise in certain regions of the input PCM signal spectrum.

For 16-bit PCM input data the standard assumption implies that a sinusoid with amplitude equal to the overload level of the quantizer would have an SPL of about 96 dB (6 dB/bit * 16 bits). If we define our quantizer overload level x_{max} to be equal to 1, this assumption implies that the SPL of a sinusoidal input with amplitude A is equal to:

$$SPL = 96 \text{ dB} + 10 \log_{10}(A^2)$$

Notice how this formula correctly has an SPL of 96 dB when the input amplitude reaches its maximum for A = 1.

Having made an assumption that allows us to define absolute SPLs in our input signals, we need to be able to translate our frequency-domain representation into units of SPL. Since our SPLs are defined in terms of the amplitudes of input sinusoids, translating the frequency-domain representation into SPL implies being careful with normalization in our time-to-frequency mappings. This care is needed in both the DFT used for the psychoacoustic modeling and in the coder main path's time-to-frequency mapping. In both cases, the choice of window affects the gain of the transform. Knowing the SPL of the maximum sinusoid (for example 96 dB for 16 bit PCM), however, allows you to define the correct translation factor for any particular case.

The basic approach to calculating the translation factor is to use Parseval's Theorem to relate the spectral density integrated over a frequency peak to the power of the input sinusoid. For example, by utilizing Parseval's Theorem for the DFT we have:

$$< x^2 > \equiv \tfrac{1}{N} \sum_{n=0}^{N-1} x[n]^2 = \tfrac{1}{N^2} \sum_{n=0}^{N-1} |X[k]|^2$$

For a sinusoid with amplitude A the average signal power is $\tfrac{1}{2}A^2$. However, a sinusoid with amplitude A that is windowed with a window $w[n]$ has an average signal power approximately equal to $\tfrac{1}{2}A^2 <w^2>$, assuming that the window function varies much more slowly in time than the sinusoid itself. Such a signal has a DFT containing two main peaks with equal spectral density: one at positive frequencies k_1 ($k_1 \in [0, N/2-1]$) and one at negative frequencies k_2 ($k_2 \in [N/2, N-1]$). We can use Parseval's Theorem to relate the sum of spectral density over a single positive frequency peak to the input signal amplitude as:

$$\frac{1}{2} A^2 < w^2 > = \tfrac{2}{N^2} \sum_{peak} |X[k]|^2$$

or equivalently:

$$A^2 = \tfrac{4}{N^2 <w^2>} \sum_{peak} |X[k]|^2$$

We can use this formula to substitute for A^2 in the SPL formula above to find:

$$SPL_{DFT} = 96 \, dB + 10 \log_{10} \left(\tfrac{4}{N^2 <w^2>} \sum_{peak} |X[k]|^2 \right)$$

where $|X[k]|^2$ is the computed power spectral density of the input signal.

For a second example, we consider how to estimate SPLs for an MDCT. The challenge here is the fact that the time-domain aliasing in the transform does not allow for an exact Parseval's Theorem. However, an approximate solution can be derived in which:

$$< x^2 > = \frac{4}{N^2} \sum_{n=0}^{N/2-1} X[k]^2 + \frac{1}{N} \sum_{n=0}^{N/2-1} \left(x[n]x[N/2-1-n] - x[N-1-n]x[N/2+n] \right)$$

$$\approx \frac{4}{N^2} \sum_{n=0}^{N/2-1} X[k]^2$$

In this case, there is only a single frequency peak for a sinusoid in the frequency range of $k \in [0, N/2-1]$ so we find that this approximate solution relates the amplitude to the sum of spectral density over a peak through:

$$A^2 \approx \frac{8}{N^2 <w^2>} \sum_{peak} |X[k]|^2$$

Again, we can substitute into the SPL formula to find:

$$SPL_{MDCT} \approx 96\,dB + 10\log_{10}\left(\frac{8}{N^2 <w>^2} \sum_{peak} |X[k]|^2 \right)$$

where X[k] represents the output of the MDCT.

The translation of frequency-domain representation into absolute SPLs depends on the choice of window utilized in the mapping onto the frequency domain, since the window choice affects the overall gain of the frequency representation. The gain factor for any specific window can be computed using the following definition:

$$< w^2 > \equiv \frac{1}{N} \sum_{n=0}^{N-1} w[n]^2$$

For completeness, we note here the appropriate gain adjustment for some of the more common windows. Since N is typically fairly large, the gain adjustments can be calculated as averages over the continuous time versions of the window. The rectangular window has $<w^2>=1$ assuming that w[n] is equal to 1 over its entire length. The sine window has $<w^2>=\frac{1}{2}$ as can be easily seen since $\sin^2(x)$ averages to $\frac{1}{2}$ over a half-integral number of periods. The Hanning window has $<w^2>=3/8$. The gain factor for the Kaiser-Bessel window depends on α but can be computed for any specific α value using the above definition.

4.   BITSTREAM FORMAT

The encoded bitstream is the means by which the encoder communicates to the decoder. This means that the encoded bitstream needs to be able to tell the decoder both how to decode the data and what the data is. Any encoded bitstream therefore includes both control data (telling the decoder what to do) and coded audio data (the signal to be decoded). A bitstream format needs to be defined in such a way that the decoder knows how to extract this data from the bitstream.

Normally, a bitstream format begins with a header. The header typically starts with a code that establishes synchronization of the bitstream and then it passes the decoder some overall information about the encoded data, for example sampling rate, data rate, copyrights, etc. To the degree that the codec has coding options, for example, input/output bits per sample, number of audio channels, algorithm used, etc., this also needs to be passed to the decoder.

After the header establishes what needs to be done, the bitstream includes the coded audio data. Each block of coded data needs to include 1) bit allocation information (when applicable) to know how many bits are used to encode signal mantissas, 2) scale factors defining the overall scale of the mantissa values, and 3) the mantissas themselves. The bitstream format defines the layout and the number of bits used for the bit allocation and scale factors. It also defines the layout of the mantissas. Entropy coding methods can be used to reduce the bits needed to write out this information. For example, masking might lead to many frequency lines using zero mantissa bits – knowing that zero bits is a common bit allocation implies that a short code should be used to denote this result. Typically, the codebook is predefined based on "training" the coder on a variety of input signals so that a decoding table doesn't need to be passed in the bitstream, but it can be simply stored in the decoder ROM.

In the case of multichannel audio, for example, stereo channels, the bitstream also needs to define how the different audio channels are laid out relative to each other in each data block. Sometimes the channels are interleaved so you get the data for each channel at a given frequency line before reading the next frequency line's data. Sometimes, however, channel transformations are made to allow for bit reduction based on similarities between channels. For example, stereo is sometimes transformed from L (left) and R (right) channels into sum (M = L + R "Mid") and difference (S = L - R "Side") channels so the knowledge that S is typically small can be leveraged into allocating it fewer bits. Likewise, correlations between channels can be exploited in cases with larger numbers of channels so various channel matrixing transformations are defined to allow for channel coding opportunities to save on bits. Note that control data need to be passed telling the decoder what format the channel data is in if it allows different options.

The individual blocks of coded data are usually bundled in larger chunks often called "frames". If the signal is fairly stationary, we would expect subsequent blocks of data to be fairly similar. Cross-block similarity can be exploited by sharing scale factors across blocks and/or by only passing differences in data between subsequent blocks in a frame. The header and some control data are typically passed on a frame-by-frame basis rather than on a block-by-block basis, telling the decoder any dynamic changes it needs to make in decoding the data. If the encoder detected a transient and shifted to shorter data blocks the decoder needs to be told. In this case, because of the non-stationary nature of the signal, scale factors are transmitted on a block-by-block basis.

To render the bitstream format issue more tangible, *Figure 3* provides an example of both a PCM data file format and a file format for a simple perceptual coder that works on files in batch mode. The PCM data file begins with a 4-byte code equal to the string "PCM " to make sure that it is a PCM file. It then includes a 4-byte integer representing the sample rate of the signal measured in Hz, for example, 44.1 kHz would be equal to 44,100. Then it has a 2-byte integer representing the number of channels (1 for mono, 2 for stereo, etc.). The header finishes with a 2-byte integer representing the number of bits per PCM data sample (8 bits, 16 bits, etc.) and a 4-byte integer representing the number of samples in the file. Following the header, the PCM file contains the signal data samples interleaved by channel, each sample being represented using nSampleBits bits as a PCM quantization code.

The coded data file in *Figure 3* represents a simple perceptual audio coder. This coder takes PCM input data, loads each channel into data blocks 2*BlockSize long (with BlockSize new data samples for each block), performs an MDCT for each channel to convert the data block into BlockSize frequency components. It uses a perceptual model that computes SMR for each of 25 critical band-based frequency bands, allocates mantissa bits for each frequency bands, block floating point quantizes each of the frequency bands using one scale factor per critical band and the allocated number of mantissa bits per sample, and finally writes each block's result into a coded file. The coded file format begins with the header, which includes a 4-byte code equal to the string "CODE". It then includes a 4-byte integer for the sample rate in Hz, a 2-byte integer for the number of channels, and a 2-byte integer for the number of PCM bits per sample when decoded. The control parameters passed in the bitstream include a 2-byte integer representing the number of scale factor bits used by each of the 25 scale factors, and then has a 2-byte integer representing the number of bits used to define the bit allocation for each of the 25 frequency bands. A 4-byte number representing the number of frequency samples in each block of frequency data and a 4-byte number representing the number of data blocks in the file is also passed. Following the control parameters, the coded audio file then has the signal data grouped by data blocks. Each data block starts with the 25 scale factors (nScaleBits each) and the 25-frequency-band bit allocations (nBitAllocBits each). Finally, the BlockSize mantissa values for each channel are interleaved, each one using the number of bits defined for its frequency band in the bit allocation information.

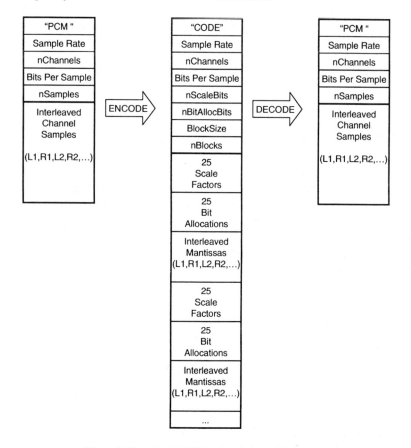

Figure 3. Very simple PCM and coded data file formats

The coded file format in *Figure 3* makes clear how simple the underlying coder is. For example, it doesn't allow for changing block size or to detect transients and it doesn't employ any cross-channel or cross-block coding tricks to squeeze out extra bits. The coder doesn't even use entropy coding-based codebooks to reduce redundancy in how it writes out the scale factors, or mantissas – both pretty easy to implement. However, it does make use of a perceptual model to allocate bits based on hearing threshold and masking models, so it quite possibly still does a reasonable job of reducing bitrate without too many audible artifacts. In subsequent chapters, we study a number of coders out in the market. In the cases where the bitstream formats are publicly available, studying the format definition gives a lot of information about the techniques employed in the encoder to squeeze bits out of the signal.

5.   BUSINESS MODELS AND CODING SECRETS

Once a coder has been developed, the goal is to get it deployed in the market. At this point, the coder developer needs to decide what is the best means to achieve market share in the target market space. A variety of business models have been used to gain market acceptance of perceptual audio coders.

The most basic business model is to create and sell customer-friendly encoding and decoding tools. Depending on the application, such tools could be hardware-based (for example built into a chip) or software-based. In such cases, the details of the inner workings of the codec (including the coded file format) are likely to be considered business secrets and details are going to be kept fairly proprietary (other than what's needed for marketing purposes). Much effort in such a business is going to be spent in sales efforts for coding tools and in keeping secret or protecting the intellectual property behind the coder.

A more recent business model that has arisen is a model wherein money is made on the encoders while the decoders are free or extremely cheap. The idea in this business model is to make your decoder ubiquitous in the target market. In this case, you'd like as many users as possible to be using your decoder and so you find ways to make that happen. For example, you might give the decoder away free over the internet or aggressively license your decoding technology to companies making players for the type of content you are encoding (for example satellite television receivers, cd/dvd/mp3 players, video game consoles).

Another recent business model that has developed is based on the idea that better technology can be made by combining the efforts of several companies in a related field. In this business model, several companies pool their efforts to develop a joint coding standard. The hope is that the technology that results is so much better than anything else in the market that it creates enough profits for each participant to have been better off than doing it alone. Although far from universally accepted, this last business model has become an increasingly important one in the world of coders. One of the first very successful examples of such approach was applied to the MPEG-2 video standard (see for example [Mitchell, Pennebaker, Fogg and LeGall 97]). Many of the most popular coders in the market today (MP3 as a notable example), are the result of setting up a standardization committee and defining an industry-standard codec for certain applications.

In the standards process, participating companies offer up technology to become part of the standard coder. For example, one company might provide the structure of the psychoacoustic model/bit allocation routine, another might provide the transform coding kernel, and yet a third company might provide the entropy coding codebook for the bit allocations and scale factors. The specifications of the resulting decoder would then become publicly available and steps taken so potential users could easily license the standard coder technology. If a patent pool is set up, typically the resulting royalties would be shared by the participating companies in some allocation mutually agreed upon, but in general related to the share of the intellectual property provided.

Usually only the bitstream format and decoding process become standardized – the encoder remaining proprietary so that companies can still compete on having the best sounding coder. An example encoder is described in the informative part of the standard, but companies can put together encoders that perform very differently while still conforming with the mandatory standard specifications. This is not the case with decoders where, to be compliant with the standard, a decoder must behave exactly as specified.

Keeping a coder proprietary means that it is hard for students, academics, and others to learn what's really going on inside the coder. The fact that the encoder part of a standardized codec remains competitive often means that the standards documents remain very cryptic, again limiting an outsider's ability to understand what is going on inside. After all, if you make money based on having the best encoders it can be in your financial interests to only lay out in the standard what steps need to be taken without explaining why they must be taken. One of the goals of this book is to help demystify some of the coding "secrets" that typically remain out of reach to outsiders.

6. REFERENCES

[ISO/IEC 11172-3]: ISO/IEC 11172-3, Information Technology, "Coding of moving pictures and associated audio for digital storage media at up to about 1.5 Mbit/s, Part 3: Audio", 1993.

[Mitchell, Pennebaker, Fogg and LeGall 97]: J. Mitchell, W. B. Pennebaker, C. E. Fogg and D. J. LeGall, *MPEG Video Compression Standard*, Chapman and Hall, New York, 1997.

[Schroeder, Atal and Hall 79]: M. R. Schroeder, B. S. Atal and J. L. Hall, "Optimizing Digital Speech Coders by Exploiting Masking Properties of the Human Ear", J. Acoust. Soc. Am., Vol. 66 no. 6, pp. 1647-1652, December 1979.

7. EXERCISES

Class Project:
The class project is to build and tune an MDCT-based perceptual audio coder. We recommend that students form groups of 2-3 students per group to work together on the coder. At the end of the course, each group will present their coder to the rest of the class. The presentations should describe how each coder works, discuss some of the design choices that were made, and let the class listen to a variety of sound examples that have been encoded/decoded at various compression ratios using the group's codec.

Chapter 10

Quality Measurement of Perceptual Audio Codecs

1. INTRODUCTION

Audio coding involves balancing data rate and system complexity limitations against needs for high-quality audio. While audio quality is a fundamental concept in audio coding, it remains very difficult to describe it in objective terms. Traditional quality measurements such as the signal to noise ratio or the total block distortion provide simple, objective measures of audio quality but they ignore psychoacoustic effects that can lead to large differences in perceived quality. In contrast, perceptual objective measurement schemes, which rely upon specific models of hearing, are subject to the criticism that the predicted results do not correlate well with the perceived audio quality. While neither simple objective measures nor perceptual measures are considered fully satisfactory, audio coding has traditionally relied on formal listening tests to assess a system's audio quality when a highly accurate assessment is needed. After all, human listeners are the ultimate judges of quality in any application.

The inadequacy of simple objective quality measures was made dramatically clear in the late eighties when J. Johnston and K. Brandenburg, then researchers at Bell Labs, presented the so-called "13 dB Miracle". In that example, two processed signals with a measured SNR of 13 dB were presented to the audience. In one processed signal the original signal was injected with white noise while in the other the noise injection was perceptually shaped. In the case of injected white noise, the distortion was a quite annoying background hiss. In contrast, the distortion in the perceptually shaped noise case varied between being just barely noticeable to being inaudible (i.e., the distortion was partially or completely masked by the signal components). Although the SNR measure was the same for both processed signals the perceived quality was very different, the second signal being judged as a very good quality signal (see also [Brandenburg and Spörer 92]). This example made clear to the audio community that quality measurements that reflect perceptual effects were needed to assess modern audio coders.

Throughout this chapter it is important to keep in mind that the perceived quality of a specific coder depends on both the type of material being coded and the data rate being used. Different material stresses different aspects of a coder. For example, highly transient signals such as percussive instruments will test the coder's ability to reproduce transient sounds effectively. In contrast, the closeness of spectral lines in a harpsichord piece will test the frequency resolution of a coder. Because of this dependence on source material, any quality assessment needs a good set of critical material for the assessment. Moreover, coding artifacts will become more pronounced as the coder's data rate is reduced. Any quality assessment comparing one coder against another needs to take into consideration the data rates used for each coder when ranking different coding systems.

Quality measurement is not only essential in the final assessment of an audio coder, but it is also critical throughout the design and fine-tuning of the different stages of the coding system. Designing an audio coder requires many decisions and judgement calls along the way, and it is very common to test and refine coding parameters by performing listening tests. Audio coding engineers have spent many long hours performing this important but arduous task! For example, in the development of MPEG-2 Advanced Audio Coding, AAC (see also Chapter 13), a number of experiments were carried out to compare technology alternatives by conducting listening tests in different sites. The results of these experiments were then analyzed and used to determine which technology was to be incorporated in the standard. Familiarity with audio coding artifacts and the ability to perform listening tests are important tools of the trade for anyone interested in developing audio coding systems.

In this chapter, we present an overview of the methods for carrying out listening tests. As we shall see in the next sections, formal listening tests require both sophistication and care to be useful. They require large numbers of trained subjects listening in a controlled environment to carefully choreographed selections of material. Although no substitute for formal listening tests has been found for most critical applications, the difficulty in doing it well has created great pent-up demand for acceptable substitutes in more forgiving applications. Coder design decisions are often made based on simple objective measurements or informal listening tests carried out with just a few subjects, and objective measures of perceptual quality are a hot topic for many researchers. In the second part of this chapter we discuss the principles behind objective perceptual quality measurements. The recent successes of the PEAQ (perceptual evaluation of audio quality) measurement system provide insurance that objective measurements can be used for informal assessment and in conjunction with formal listening tests.

Finally we briefly describe what we are listening for during listening tests and introduce the most commonly found artifacts in perceptual audio coding.

2. AUDIO QUALITY

The audio quality of a coding system can be linked to the perceived difference between the output of a system under test and a known reference signal. These differences are sometimes referred to as impairments. In evaluating the quality of a system, we need to be prepared for test signals that can range between perfect replicas of the reference signal (for example a lossless compression scheme) to test signals that bear very little resemblance to the reference. Depending where we are in this range, different strategies will be used to assess quality.

A very useful concept in quality assessment is that of "transparency". When even listeners expert in identifying coding impairments cannot distinguish between the reference and test signals, we refer to the coding system under test as being transparent. One way of measuring whether or not the coding system is transparent is to present both the test and reference signals to the listener in random order and to have them pick out which is the test signal. If the coding system is truly transparent, listeners will get it wrong roughly 50% of the time.

The questions we will want answered about coder quality will differ greatly depending on whether or not we are in the region of transparency. When we are in the region of transparency, the "coding margin" of the coder is an attribute that the test can assess. Coding margin refers to a measure of how far the coder is from the onset of audible impairments. Normally, we estimate coding margin using listening tests to find out how much we can reduce the coder's data rate before listeners can detect the test signal with statistically significant accuracy. To the degree that perceptual objective measures can assess how far below the masking curves the coding errors are positioned, they also can provide estimates of coding margin. For example, if the objective measure can report the worst-case noise-to-mask ratio in the signal (where the NMR represents the difference between the signal to mask ratio, SMR, and the SNR), we can estimate how many fewer bits would start making the impairments audible.

When we are below the region of transparency, we are interested in knowing how annoying the audible impairments are for different types of test signals. In this manner we can determine whether or not the coder is adequate at the tested data rate for a specific target application. In most cases, we are most interested in using the coder in or near the transparent region. In such cases we are concerned with identifying and rating impairments that are very small. It is exactly for such situations that the experts in the International Telecommunication Union, Radiocommunication Bureau, ITU-R, formerly know as International Radio Consultative Committee, CCIR, designed the five-grade impairment scale and formal listening test process we present in the next sections.

3.   SYSTEMS WITH SMALL IMPAIRMENTS

In this section we review the major features of carrying out a listening test to evaluate an audio codec producing signals with small impairments with respect to the reference signal [ITU-R BS. 1116]. The goal is to gain an appreciation of what's involved in carrying out such a test. For readers interested in further exploration of this topic, reading the ITU-R reference material [ITU-R BS.1116 and ITU-R BS. 562-3] is highly recommended.

3.1   Five-Grade Impairment Scale

The grading scale used in BS. 1116 listening tests is based on the five-grade impairment scale as defined by [ITU-R BS.562-3] and shown in *Figure 1*. According to BS.562-3, any perceived difference between the reference and the systems under test output should be interpreted as an impairment and the discrete five-grade scale measures the degree of perceptibility of the impairment. In BS.1116, the ratings are represented on a continuous scale between grades of 5.0 for transparent coding down to 1.0 for highly annoying impairments. The five-grade impairment as defined by BS.562-3 is related to the five-grade quality scale as shown in *Table 1*.

Table 1. Relationship between quality and impairment scale [ITU-R BS.562-3]

Quality	Impairment
5 Excellent	5 Imperceptible
4 Good	4 Perceptible, but not annoying
3 Fair	3 Slightly annoying
2 Poor	2 Annoying
1 Bad	1 Very annoying

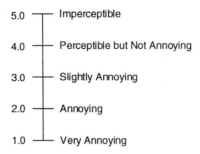

Figure 1. ITU-R five-grade impairment scale

Very often, to facilitate the data analysis, the difference grade between the listener's rating of the reference and coded signal is considered. This value, called the subjective difference grade, SDG, is defined as follows:

$$SDG = Grade_{coded\ signal} - Grade_{reference\ signal}$$

The SDG has a negative value when the listener successfully distinguishes the reference from the coded signal and it has a positive value when the listener erroneously identifies the coded signal as the reference. An SDG of zero means we are in the transparency region and any impairments are imperceptible, while an SDG of –4 indicates a very annoying impairment. *Table 2* shows the relationship between the five-grade impairment scale and the subjective difference grades.

Table 2. Subjective Difference Grades (SDGs) and their relationship with the ITU-R 5-grade impairment scale (assuming that the reference signal is identified correctly).

Impairment Description	ITU-R Grade	SDG
Imperceptible	5.0	0.0
Perceptible, but not annoying	4.0	-1.0
Slightly annoying	3.0	-2.0
Annoying	2.0	-3.0
Very annoying	1.0	-4.0

3.2   Test Method

The test method most widely accepted for testing systems with small impairments is the so-called "double-blind, triple-stimulus with hidden reference" method. In this method the listener is presented with three signals ("stimuli"): the reference signal, R, and then the test signals A and B. One of the two test signals will be identical to the reference signal and the other will be the coded signal. The test is carried out "double blind" in that neither the listener nor the test administrator should know beforehand which test signal is which. The assignments of signals A and B should be done randomly by some entity different from the test administrator entity so that neither the test administrator nor test subject has any basis for predicting which test signal is the coded one.

The listener is asked to assess the impairments of A compared to R, and of B compared to R according to the grading scale of *Figure 1*. Since one of the stimuli is actually the reference signal, one of them should be receiving a grade equal to five while the other stimulus may receive a grade that describes the listener's assessment of the impairment. If the system under test produces an output whose quality is in the transparency region, the listener will perceive no differences between the stimuli. In this case, one may decide to vary the data rate of the system to derive an estimate of the coding margin of the system. In addition to the basic quality assessment, the listener may be asked to grade spatial attributes such as stereophonic image, front image, and impression of surround quality separately for stereo and other multichannel material.

The double-blind, triple-stimulus with hidden reference method has been implemented in differt ways. For example, the system under test can be a real-time hardware implementation or a software simulation of the system. The stimuli can be presented with a tape-based reproduction or with a play-back system from computer hard disk. Preferably, only one listener is performing the test at one time. The listener is allowed to switch at will between R, A or B and to loop through the test sequences. In this fashion, the cognitive limitation of utilizing only echoic and short-term memory for judging the impairments in the relatively short sequence are mitigated (see also the description of the selection of critical material later in this chapter). The inclusion of the hidden reference in each trial provides an easy mean to check that the listener does not consistently make mistakes and therefore provides a control condition on the expertise of the listener.

The double-blind, triple-stimulus with hidden reference method has been employed worldwide for many formal listening tests of perceptual audio codecs. The consensus is that it provides a very sensitive, accurate, and stable way of assessing small impairments in audio systems.

3.3   Training and Grading Sessions

A listening test usually consists of two separate parts: a training phase and a formal grading phase. The training phase or "calibration" phase is carried out prior to the formal grading phase and it allows the listening panel to become familiar with the test environment, the grading process, and the codec impairments. It is essential for the listening panel to be familiar with the artifacts under study. A small unfamiliar distortion is much more difficult to assess than a small familiar distortion. This phenomenon is also known as informational masking, where the threshold of a complex maskee masked by a complex masker can decrease on the order of 40 dB after training [Leek and Watson 84]. Although the effects of the training phase in the assessment of perceptual audio coding have not been quantified, it is believed that this phase considerably reduces the informational masking that might occur.

Since the tests present the listener with the rather difficult task of recognizing very small impairments, it is common practice to introduce a "low anchor". A low anchor is an audio sequence with easily recognizable artifacts. The purpose of the low anchor is to help the listener in identifying artifacts.

An example of a test session grading sheet is shown in *Figure 2*. The sheet shown comes from one of the listening tests carried out during the development of the MPEG AAC coder. This particular example was used in the core experiment to assess the quality of reference model three (RM3) in 1996. The same core experiment was conducted in several test sites worldwide, including AT&T and Dolby Laboratories in the US, Fraunhofer Gesellschaft in Germany, and Sony in Japan. The particular core experiment described by the grading sheet of *Figure 2* was carried out through STAX headphones at Dolby Laboratories in San Francisco. The test material was presented to the subject via tape and consisted of two sessions of nine trials each. In Tape 1, Session 1, Trial 1, for example, the subject recognized A as being the hidden reference and B being the better than "perceptible but not annoying" system under test output.

MPEG-2 Audio NBC RM3 Test

Test site	Date	Loudspeaker/ headphone	Name of subject	Age profile e.g. 20-25,25-30, 30-35, 35-40 etc	Male/female M / F	Subject No	Tape No	Session No	Trial No	Grade for A	Grade for B	Remarks or comments (of test subject)
Dolby	14-Feb-96	STAX		30-35	F	1	1	1	1	5	4.5	
							1	1	2	5	4	
							1	1	3	3.5	5	
							1	1	4	5	4.7	
							1	1	5	4.9	5	
							1	1	6	4.8	5	
							1	1	7	5	4.8	
							1	1	8	5	4.2	
							1	1	9	4.5	5	
							1	1	10	4.4	5	
							1	2	11	4.4	5	
							1	2	12	5	4.9	
							1	2	13	5	4.9	
							1	2	14	5	2.5	
							1	2	15	4.5	5	
							1	2	16	2.7	5	
							1	2	17	5	4.2	
							1	2	18	5	4.8	

Figure 2. Example of a grading sheet from a listening test

3.4   Expert Listeners and Critical Material

The demanding nature of the test procedures is justified by the fact that the aim is to reveal any impairment in the system under test. These impairments may be recognized initially as very subtle, but may become more obvious after extensive exposure under different conditions once the system has been introduced to the general public. In general, a test is successfully designed if it can isolate the worst-case scenario for the system under study. In order to be able to accomplish this goal, only expert listeners and critical material that stresses the system under test are employed in formal listening tests.

The term expert listener applies to listeners who have recent and extensive experience of assessing impairments of the type being studied in the test. Even in cases where professional listeners are available, the training phase is very important. The expert listener panel is typically selected by employing pre-screening and post-screening procedures. An example of pre-screening procedures is given by an audiometric test. Post-screening is employed after the resulting data from the test are collected. Post-screening is based on the ability of the listener to consistently identify the hidden reference versus the system under test output sequence. There has been a long debate on the benefits versus the drawbacks of applying pre and post-screening procedures (see also [Ryden 96]). A demanding screening procedure may lead to the selection of a small number of expert listeners limiting the relevance of the results. On the other hand, the efficiency of the test may increase in doing so. In general, the size of the panel depends on the required resolution of the test, the desired representativity, etc. Typically, the number of expert listeners involved in a formal test varies between twenty and sixty.

The selection of critical material is an important aspect of the test procedure. While a database of difficult material for perceptual audio codecs has been collected over the past ten years with the work of MPEG and ITU-R (see also [Soulodre et al. 98] and [Treurniet and Soulodre 00]), it is impossible to create a complete list of such material. Critical material must be sought for each codec to be tested. Typically, an exhaustive search and selection by an expert group is conducted prior to the formal presentation of the test. If truly critical material cannot be found, the test fails to reveal differences among the systems and therefore is inconclusive. Generally, other than synthetic signals that deliberately break the system under test, any potential broadcast material or dedicated recordings that stresses the system under test is examined during the critical material selection stage. If more than one system is studied, then it is recommended to have an average of at least 1.5 audio excerpts for each codec under test with a minimum of five excerpts. Each excerpt should be relatively short, typically lasting about 10 seconds.

3.5   Listening Conditions

In order to be able to reliably reproduce the test, the listening conditions and the equipment need to be precisely specified. In [ITU-R BS.1116] the listening conditions include the characteristics of the listening room (such as its geometric properties, its reverberation time, early reflections, background noise, etc.), the characteristics and the arrangement of the loudspeakers in the listening room, and the reference listening area. In *Table 3* a summary of the characteristics of reference monitors and the listening room as per BS.1116 is shown. In *Figure 3* the multichannel loudspeaker configuration and the reference and worst case listening positions are shown. Typically, for mono and stereo program material, testing with both headphones and loudspeakers is recommended. Experience has shown that headphones highlight some types of artifacts better than loudspeakers and vice versa.

In addition, in [ITU-R BS.1116] specific listening levels are defined. Some listeners strongly prefer to have direct control on the absolute listening level. In general, arbitrary variations in the listening levels are not recommended since they may introduce unpredictable offsets in the masked thresholds and therefore increase the variance.

Finally, it should be noted that one of the most difficult criteria to meet in the ITU-R BS.1116 room specifications is the background noise. The Dolby listening room utilized in the MPEG-2 AAC core experiments exhibits a background noise defined by the NC 20 curve, while ITU-R BS.1116 requires the background noise to be contained between NR 10 and NR 15, in any case not to exceed NR 15.[4]

---

[4] Noise criterion, NC [Beranek 57], and noise rating, NR [Kosten and Van Os 62 and ISO 1996-1, ISO 1996-2, ISO 1996-3] are standardized curves of maximum permissible noise as a function of frequency.

Table 3. Reference monitor and room specifications as per ITU-R BS.1116

Parameter	BS.1116 Specifications
Reference loudspeaker monitors amplitude vs. frequency response	40 to 16 kHz (1/3 octave, free-field) ± 10° frontal axis ±3 dB re 0° ± 30° frontal axis ±4 dB re 0°
Reference loudspeaker monitors directivity index	3.5.1.1 0 dB ≤directivity index ≤12 dB 40 – 10000 Hz
Reference loudspeaker monitors non-linear distortion at 90 dB SPL	< −30 dB @ 40 to <250 Hz < −40 dB @ 250 to 16 kHz
Reference monitors time delay	< 100 μs between channels < 20 μs for headphones
Height and orientation of loudspeakers	1.10 m above floor reference axis at listener's ears
Loudspeaker configuration	Distance between loudspeakers 2 to 3 m Angle to loudspeakers 0°, ± 30°, ± 110°, Distance from walls > 1 m
Room dimensions and proportions	20 to 60 m^2 area for mono/stereophonic reproduction 30-70 m2 for multichannel reproduction 1.1 w/h ≤ l/h ≤ (4.5 w/h-4) l/h < 3 w/h < 3 where: l is length, w is width, h is height
Room reverberation time (T_m)	$T_{m/s} = 0.3(V/100)^{1/3}$ for 200Hz≤frequency≤4kHz where: V = volume of the room the following limits apply: 0 to 578 ms @ 63 Hz 428 to 228 ms @ 125 Hz 328 to 228 ms @ 200 to 4000 Hz 178 to 378 ms @ >8000 Hz
Room early reflections	< −10 dB for t ≤ 15 ms
Operational room response	≤ +3 dB, −7 dB @ 50 Hz ≤ +3 dB, −5 dB @ 125 Hz ≤ +3 dB, @ 250 – 2000 Hz ≤ +3 dB, −4.5 dB @ 4000 Hz ≤ +3 dB, −6.0 dB @ 8000 Hz ≤ +3 dB, −7.5 dB @ 16000 Hz
Background noise (equipment & HVAC on)	< NR15

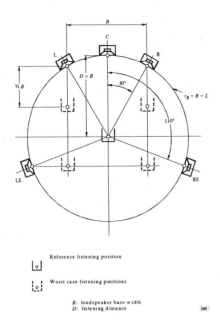

Figure 3. Multichannel loudspeakers configuration from [ITU-R BS.1116]

3.6 Data Analysis

After the expert listeners grades are collected, the data analysis starts. It is important to stress that the grades do not represent a physical measurement, but rather the individual interpretation of the grading scale. For example, it appears that the distances between steps of the five-grade impairment scale are different for different languages [Ryden 96]. Some experts have argued that given the nature of the quality of the scale employed, only non-parametric methods should be applied. On the other hand, provided that the assumptions underlying the parametric methods are met, these methods are considered very sensitive and powerful. Assuming that the tests have been conducted according the strict rules specified in [ITU-R BS.1116], then it is likely that each step of the grading scale is approximately of equal size to all the others. A variance model such as the ANOVA (ANalysis Of VAriance) method is most commonly used. The appropriate basis for a detailed statistical analysis is the difference grade (SDG) as defined earlier in this chapter, not the absolute grade because of the interdependence of any pair of observations. The results of the data analysis should be able to give a measure of the average performance of the system under test and, if more than one system is examined, the differences between the systems under tests.

The resolution achieved by the listening test is reflected in the confidence interval. This interval contains the SDG values with a specified degree of confidence, 1- α, where α represents the probability that inaudible differences are labeled as audible. In practice a value of 0.05 is chosen for α, which corresponds to a 95% confidence interval. In *Figure 4* an example of test results presentation is shown from [ISO/IEC MPEG N1420]. The specific test result shown corresponds to formal listening tests of MPEG-2 NBC (non backward compatible coder, later renamed MPEG-2 AAC) in the multichannel configuration at a data rate of 320 kb/s per five channel carried out at the BBC, UK in June 1996. *Figure 4* shows an assessment of the average quality of MPEG-2 NBC for ten critical items. The data were derived from the grades relative to the ten critical sequences analyzed with ANOVA and a confidence interval of 95%.

Finally, it should be mentioned that someone could argue that the strictly controlled test environment as described in this section not only is far from the reality of our living rooms, but also, in some cases a bit more forgiving. For example, consumer type of reproduction devices could reveal more artifacts than may be detected by professional equipment. Due to imperfections in the reproduction devices, such as notches in some frequency ranges, unmasking effects may occur. A listening area different from the one specified in [ITU-R BS.1116], may also enable the listener to distinguish distortion that was originally masked if the listener was at a certain distance and angle from the loudspeakers. Although these and other issues were debated by a number of experts, the general consensus was that one of the fundamental objectives in the design of listening test procedures was the reproducibility of the results, and this can only be obtained by a well-controlled procedure.

In general formal listening tests have shown very good reliability in the evaluation of audio coding systems and high correlation in their results, see for example [ISO/IEC MPEG 94/063, ITU-R 10/51, ITU-R 10/2-23, ISO/IEC MPEG 91/010]. In general they have proven to be a very effective tool in evaluating high-quality audio systems with small impairments.

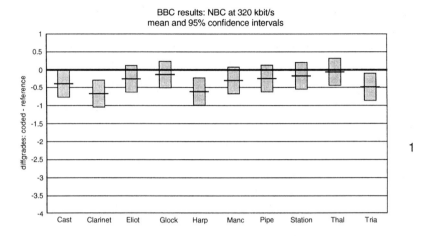

Figure 4. Example of formal listening test results from [ISO/IEC MPEG N1420]

3.7   The MUSHRA Method

While ITU-R BS.1116 is very effective in evaluating high quality audio systems with small impairments, other methods can be used for systems with intermediate quality.   For example, for speech signals in telephone environments recommendations [ITU-T P.800, ITU-T P.810 and ITU-T P.830] provide guidelines for assessment.   If one wishes to provide relative ranking between two systems in the region far from transparency, then [ITU-R BS.1284] provides appropriate guidelines.   In this case the seven-grade comparison scale is also recommended (see also *Table 4*).

Table 4. Seven-grade comparison scale

Grade	Comparison
3	Much better
2	Better
1	Slightly better
0	The same
−1	Slightly worse
−2	Worse
−3	Much worse

For systems where limitations are known a priori, such as, for example, digital transmission with reduced bandwidth, internet and mobile multimedia, etc., a new method, nicknamed MUSHRA, (MUltiple Stimulus with Hidden Reference and Anchors) was recently recommended by the ITU-R [ITU-R BS. 1534].

MUSHRA is a double-blind multi-stimulus test method with hidden reference and one or more hidden anchors as opposed to BS.1116's "double-blind triple-stimulus test method with hidden reference" test method. At least one of the anchors is required to be a low-passed version of the reference signal. The presence of the anchor(s) is meant as an aid in the task of weighing the relative annoyance of the various artifacts.

While there are common requirements with BS.1116 such as the selection of expert listeners, training phase, pre and post-screening of the listeners, listening conditions, in BS.1534 the subject is allowed to adjust the play-back level and the grading scale is modified since the grading of systems of intermediate quality would tend to cover mostly the lower half of the five-grade impairment scale.

According to the MUSHRA guidelines the subjects are required to score the stimuli according to a continuous quality scale divided in five equal intervals labeled, from top to bottom, excellent, good, fair, poor and bad (see for example [ITU-R BT.710]). The scores are then normalized in the range between 0 and 100, where 0 corresponds to the bottom of the scale (bad quality).

The data analysis is performed as the average across subjects of the differences between the score associated to the hidden reference and the score associated to each other stimulus. Typically a 95% confidence interval is utilized. Additional analysis, such as ANOVA etc., may also be calculated depending on the goal of the tests. The interested reader should consult [ITU-R BS 1534] for further details.

While listening tests have shown very good reliability in the evaluation of audio codecs, their cost can be high and sometimes the required level of effort might be impractical. Perceptual objective measurements have been studied since the late seventies and successfully applied to speech coding systems (see for example [ITU-T P.861] and [ITU-T P.862]). In recent years perceptual objective measurements for audio coding systems have reached a level of reliability and correlation with subjective listening tests that makes them an important complement in the assessment of audio coding systems. We turn next to the description of the underlying principles in perceptual objective measurements and the description of PEAQ, the ITU-R standard for such measurements.

4. OBJECTIVE PERCEPTUAL MEASUREMENTS OF AUDIO QUALITY

The aim of objective perceptual measurements is to predict the basic audio quality by using objective measurements incorporating psychoacoustics principles. Objective measurements that incorporate perceptual models have been introduced since the late seventies [Schroeder 79] for speech applications. More recently, psychoacoustics models have been exploited in the measurements of perceived quality of audio coding systems, see for example [Karjalainen 85], [Brandenburg and Spörer 92], [Beerends and Stemerdink 92], [Paillard, Mabilleu, Morissette and Soumagne 92], and [Colomes, Lever, Rault and Dehéry 93]. The effectiveness of objective quality measurements can only be assessed by comparison with corresponding scores obtained from subjective listening tests. One of the first global opportunities for correlating the results of these different audio objective subjective evaluations with informal subjective listening test results arose in 1995 in the early stages of the development of the MPEG-2 AAC codec. The need to test different reference models in the development of MPEG-2 AAC led to the study of objective subjective tests as a supplement and as an alternative to listening tests. Unfortunately, none of the objective subjective techniques under examination at that time showed reliable correlation with the results of the listening tests [ISO/IEC MPEG 95/201]. Similar conclusions were reached at the time within the work of ITU-R.

The recent adoption by ITU-R of PEAQ in BS.1387 [ITU-R BS.1387, Thiede et al.00] came in conjunction with data that corroborated the correlation between PEAQ objective difference grades, ODGs, with the SDGs obtained averaging the results of previous formal subjective listening tests [Treurniet and Soulodre 00]. While PEAQ is based on a refinement of generally accepted psychoacoustics models, it also includes new cognitive components to account for higher-level processes that come to play a role in the judgment of audio quality.

4.1 Different Approaches in Perceptual Objective Measurements

Before describing PEAQ, it is interesting to briefly review the two basic approaches used in perceptual objective measurements: the masked threshold method [Schroeder, Atal and Hall 79, Brandenburg and Spörer 92] and the internal representation method [Karjalainen 85, Beerends and

Stemerdink 92, Paillard, Mabilleu, Morissette and Soumagne 92, Colomes, Lever, Rault and Dehéry 93].

In the masked threshold method the error signal, computed as the difference between the original and the processed signal, is compared to the masked threshold of the original signal (see *Figure 5*). The error at a certain time and frequency is labeled as inaudible if its level falls below the masked threshold. Key to use of this method is an accurate model of masking.

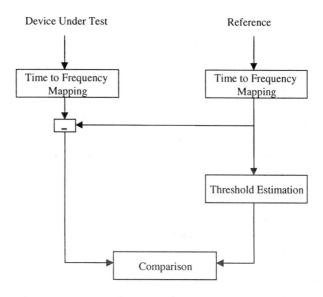

Figure 5. Block diagram of the masked threshold method

In the internal representation method, excitation patterns of the cochlea are estimated by modeling the signal transformations that take place in the human ear. The excitation patterns of the reference and of the output of the device under test are then compared to see if any differences in the excitation pattern can be discerned by the auditory system (see *Figure 6*). The internal representation method seems to be closer to the physiology of human perception than the masked threshold method previously described and it has the capacity of modeling more complex auditory phenomena. Key to the use of this method is a good description of the ability of the auditory system to discern changes in cochlear excitation patterns.

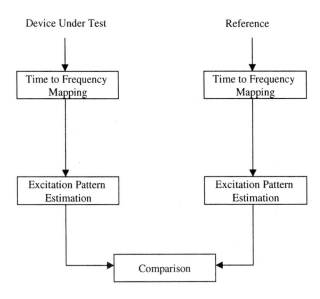

Figure 6. Block diagram of the internal representation method

4.2 Perceptual Evaluation of Audio Quality, PEAQ

PEAQ takes advantage of both masked threshold and internal representation methods [Thiede et al. 00]. In PEAQ's advanced version the peripheral ear is modeled both through a DFT and a bank of forty pairs of linear-phase filters with center frequencies and bandwidths corresponding to the auditory filters bandwidths. The model output values (MOVs) are based partly on the masked threshold method and partly on the internal representation method. The cognitive model compares the internal representations and calculates variables that summarize the behavior of the psychoacoustic activity over time. The MOVs include partial loudness of linear and non-linear distortion, noise to mask ratios, alteration of temporal envelopes, harmonic errors, probability of error detection, and proportion of signal frames containing audible distortions. Selected MOVs are used to predict the subjective quality rating (e.g., SDG) that would be assigned to the systems under test through formal listening tests. The MOVs are mapped to an objective difference grade (ODG) via an artificial neural network. The ODGs represent a prediction of the SDG values. The mapping of the ODGs derived from the MOVs was optimized by minimizing the difference between the ODG distribution and the corresponding distribution of mean SDGs from a number of formal listening tests.

In *Figure 7* the block diagram for the advanced version of PEAQ is shown. In contrast, the basic version utilizes the DFT-based peripheral ear model only. In general the correlation between subjective and objective quality evaluations are slightly higher for the advanced model than for the basic version. The pattern for the two versions, however, is similar [Treurniet and Soulodre 00].

PEAQ was used to generate objective quality measurements for audio data previously utilized in formal listening tests of state-of-the-art perceptual audio codecs. The performance of PEAQ was evaluated in different ways. The objective and mean subjective ratings were compared for each critical audio item used in formal tests. Then, the objective and subjective overall system quality measurements were compared by averaging codec quality measurements over critical items. The correlation between subjective and objective results proved very good and analysis of SDG and ODG showed no significant statistical differences [Treurniet and Soulodre 00]. The accuracy of the ODG demonstrated the capacity of PEAQ to correctly predict the outcome of the formal listening tests including the ranking of the codecs in terms of measured quality. PEAQ was also tested as a tool in aiding the selection of critical material for formal listening tests. On the basis of quality measurement, the PEAQ set of critical material included more than half the critical sequences used in the formal listening test under exam [Treurniet and Soulodre 00].

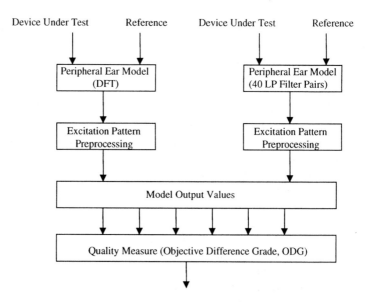

Figure 7. Block diagram of the advanced version of PEAQ [Thiede et al.00]

5. WHAT ARE WE LISTENING FOR?

In the previous sections, we have described how we can assess perceptual audio codecs. Formal listening tests and perceptual objective measurements are the most appropriate tools to assist us in this task. In this section we now address the question: "What is that we are listening for?". To inexperienced ears different versions of a codec may sound equally good. The more familiar one becomes with coding artifacts, the easier it is to recognize the codec impairments and to distinguish between different versions. In addition to general distortion due to bit starvation, there are a number of less obvious artifacts commonly encountered in audio coding. In this section, we briefly describe some of the most common coding artifacts that one may expect when listening to perceptual audio coding systems. For detailed sound examples, the reader can refer to [AES CD-ROM On Perceptual Audio Coders 2001].

5.1 Pre-echo

We saw in Chapter 9 that the first stage in perceptual audio coding is typically a time to frequency mapping stage. In this stage, one would like to maximize the time-frequency resolution of the signal representation. Block size values go up to 2048 time samples in state-of-the-art audio coders. In Chapter 6, we described how temporal masking effects cover a range of the order of few ms before the onset of the signal (backward or pre-masking) and few 100 ms after the onset of the masker (forward or post –masking). In the case of signals with sharp attacks, like for example castanets, some of the quantization noise may spread before the onset of the masker through the input block length in a time region where it is not masked (see also *Figure 8* in Chapter 6). In this case, the spreading in time of quantization noise results in the artifact known as pre-echo. Pre-echo effects dampen the sharpness and clarity of the attacks, resulting in what some call "double attacks". As mentioned in Chapter 5, pre-echo can be mitigated by trading-off frequency resolution for time resolution of the filter bank, that is by applying block switching.

5.2 Aliasing

If the filter bank is implemented as a set of sub-band filters (see also Chapter 4), like for example the PQMF utilized in the MPEG Audio coders, one may expect that aliasing effects due to the nature of these filters may introduce artifacts. It appears that, in normal conditions, this artifact is hardly audible [Erne 01]. Analogously, in the MDCT approach, although the overall system is a perfect reconstruction system in absence of quantization, coarse quantization may impede full time-domain aliasing cancellation resulting in audible artifacts. In general, this is not a problem in normal conditions.

5.3   "Birdies"

This artifact arises when, at low data rate for spectrally demanding signals, the highest frequency bands bit allocation changes from block to block. Consequently, some spectral coefficients may temporarily appear and disappear. The resulting effects cause a very noticeable change in timbre at high frequencies, sounding almost like a chirp, therefore the name "birdies". A potential solution to this problem is to low-pass the signal prior to coding in order to prevent bit allocation in this region. The resulting signal will sound band-limited, but this effect is in general much less disturbing than the birdies artifact. Even when the signal is band-limited, however, there is still the possibility that this artifact may occur. Ideally, a higher data rate should be selected in order to maintain high quality.

5.4   Speech Reverberation

Typically, audio coders are not tuned to any specific sound source, like for example speech coders, but, on the contrary, try to address general wide-band audio signals. For general audio coders speech is a very demanding signal since it requires both high frequency resolution, for example for highly tonal segments like vowels, and high time resolution for fricatives and plosives. If a large block size is employed for the filter bank at low data rates, the speech may sound unnaturally reverberant with a "metallic" quality to it. This artifact sometimes referred to as "speech reverberation" can be mitigated by adopting a filter bank which dynamically adapts its resolution to the characteristics of the input signal.

5.5   Multichannel Artifacts

Multichannel artifacts arise from differences in the perceived sound field of the coded signal. Spatial attributes such as stereophonic image, front image, and impression of surround quality may exhibit differences in the coded version. Some of the most common artifacts include a loss or a shift in the stereo image and changes in the signal envelope at high frequencies, a phenomenon related to the effects of binaural masking. Joint stereo coding strategies such as M/S coding and intensity stereo coding are currently employed for multichannel coding. Of the two approaches, M/S tends to be lossless or nearly lossless, while intensity stereo coding may introduce quite noticeable artifacts at low data rates. Intensity stereo coding reconstructs the output multichannel signal above a certain frequency from a single channel by appropriately scaling it. If the signal is not stationary for the duration of the input block and has different envelopes in different channels the recovery will introduce artifacts. A particularly revealing excerpt for these types artifacts is the applause sample in [AES CD-ROM On Perceptual Audio Coders 2001].

6. SUMMARY

In this chapter, we discussed the importance of subjective listening tests in the assessment of perceptual audio coding. The more controlled are the parameters in the test, the more reliable are the test results. The double blind, triple stimulus with hidden reference method as per the ITU-R BS.1116 specifications has proven to generate reliable results. Although test sites for formal listening tests need to be fully compliant with BS.1116, the basic guidelines are also useful for carrying out informal listening tests. Performing listening tests plays a central role not only in the final assessment of an audio coder, but also during its development by providing invaluable feedback for the fine-tuning of different parameters. Subjective listening test results provide a measure of the degree of transparency of the perceptual codecs under tests and the reliability of differences between the different codecs. In recent years, perceptual objective measurements have also been developed such as PEAQ that show good correlation with subjective tests results. These also represent an important tool in the development of audio coders.

This chapter concludes the first part of this book devoted to a discussion of the underlying principles and implementation issues in perceptual audio coding. In the remaining chapters we review how these basic principles are applied in state-of-the-art perceptual audio coders such as the MPEG and the Dolby families of audio coders, and how different implementation strategies have affected the final results.

7. REFERENCES

[AES CD-ROM On Perceptual Audio Coders 2001]: "Perceptual Audio Coders: What to Listen For", AES 2001.

[Beerends and Stemerdink 92]: J. G. Beerends and J. A. Stemerdink, "A Perceptual Audio Quality Measure Based on a Psychoacoustic Sound Representation", J. Audio Eng. Soc., Vol. 40, no. 12, pp. 963-978, December 1992.

[Beranek 57]: L. L. Beranek, "Revised Criteria for Noise in Buildings", Noise Control, Vol. 3, pp. 19-27, 1957.

[Brandenburg and Spörer 92]: K. Brandenburg and T. Spörer, "NMR and Masking Flag: Evaluation of Quality Using Perceptual Criteria," Proc. of AES 11th Intl. Conf., Portland, May 1992.

[Colomes, Lever, Rault and Dehéry 93]: C. Colomes, M. Lever, J.B. Rault and Y.F. Dehéry, "A Perceptual Model Applied to Audio Bit Rate Reduction," presented at the 95th AES Convention, Preprint 3742, New York, October 1993.

[Erne 01]: M. Erne, " Perceptual Audio Coders: What to Listen For," presented at the 111th AES Convention, New York, November 2001.

[ISO 1996-1]: ISO 1996, "Acoustics – Description and Measurement of Environmental Noise – Part 1: Basic Quantities and Procedures", Geneva 1982.

[ISO 1996-2]: ISO 1996, "Acoustics – Description and Measurement of Environmental Noise – Part 2: Acquisition of Data Pertinent to Land Use", Geneva 1987.

[ISO 1996-3]: ISO 1996, "Acoustics – Description and Measurement of Environmental Noise – Part 3: Applications to Noise Limits", Geneva 1987.

[ISO/IEC MPEG 91/010]: ISO/IEC JTC 1/SC 29/WG 11 MPEG 91/010, "The MPEG/AUDIO Subjective Listening Test" Stockholm, April/May 1991.

[ISO/IEC MPEG 94/063]: ISO/IEC JTC 1/SC 29/WG 11 MPEG 94/063, "Report on the MPEG/Audio Multichannel Formal Subjective Listening Tests", 1994.

[ISO/IEC MPEG 95/201]: ISO/IEC JTC 1/SC 29/WG 11 MPEG 95/201, "Chairman's Report on the Work of the Audio ad Hoc Group on Objective Measurements" Tokyo, July 1995.

[ISO/IEC MPEG N1420]: ISO/IEC JTC 1/SC 29/WG 11 N1420, "Overview of the Report on the Formal Subjective Listening Tests of MPEG-2 NBC Multichannel Audio Coding", 1996.

[ITU-R 10/2-23]: International Telecommunications Union, Radiocommunication Sector 10/2-23-E, "Chairman Report of the Second Meeting of the Task Group 10/2", Geneva 1992.

[ITU-R 10/51]: International Telecommunications Union, Radiocommunication Sector 10/51-E, "Low Bit Rate Multichannel Audio Coder Test Results", Geneva 1995.

[ITU-R BS.1116]: International Telecommunications Union, Radiocommunication Sector BS.1116 (rev. 1), "Methods for the Subjective Assessment of Small Impairments in Audio Systems Including Multichannel Sound Systems ", Geneva 1997.

[ITU-R BS.1284]: International Telecommunications Union, Radiocommunication Sector BS.1284, "Methods for the Subjective Assessment of Sound Quality - General Requirements", Geneva 1997.

[ITU-R BS.1387]: International Telecommunications Union, Radiocommunication Sector BS.1387, "Method for the Objective Measurements of Perceived Audio Quality ", Geneva 1998.

[ITU-R BS.1534]: International Telecommunications Union, Radiocommunication Sector BS.1534, "Method for the Subjective Assessment of Intermediate Quality Level Coding Systems - General requirements", Geneva 2001.

[ITU-R BS.562-3]: International Telecommunications Union, Radiocommunication Sector BS.562-3, "Subjective Assessment of Sound Quality", Geneva 1978-1982-1984-1990.

[ITU-R BT.710]: International Telecommunications Union, Radiocommunication Sector BT.710, "Subjective Assessment Methods for Image Quality in High-Definition Television", Geneva 1998.

[ITU-T P.800]: International Telecommunications Union, Telecommunication Sector P.800, "Methods for Subjective Determination of Transmission Quality", Geneva 1996.

[ITU-T P.810]: International Telecommunications Union, Telecommunication Sector P.810, "Modulated noise reference unit (MNRU)", Geneva 1994.

[ITU-T P.830]: International Telecommunications Union, Telecommunication Sector P.830, "Subjective Performance Assessment of Telephone Band and Wide Band Digital Codecs", Geneva 1996.

[ITU-T P.861]: International Telecommunications Union, Telecommunication Sector P.861, "Objective Quality Measurement of Telephone Band (300-3400 Hz) Speech Codecs", Geneva 1998.

[ITU-T P.862]: International Telecommunications Union, Telecommunication Sector P.862, "Perceptual Evaluation of Speech Quality (PESQ), an Objective

Method for End-to-End Speech Quality Assessment of Narrowband Telephone Networks and Speech Codecs", Geneva 2001.

[Karjalainen 85]: M. Karjalainen, "A New Auditory Model for the Evaluation of Sound Quality of Audio Systems", Proc. of ICASSP, pp. 608-611, March 1985.

[Kosten and van Os 62]: Kosten and van Os, "Community Reaction Criteria for External Noises", National Physical Laboratory Symposium No. 12, P. 377, London H. M.S.O. 1962.

[Leek and Watson 84]: M. R. Leek and C. S. Watson, "Learning to Detect Auditory Pattern Components", J. Acoust. Soc. Am., Vol. 76 no. 4, pp. 1037-1044, October 1984.

[Paillard, Mabilleu, Morissette and Soumagne 92]: B. Paillard, P. Mabilleu, S. Morissette and J. Soumagne, "PERCEVAL: Perceptual Evaluation of the Quality of Audio Signals," J. Audio Eng. Soc., pp. 21-31, vol. 40, January/February 1992.

[Ryden 96]: T. Ryden, "Using Listening Tests to Assess Audio Codecs", in *Collected Papers on Digital Audio Bit-Rate Reduction*, N. Gilchrist and C. Gerwin (ed.) pp. 115-125, AES 1996.

[Schroeder, Atal and Hall 79]: M. R. Schroeder, B. S. Atal and J. L. Hall, "Optimizing Digital Speech Coders by Exploiting Masking Properties of the Human Ear", J. Acoust. Soc. Am., Vol. 66 no. 6, pp. 1647-1652, December 1979.

[Soulodre et al. 98]: G. A. Soulodre, T. Grusec, M. Lavoie, and L. Thibault, "Subjective Evaluation of State-of-theArt Two-Channel Audio Codecs", J. Audio Eng. Soc., Vol. 46, no. 3, pp. 164-177, March 1998.

[Thiede et al. 00]: T. Thiede, W. Treurniet, R. Bitto, C. Schmidmer, T. Sporer, J. Beerends, C. Colomes, M. Keyhl, G. Stoll, K. Brandenburg and B. Feiten,, "PEAQ-The ITU Standard for Objective Measurement of Perceived Audio Quality", J. Audio Eng. Soc., Vol. 48, no. 1/2, pp. 3-29, January/February 2000.

[Treurniet and Soulodre 00]: W. C. Treurniet and G. A. Soulodre, "Evaluation of the ITU-R Objective Audio Quality Measurement Method", J. Audio Eng. Soc., Vol. 48, no. 3, pp. 164-173, March 2000.

8. EXERCISES

Listening Test:
In this exercise you will perform a listening test to compare the coders you built in Chapters 2 and 5 on a variety of test samples. You will rate the coders using the ITU-R five-grade impairment scale.

1. Prepare a set of short test signals to be used for your listening test. Make sure that the set includes 1) human speech, 2) highly tonal music (e.g., flute), and 3) music with sharp attacks (e.g., drum solo).
2. Encode/decode each of your test signals using 1) your coder from Chapter 2 with 4-bit midtread uniform quantization, 2) your coder from Chapter 5 with three scale bits and five mantissa bit midtread floating point quantization, 3) your coder from Chapter 5 with N = 2048 and 4-bit midtread uniform quantization, 4) your coder from Chapter 5 with N = 2048 and three scale bit and five mantissa bit floating point quantization, and 5) your coder from Chapter 5 with N = 256 and three scale bit and five mantissa bit floating point quantization.
3. Grade each of your encoded/decoded test signals using the ITU-R five-grade impairment scale. Summarize the performance of your coders.
4. Team with a classmate to perform a double-blind, triple-stimulus with hidden reference listening test using several of your friends and classmates as test subjects to evaluate your encoded/decoded test signals. For each coder, prepare a graphical summary of the results showing the highest/lowest/mean SDG score for each sound test signal. Summarize the results. Do your classmates (who are hopefully trained listeners at this point) give significantly different ratings than your other (untrained) friends?

PART II: AUDIO CODING STANDARDS

Chapter 11

MPEG-1 Audio

1. INTRODUCTION

After the introduction of digital video technologies and the CD format in the mid eighties, a flurry of applications that involved digital audio/video and multimedia technologies started to emerge. The need for interoperability, high-quality picture accompanied by CD-quality audio at lower data rates, and for a common file format led to the institution of a new standardization group within the joint technical committee on information technology (JTC 1) sponsored by the International Organization for Standardization (ISO) and the International Electrotechnical Commission (IEC). This group, the Moving Picture Experts Group (MPEG), was established at the end of the eighties with the mandate to develop standards for coded representation of moving pictures, associated audio, and their combination [Chiariglione 95].

MPEG-1 was the initial milestone achieved by this committee after over three years of concurrent work. MPEG-1 Audio represents the first international standard that specifies the digital format for high quality audio, where the aim is to reduce the data rate while maintaining CD-like quality. Other compression algorithms standardized prior to MPEG-1 addressed either speech-only applications or provided only medium-quality audio performance. The success of the MPEG standard enabled the adoption of compressed high-quality audio in a large range of applications from digital broadcasting to internet applications. Everyone is now familiar with the MP3 format (MPEG Layer III). The introduction of MPEG Audio technology radically changed the perspective of digital distribution of music, touching diverse aspects of it, including copyright protection, business models, and ultimately our every-day life.

In this chapter and Chapters 12, 13, and 15, we discuss different audio coding algorithms standardized by MPEG. In this chapter, after presenting a brief history of the MPEG standards with emphasis on the MPEG Audio goals and objectives, we discuss in depth the layered approach and attributes of MPEG-1 Audio.

2.   BRIEF HISTORY OF MPEG STANDARDS

The Moving Pictures Experts Group (MPEG) was established with the mandate to develop standards for coded representation of moving pictures, associated audio, and their combination. The original group of about 25 people met for the first time in 1988. Later MPEG become working group 11 of ISO/IEC JTC 1 sub-committee 29. Any official document of the MPEG group can be recognized by the ISO/IEC JTC 1/SC 29/WG 11 header. There were originally three work items approved for MPEG:

- The MPEG-1 standard [ISO/IEC 11172] coding of synchronized video and audio at a total data rate of about 1.5 Mb/s was finalized in 1992.
- The MPEG-2 standard [ISO/IEC 13818] coding synchronized video and audio at a total data rate of about 10 Mb/s was finalized in 1994.
- The third work item, MPEG-3, addressing coding of synchronized video and audio at a total data rate of about 40 Mb/s was dropped in July 1993, after being deemed redundant since its attributes were incorporated in the MPEG-2 specifications.

After the initial work started, a proposal for audiovisual coding at very low data rates with additional functionalities, such as scalability, 3-D, synthetic/natural hybrid coding, was first discussed in 1991 and then proposed in 1992 [ISO/IEC MPEG N271]. This phase of MPEG standardization was called MPEG-4 giving origin to the somewhat disconnected numbering of subsequent phases of MPEG. MPEG-4 was finalized in 1998 as [ISO/IEC 14496]. The MPEG-1, 2, 4 standards address video, audio compression as well as synchronization, compliance, and reference software issues. Although MPEG Audio is often utilized as a stand-alone standard, it is one component of a multi-part standard, where typically "part one" describes the system elements (i.e. synchronization of video and audio stream, etc.) of the standard, "part two" the video coding elements, and "part three" the audio coding elements. After MPEG-4 the work of MPEG started focusing more and more towards coding-related technology rather than coding technology per se. MPEG-7, whose completion was reached in July 2001 [ISO/IEC 15938], addresses the description of multimedia content for multimedia database search. Currently in the developmental stage (only three parts of the standard have been approved), MPEG-21 is addressing the many elements needed to build an infrastructure for the usage of multimedia content, see for example [ISO/IEC MPEG N4318].

The goal of MPEG-1 Audio was originally to define the coded representation of high quality audio for storage media and a method for decoding high quality audio signals. Later the algorithms specified by MPEG were tested within the work of ITU-R for broadcasting applications and recommended for use in contribution, distribution, commentary, and emission channels [ITU-R BS.1115]. Common to all phases of MPEG was the standardization of the bitstream and decoder specifications only, but not of the encoder. A sample encoder algorithm is described in an "informative" part of the standard, but following the sample algorithm is not required to be compliant with the standard. This approach, while allowing for interoperability between implementation from different manufacturers, also allowed encoder manufactures to retain control on the core intellectual property and know-how that contributed to the success of the coding system. The input of the MPEG-1 audio encoder and the output of the decoder are compatible with existing PCM standards such as the CD and the digital audio tape, DAT, formats. MPEG-1 audio aimed to support one or two main channels, depending on the configuration (see more details on channel configuration in the next sections) and sampling frequencies of 32 kHz, 44.1 khz, and 48 kHz.

In MPEG-2 Audio the initial goal was to define the multichannel extension to MPEG-1 audio (MPEG-2 BC, backwards compatible) and to define audio coding systems at lower sampling rates than MPEG-1, namely at 16 kHz, 22.5 kHz and 24 kHz. This phase of the work of the MPEG audio sub-group was partially motivated by the debut of multichannel de-facto standards in the cinema industry such as Dolby AC-3 (currently known also as Dolby Digital, see also Chapter 14) and the need for lower data rates for the emerging internet applications. After a call for proposals in late 1993, the work on a new aspect of multichannel audio, the so-called MPEG-2 non-backwards compatible, NBC (later renamed MPEG Advanced Audio Coding, AAC), was started in 1994. The objective was to define a higher-quality multichannel standard than achievable with MPEG-1 extensions. A number of studies highlighted the burden in terms of quality, or equivalently in terms of increased data rates demands, suffered by the design of a multichannel audio system when the backwards compatibility requirement was enforced (see for example [Bosi, Todd and Holman 93 and ISO/IEC MPEG N1229] and see also next chapter for a detailed discussion on this issue). As a result of this phase of work, MPEG-2 AAC was standardized in 1997 [ISO/IEC 13818-7]. In a number of subjective tests MPEG-2 AAC shows comparable or better audio quality than MPEG-2 Layer II BC operating at twice the data rate, see for example [ISO/IEC MPEG N1420].

The MPEG-4 Audio goals were to provide a high coding efficiency, where the data rates introduced ranging from 200 b/s to 64 kb/s reach lower values than the data rates defined in MPEG-1 or 2. In addition to general audio coding technology MPEG-4 also accommodates:
– speech coding technology
– error protection
– content-based interactivity such as flexible access and manipulation, for example pitch/speed modifications;
– universal access, for example access to a subset of data or scalability
– support for synthetic audio and speech, such as in structured audio, SA, and text to speech, TTS, interfaces;
– additional effects such as post-processing (reverberation, 3D, etc.) and scene composition.

From its onset, the MPEG standardization process played a very relevant role in promoting technology across the boundaries of a single organization or country. As a result, teams around the world joined forces and expertise to design algorithms that incorporated the most advanced technology available given a certain range of applications. In the first phases of the MPEG work, the focus was centered on coding technologies. In this chapter and in Chapters 12 , 13 and 15 a detailed description of the audio coding algorithms developed during the MPEG-1 through 4 phases are presented. In particular, the next sections of this chapter present the details of the MPEG-1 Audio algorithms.

3.   MPEG-1 AUDIO

MPEG-1 is a compression standard that addresses the compression of synchronized video and audio at a total data rate of 1.5 Mb/s. It includes systems, video, and audio specifications. MPEG-1 became a standard in 1992, and is also known as [ISO/IEC 11172]. [ISO/IEC 11172-3] specifies the audio portion of the MPEG-1 standard. It includes the syntax of the audio coded bitstream and a description of the decoding process. In addition, reference software modules and a set of test vectors for assessing the compliance of the decoder are also provided by the standard specifications. The MPEG-1 audio encoder structure is not a mandatory part of the standard specifications and its description is an informative annex to the standard. While the mandatory nature of the syntax and decoding process ensures interoperability, the encoder implementation is left to the designers of the system, leaving a large degree of differentiation within the boundaries of the standard specifications. The MPEG-1 standard describes a perceptual audio coding algorithm that is designed for general audio signals. There is no specific source model applied as, for example, in speech codecs. It is simply assumed that the statistics of the input signal are quasi-stationary. The audio signal is then represented by its spectral components on a frame-by-frame basis and encoded exploiting perceptual models. The aim of the algorithm is to provide a perceptually lossless coding scheme. The MPEG-1 Audio standard specifications were derived from two main proposals: MUSICAM [Dehéry, Stoll and Kerkhof 91] presented by CCETT, IRT and Philips, which is the basis for the low-complexity first two layers (see also next sections), and ASPEC (see [Brandenburg and Johnston 90] and [Brandenburg et al. 91]) presented by AT&T, FhG, and Telefunken which is the basis for layer III. The quality of the audio standard was tested by extensive subjective listening tests during its development. The resulting data, see for example [ISO/IEC MPEG 91/010], showed that, under strictly controlled listening conditions, experts listeners were not able to distinguish between coded and original sequences with statistical significance at typical codec data rates. Typical data rates for the coded sequences were 192 kb/s per channel for MPEG Layer I, 128 kb/s per channel for Layer II and Layer III (see detailed description of the different MPEG-1 Audio Layers later in this chapter and also the MPEG public documents at [MPEG]).

3.1   Main Features of MPEG-1 Audio

The sampling rates supported by MPEG-1 are 32, 44.1, and 48 kHz. The channel configurations encompass one or two channels. In addition to a monophonic mode for a single audio channel configuration, a dual monophonic mode for two independent channels is included. A stereo mode for stereophonic channels, which shares the available bit pool amongst the two channels but does not exploit any other spatial perceptual model, is also covered. Moreover, joint stereo modes that take advantage of correlation and irrelevancies between the stereo channels are described in the standard. The data rates vary between 32 and 224 kb/s per channel allowing for compression ratios ranging from 2.7 to 24:1 depending on the sampling rate. In addition to the pre-defined data rates, a free format mode can support supplementary, fixed data rates.

MPEG-1 Audio specifies three layers. The different layers offer increasingly higher audio quality at slightly increased complexity. While

Layers I and II share the basic structure of the encoding process having their roots in an earlier algorithm also known as MUSICAM [Dehéry, Stoll and Kerkhof 91], Layer III is substantially different. The Layer III algorithm was derived from the merge of ASPEC [Brandenburg et al. 91] with the Layer I and II filter bank, the idea being that a Layer III decoder should be able to decode Layer I and II bitstreams. Layer I is the simplest layer and it operates at data rates between 32 and 224 kb/s per channel. The preferred range of operation is above 128 kb/s. Layer I finds an application, for example, in the digital compact cassette, DCC, at 192 kb/s per channel. Layer II is of medium complexity and it employs data rates between 32 and 192 kb/s per channel. At 128 kb/s per channel it provides very good audio quality. A number of applications take advantage of Layer II including digital audio broadcasting, DAB, [ETS 300 401 v2] and digital video broadcasting, DVB [ETS 300 421, ETS 300 429, ETS 300 744]. Layer III exhibits the highest quality of the three layers at an increased complexity. The data rates for Layer III are lower than the rates for Layers I and II and they vary between 32 and 160 kb/s per channel. Layer III displays very good quality at rates below 128 kb/s per channel. Applications of Layer III include transmission over ISDN lines and internet applications. A modification of the MPEG Layer III format at lower sampling frequencies gave origin to the ubiquitous MP3 file format.

In spite of the differences in complexity, single-chip, real-time decoder implementations exist for all three layers. It should be noted that, in addition to the main audio data, all three layers provide a means of including auxiliary data within the bitstream syntax. Finally it should be mentioned that, MPEG-1 Layers II and III were also selected by ITU-R task group, TG, 10/2 for broadcasting applications in recommendation BS.1115. In ITU-R BS.1115, Layer II is recommended for emission at the data rate of 128 kb/s per channel, and for distribution and contribution at data rates above 180 kb/s per channel. Layer III is also recommended in BS.1115 for commentary broadcasting at data rates of about 60 kb/s per channel.

The main building blocks of the MPEG-1 audio coding scheme are shown in *Figure 1* and *Figure 2*. The basic building blocks include a time to frequency mapping stage followed by a bit or noise allocation stage. The input signal also feeds a psychoacoustic model block whose output determines the precision of the allocation stage. The bitstream formatting stage interleaves the representation of the quantized data with side information and optional ancillary data. The decoder interprets the bitstream, restores the quantized spectral components of the signal and finally reconstructs the time domain representation of the audio signal from its frequency representation.

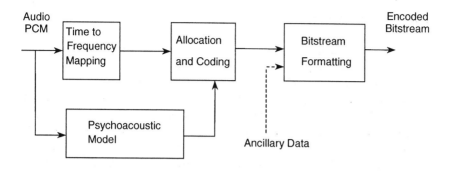

Figure 1. MPEG-1 Audio encoder basic building blocks

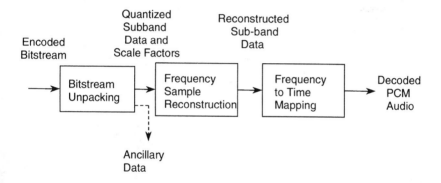

Figure 2. MPEG-1 Audio decoder basic building blocks

3.2   Different Layers Coding Options

The general approach to the coded representation of audio signals is the same for all layers. Based on the time to frequency mapping of the signals with a source model design based on statistics of generic audio signals, they share the basic building blocks and group the input PCM samples into frames of samples for analysis/synthesis. There are, however, a number of differences in the different layers' algorithms going from the simple approach of Layer I to the more sophisticated approach of Layer III at increased complexity. In *Figure 3* and *Figure 4* the block diagrams of Layers I, II, and III in single channel mode are shown.

3.2.1   Layers I and II

For Layers I and II, the time to frequency mapping is performed by applying a 32-band PQMF (see also Chapter 4) to the main audio path data. The frequency representation of the signal is scaled and then quantized with a uniform midtread quantizer (see also Chapter 2) whose precision is determined by the output of the psychaocustic model. Typically, Psychoacoustic Model 1 (see also next sections) is applied, where the psychoacoustic analysis stage is performed with a 512-point FFT (Layer I) or 1024-point FFT (Layer II). In order to further reduce the data rate, Layer II applies group coding of consecutive quantized samples certain levels (see also next sections).

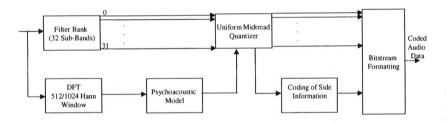

Figure 3. Block diagram of Layers I and II (single channel mode)

3.2.2   Layer III

For Layer III the output of the PQMF is fed to an MDCT stage (see also Chapter 5). In addition, the Layer III filter bank is not static as in Layers I and II, but it is signal adaptive (see next sections). The output of this hybrid filter bank is scaled and then non-uniformly quantized with a midtread quantizer. Noiseless coding is also applied in Layer III. In an iterative loop that performs the synthesis of the Huffman-encoded, quantized signal and compares its relative error levels with the masked thresholds levels, the quantizer step size is calculated for each spectral region. The quantizer step is once again determined by the output of the psychoacoustic model, however, the nature of the psychoacoustic model (Model 2, see next sections) applied to Layer III is substantially different from the model applied for Layers I and II. *Figure 4* highlights one of the differences, the analysis stage, which is performed by applying two, 1024-point FFTs. In all layers the audio data together with the side information such as bit allocation and control parameters are multiplexed with the optional ancillary data and then stored or transmitted.

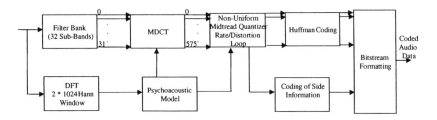

Figure 4. Block diagram of Layer III (single channel mode)

In the next sections, we describe the common characteristics of the audio coding algorithms in the three layers.

4. TIME TO FREQUENCY MAPPING

A PQMF filter bank (see also Chapter 4) is part of the time to frequency mapping stage for all three MPEG layers. This filter divides the frequency spectrum into 32 equally spaced frequency sub-bands. For Layers I and II the output of the PQMF represents the signal spectral data to be quantized. The frequency resolution of the Layer I and II filterbank is 750 Hz at a 48 kHz sampling rate. For Layer III, the PQMF is cascaded with an 18 frequency-line MDCT for a total of 576 frequency channels in order to increase the filter bank resolution.

4.1 Layer III Hybrid Filter Bank

The block diagram of Layer III filter bank analysis stage is shown in *Figure 5*. After the 32-band PQMF filter, blocks of 36 sub-band samples (for steady state conditions) are overlapped by 50 percent, multiplied by a sine window and then processed by the MDCT transform (see also Chapter 5). It should be noted that, in addition to the potential frequency aliasing introduced by the PQMF, the OTDAC transform introduces also time aliasing that cancels out between adjacent time-blocks in absence of quantization in the overlap-add stage of the decoder process. In order to lessen some of the artifacts potentially introduced by the overlapping bands of the PQMF, for long blocks (steady state conditions) the Layer III filter bank multiplies the MDCT output by coefficients that reduce the signal aliasing [Edler 92].

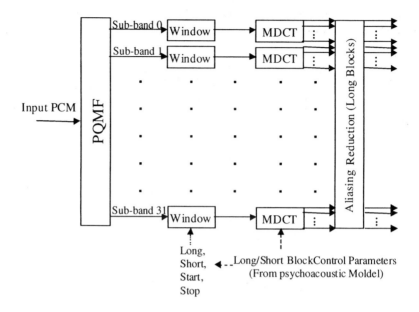

Figure 5. MPEG Audio Layer III analysis filter bank structure

In the decoder, the inverse aliasing reduction process is applied prior to the IMDCT in order to provide the correct sub-band samples for the PQMF synthesis stage for aliasing cancellation (see *Figure 6*). A pure sine wave signal processed by the hybrid PQMF/MDCT filter bank without aliasing reduction can present a spurious component as high as −12 dB with respect to the original signal. After the aliasing reduction process, the spurious component magnitude is reduced significantly. It should be noted, however, that, although the aliasing reduction process greatly improves the frequency representation of the signal, residual aliasing components might still be present. In the synthesis stage, the IMDCT is applied prior to the reconstruction PQMF. After the de-quantization of the spectral components and, when applicable, the joint stereo processing of the signal, the inverse aliasing reduction is applied. Next, the IMDCT is employed followed by the windowing process, where the windows applied are defined in the same manner as the analysis windows. The first half of the current windowed block is overlapped and added to the second half of the windowed samples of the previous block. For the long block the output of the overlap and add stage consists of 18 samples for each of the 32 synthesis PQMF sub-bands.

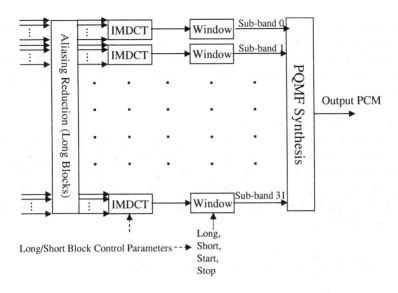

Figure 6. MPEG Audio Layer III synthesis filter bank structure

4.1.1 Block Switching

The total block size processed by the Layer III filter bank is given by 32 * 36 = 1152 time-samples. This block length ensures a frequency resolution of about 41.66 Hz at 48 kHz sampling rate. The increased frequency resolution for Layer III is much better suited to accommodate allocation of the bit pool based on psychoacoustic models. One drawback of this approach is that quantization errors can now be spread over a block of 1152 time-samples. For signals containing transients, such as castanet excerpts, this translates into unmasked temporal noise, specifically pre-echo. In the case of transient signals, the Layer III filter bank can switch to a higher time resolution in order to avoid pre-echo effects. Namely, during transients, Layer III utilizes a shorter block size of 32 * 12 = 384 time samples, reducing the temporal spreading of quantization noise for sharp attacks. The short block size represents a third of the long block length. During transients, a sequence of three short windows replaces the long window, maintaining the same total number of samples per frame. In order to ensure a smooth permutation between long and short blocks and vice versa, two transition blocks, long-to-short and short-to-long, which have the same size as the long block are employed. This approach was first presented by Edler [Edler 89] and, based on the shape of the windows and overlap regions, it maintains the time domain aliasing cancellation property of the MDCT (see also Chapter 5). In addition, the frame size is kept constant during the allowed window size sequences. This characteristic ensures the subsistence of a simple overall structure of the algorithm and bitstream formatting routines.

4.1.1.1 Window Sequence

In *Figure 7* a typical window sequence from long to short windows and from short to long windows is shown together with the corresponding amount of overlapping between adjacent windows in order to maintain the time-domain aliasing cancellation for the transform. The basic window, w[n], utilized by Layer III is a sine widow. The different windows are defined as follows:

$$w[n] = \sin\left(\frac{\pi}{36}\left(n + \frac{1}{2}\right)\right) \qquad n = 0,...,35 \quad \text{(long window)}$$

$$w[n] = \sin\left(\frac{\pi}{12}\left(n + \frac{1}{2}\right)\right) \qquad n = 0,...,11 \quad \text{(short window)}$$

$$w[n] = \begin{cases} \sin\left(\frac{\pi}{36}\left(n + \frac{1}{2}\right)\right) & n = 0,...,17 \\ 1 & n = 18,...,23 \\ \sin\left(\frac{\pi}{12}\left(n - 18 + \frac{1}{2}\right)\right) & n = 24,...,29 \\ 0 & n = 30,...,35 \end{cases} \quad \text{(start window)}$$

$$w[n] = \begin{cases} 0 & n = 0,...,5 \\ \sin\left(\frac{\pi}{12}\left(n - 6 + \frac{1}{2}\right)\right) & n = 6,...,11 \\ 1 & n = 12,...,17 \\ \sin\left(\frac{\pi}{36}\left(n + \frac{1}{2}\right)\right) & n = 18,...,35 \end{cases} \quad \text{(stop window)}$$

It should be noted that the Layer III filter bank allows for a mixed block mode. In this mode, the two lower frequency PQMF sub-bands are always processed with long blocks, while the remaining sub-bands are processed with short blocks during transients. This mode ensures high frequency resolution at low frequencies where it is most needed and high time resolution at high frequencies.

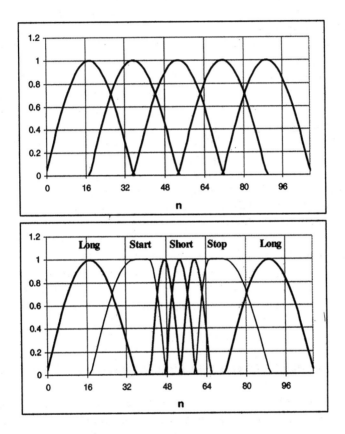

Figure 7. Typical Layer III window sequence: top for steady state signal, bottom for transients occurring in the time region between n = 45 and n = 60

4.1.2   Hybrid Filter Bank Versus PQMF Characteristics

The Layer III hybrid filter bank provides much higher frequency resolution than the Layers I and II PQMF. The time resolution, however, is decreased. At 48 kHz sampling rate, the time resolution for Layers I and II is 0.66 ms, for Layer III is 4 ms. The decreased time resolution renders Layer III more prone to pre-echo. A number of measures to reduce pre-echo are incorporated in Layer III including a detection mechanism in the psychoacoustic model and the ability to "borrow" bits from the bit reservoir (see also next sections) in addition to block switching. The inherent filter bank structure with long impulse response of 384 + 512 = 896 samples even in the short block mode, however, makes the encoding of transients a challenge for Layer III.

In summary, the Layer III hybrid filter bank approach offers advantages such as high frequency resolution, a dynamic, adaptive trade-off between time and frequency resolution, and full compatibility with Layers I and II. The shortcomings include potential aliasing effects exposed by the MDCT stage and long impulse response filters. Both shortcomings are reduced in the standard specifications by adopting procedures to mitigate them. The complexity of the Layer III filter bank is increased with respect to the complexity of Layers I and II. In addition to the PQMF, the MDCT stage contributes to its complexity. In general, fast implementations of the MDCT exploit the use of FFTs. It should be noted that the size of the MDCT is non-power of two, therefore the implementation via FFT requires a decomposition to a power-of-two length sequence if a radix-2 FFT is utilized. Considering the different stages of the filter bank implementation and assuming that the MDCT is implemented via an FFT, the complexity for a long window is given by (18 + 9 + 18) additional complex multiplications and additions per sub-band block with respect to the PQMF alone, or equivalently a little over 1 additional multiplication and addition per sub-band sample.

5.   MPEG AUDIO PSYCHOACOUSTIC MODELS

The goal of MPEG Audio is to provide perceptually lossless quality. In other words, the output of the MPEG coder should be a signal statistically indistinguishable from its input. In order to achieve this objective at relatively low data rates, MPEG Audio exploits the psychoacoustic principles and models we discussed in Chapter 7. During the encoding process, the input signal is analyzed on a frame-by-frame basis and the masking ability of the signal components is determined. For each frame, based on the computed masked thresholds, the bits available are distributed through the signal spectrum in order to best represent the signal. Although the encoder process is not a mandatory part of the MPEG standard, two psychoacoustic models are described in the informative part of its specifications. Either model works for all layers, but typically Model 1 is applied to Layers I and II and Model 2 to Layer III. There is a large degree of freedom in the psychoacoustic model implementation. At high data rates, the psychoacoustic model can be completely bypassed, leaving the task of assigning the available resource to the iterative process in the allocation routines simply based on the strength of the signal spectral components.

5.1 Psychoacoustic Models Analysis Stage

The input to the psychoacoustic model is the time representation of the audio signal over a certain time interval and the corresponding outputs are the signal to mask ratios (SMRs) for the coder's frequency partitions. Based on this information, the bit (Layers I and II) and noise (Layer III) allocation is determined for each block of input data (see Chapter 8 for a discussion on perceptual bit allocation). In order to provide accurate frequency representation of the input signal, a discrete Fourier transform is computed in parallel to the main audio path time to frequency mapping stage. One might argue that the output of the PQMF or the hybrid filter bank could be utilized for this purpose in order to simplify the structure of the algorithm. In the case of the evaluation of the masking thresholds, the aim is to have maximum accuracy in the signal representation. While issues like critical sampling etc. play a fundamental role in the design of the time to frequency mapping in the main audio path, they are irrelevant in the frequency representation of the audio signal for analysis only purposes. On the other hand, inadequate frequency resolution and potential aliasing can irremediably confound the evaluations of the psychoacoustic model. It should be noted that different approaches are found in the literature. For example, in Dolby AC-3 [Fielder et al. 96] and PAC [Sinha, Johnston, Dorward and Quackenbush 98] the output of the MDCT is employed for the psychoacoustic analysis; in the advanced PEAQ version the DFT analysis is employed along with a filter bank that mirrors the auditory peripheral filter bank [Thiede et al. 00].

The first step in both MPEG psychoacoustic models is to time-align the audio data used by the psychoacoustic model stage with the main path audio data. This process must take into account the delay through the filter bank and the time offset needed so that the psychoacoustic analysis window is centered on the current block of data to be coded. For example, in Layer I the delay through the filter bank is 256 samples and the block of data to be coded is 384 samples long (see also next section). The analysis window applied to Layer I data in Psychoacoustic Model 1 is 512 samples long. The offset to be applied for time alignment is therefore 256 + (512 − 384)/2 = 320 samples.

5.2   Psychoacoustic Model 1

The block diagram for Psychoacoustic Model 1 is shown in *Figure 8*. The first stage, the analysis stage, windows the input data and performs an FFT. The analysis window is a Hanning window of length N equal to 512 samples for Layer I and 1024 for Layers II and III. The overlapping between adjacent windows is N/16. Since Layers II and III utilize a 1152 sample frame, the 1024 sample analysis window does not cover the entirety of the audio data in a frame. If, for example, a transient occurs at the tail end of the main path audio frame, the relative sudden energy change would be undetected in the Psychoacoustic Model 1 analysis window. In general, however, the 1024 sample analysis window proved to be a reasonable compromise.

5.2.1   SPL Computation

After applying the FFT, the signal level is computed for each spectral line k, L_k as follows

$$L_k = 96\,dB + 10\log_{10}\left(4/N^2 \mid X[k] \mid^2 8/3\right) \quad \text{for k=0,...,N/2-1}$$

where X[k] represents the FFT output of the time-aligned, windowed input signal and N equals 512 for Layer I and 1024 for Layers II and III. The signal level is normalized so that the level of an input sine wave that just overloads the quantizers, here defined as being at $x[n] = \pm 1.0$, has a level of 96 dB when integrated over the peak. In this equation, the factor of $1/N^2$ comes from Parseval's theorem, one factor of 2 comes from only working with positive frequency components, another factor of 2 comes from the power of a unit amplitude sinusoid being equal to ½, and the factor of 8/3 comes from the reduction in gain from the Hanning window (see also Chapter 9). Since the description of Model 1 in the standard absorbs the factor of 8/3 into the Hanning window definition, a natural way to take the other factors into account is to include a factor of 2/N in the forward transform of the FFT.

The sound pressure level in each sub-band m, $L_{sb}[m]$ is then computed as the greater of the SPL of the maximum amplitude FFT spectral line in sub-band m and the lowest level that can be described with the maximum scale factor for that frame in sub-band m as follows:

$$L_{sb}[m] = \max\{L_k, 20\log_{10}(scf_{max}[m]\,32{,}768) - 10\}dB$$

where L_k represents the level of the kth line of the FFT in sub-band m with the maximum amplitude and scf_{max} is the maximum of the scale factors for sub-band m (see the next section for a discussion on MPEG Layers I and II "scale factors" which differ somewhat from the scale factors discussed in Chapter 2 in that these "scale factors" represent the actual factor that the signal is scaled by as opposed to the number of factors of 2 in the scaling). In Layers I and II coders, the scale factors range from a very small number up to 2.0 so the multiplication by 32,768 just normalizes the power of the scale factor so that the largest possible scale factor corresponds to a level of 96 dB. The −10 dB term is an adjustment to take into consideration the difference between peak and average levels. The reason for taking the scale factor into account in the above expression can be explained by closely examining the block floating quantization process, since block floating point quantization cannot scale the signal to lower amplitudes than can be represented by the scale factors themselves. This implies that for low amplitude frequency lines the quantization noise is of a size determined by the maximum scale factor.

5.2.2   Separation of Tonal and Non-Tonal Components

Having found the sound pressure level in the sub-band, we next compute the masking threshold in order to calculate the signal to mask ratio (SMR) for the sub-band. Since noise is a better masker than tones, a search for tonal maskers in the signal is performed in Model 1. This evaluation is based upon the assumption that local maxima within a critical band represent the tonal components of the signal. A local maximum L_k, is included in the list of tonal components if $L_k - L_{k+j} \geq 7$ dB where the index j varies with the center frequency of the critical band examined. If L_k represents a tonal component, then the index k, the sound pressure level derived from the sum of three adjacent spectral components centered at k, L_T, and a tonal flag define the tonal masker.

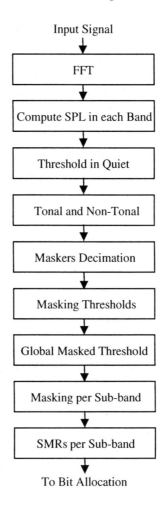

Figure 8. Block diagram of MPEG Psychoacoustic Model 1

The noise maskers in the signal are derived from the remaining spectral lines. Within a critical band, the power of the spectral components remaining after the tonal components are removed is summed to form the sound pressure level of the noise masker, L_N, for that critical band. The noise masker components for each critical band are centered at the geometric mean of the FFT spectral line indices for each critical band.

5.2.3   Maskers Decimation

Having defined the tonal and noise maskers in the signal, the number of maskers is then reduced prior to computing the global masked threshold. First, tonal and non-tonal maskers are eliminated if their levels do not exceed the threshold in quiet. Second, maskers extremely close to stronger maskers are eliminated. If two or more components are separated in frequency by less than 0.5 bark, only the component with the highest power is retained. The masking thresholds are then computed for the remaining maskers by applying a spreading function and shifting the curves down by a certain amount of dB which depends on whether the masker is tonal or noise-like and the frequency position of the masker. To keep the calculation time manageable, the masking thresholds are evaluated at only a sub-sampling of the frequency lines. The number of sub-sample lines depends on which layer is being implemented and on the sampling rate, ranging from 102 to 108 sub-sample lines in Layer I and ranging from 126 to 132 sub-sample lines in Layer II.

5.2.4   Model 1 Spreading Function and Excitation Patterns

The spreading function used in Model 1 is defined as follows:

$$B(dz, L) = -17\ dz + 0.15L\ (dz - 1)\ \theta(dz - 1) \qquad \text{for } dz \geq 0$$

$$B(dz, L) = -(6 + 0.4L)\ |dz| - (11 - 0.4L)\ (|dz| - 1)\ \theta(|dz| - 1) \qquad \text{for } dz < 0$$

where $\theta(x) = 1$ for $x \geq 0$ and 0 otherwise; L represents the sound pressure level of the masker; dz represents the distance in bark between the maskee and the masker. Notice that dz assumes positive values when the masker is located at a lower frequency than the maskee and negative values when the masker is located at a higher frequency than the maskee. The two-piece linear spreading function for upper and lower frequencies in Model 1 seems to be consistent with the masking data for tones masking tones (see also Chapter 6). The excitation patterns for a signal positioned at 8 bark with different sound pressure levels, L = 20, 40, 60, 80, 100 dB, is shown in *Figure 9*. As shown in *Figure 9* the excitation patterns are level dependent, being nearly symmetrical for levels below 40 dB and increasingly spreading towards high frequencies at higher levels.

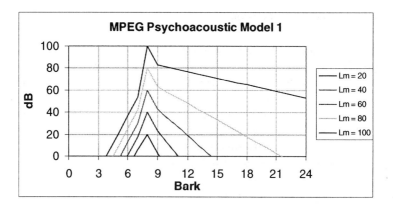

Figure 9. MPEG Psychoacoustic Model 1 excitation patterns for different SPLs

5.2.5   Masking Thresholds

In order to derive the masking threshold relative to the masker of level L, one needs to shift the excitation pattern relative to the masker by an appropriate amount. This shift depends on the tonal versus noise-like characteristics of the masker, since we know from experimental data that noise is a better masker than tones. In Model 1 the shift Δ is defined as follows:

$$\Delta_T(z) = -6.025 - 0.275 \, z \quad dB$$

$$\Delta_N(z) = -2.025 - 0.175 \, z \quad dB$$

where Δ_T represents the shift for tonal maskers and Δ_N represents the shift for noise-like maskers, and z is the frequency index of the masker in the bark scale. The masking threshold M relative to a masker of level L at a frequency equal to bark z can be expressed as:

$$M_{T,N}(L, z_i, z_j) = B(z_i\text{-}z_j,L) + \Delta_{T,N}(z_j)$$

Where z_j is the bark index of the masker and z_i is the bark index of the maskee, $\Delta = \Delta_T$ for tonal maskers and $\Delta = \Delta_N$ for noise-like maskers. In *Figure 10* the masking curve relative to a 70 dB noise-like masker centered at bark 8 is shown.

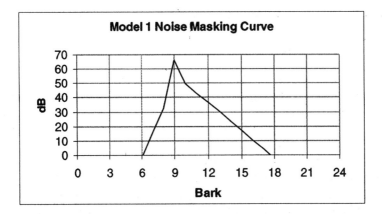

Figure 10. MPEG Psychoacoustic Model 1 masking curve for a noise-like masker centered at 8 Bark with a sound pressure level of 70 dB

5.2.6 Masked Threshold

The global masked threshold $M_G(z_i)$ at the bark location z_i of each sub-sampled frequency line is then computed in Model 1 by summing the power of the individual masking thresholds and the threshold in quiet as follows:

$$M_G(z_i) = 10\log_{10}[10^{\frac{M_q(z_i)}{10}} + \sum_{j=1}^{m}10^{\frac{M_{Tj}(L_j,z_i,z_j)}{10}} + \sum_{k=1}^{n}10^{\frac{M_{Nk}(L_k,z_i,z_k)}{10}}]$$

where M_q represents the threshold in quiet at the bark location z_i, M_{Tj} the masking threshold from the jth tonal masker, M_{Nk} the masking threshold from the kth noise-like masker, m is the total number of tonal maskers, and n is the total number of noise-like maskers.

5.2.7 SMR Computation

Since at low frequency some of the main path PQMF sub-bands span more than one critical band, the masking level $M_{Gmin}(sb)$ in each sub-band, sb, is determined based on the minimum masking level at the sub-sampled lines in that sub-band and used to determine the signal to mask ratio $SMR(sb)$ to be transmitted to the bit allocation routine:

$$SMR(sb) = L(sb) - M_{Gmin}(sb)$$

where L(sb) is the signal level for sub-band sb.

5.2.8   An Example

To help clarify the Model 1 calculations, *Figure 11* through *Figure 13* give an example taken from [Pan 95]. *Figure 11* shows the frequency spectrum of an input signal consisting of a single tonal component overlaying a shaped noise spectrum. The top part of *Figure 12* shows the signal level of each masker identified in the signal and its associated frequency location (measured in terms of the k index in the N = 1024 FFT). The tonal masker is clearly visible as are the noise maskers from each critical band. The middle part of *Figure 12* shows the maskers remaining after eliminating maskers below the threshold in quiet or too close to stronger maskers. The bottom part of *Figure 12* shows the resulting global masked threshold at each of the sub-sampled frequency locations. The top of *Figure 13* shows the resulting SMRs calculated for each of the 32 critical bands from the input signal spectrum and the masked curve. The bottom of *Figure 13* shows the resulting decoded data after being quantized using bit allocations determined by the SMRs in the top of the figure at a data rate of 64 kb/s.

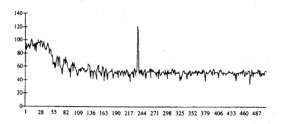

Figure 11. Low-pass filtered noise plus 11.250 kHz sine wave at a sampling rate of 48kHz; the abscissa represents the FFT frequency line indices, the ordinate the strength of the signal in dB. (From [Pan 95] © 1995 IEEE.)

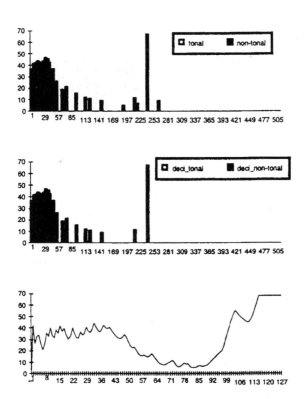

Figure 12. MPEG Psychoacoustic Model 1 processing from [Pan 95] © 1995 IEEE.

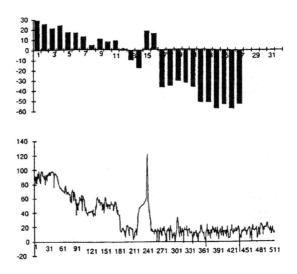

Figure 13. MPEG Psychoacoustic Model 1 SMRs (top) and coded audio data (bottom) at 64 kb/s. (From [Pan 95] © 1995 IEEE.)

5.3   Psychoacoustic Model 2

Model 2's process and calculations differ substantially from those of Model 1. A block diagram of Model 2 is shown in *Figure 14*. Notice that there are now two parallel calculation paths: that of the masking energy and that of the tonality index.   In addition, Model 2 does not have a masker decimation stage.

5.3.1   Model 2 Analysis Stage

As discussed in Model 1, Model 2 also applies a Hanning-windowed FFT to the time-aligned signal samples. The size of the FFT window is now 1024 for all Layers, however for Layers II and III this model computes two FFTs for each frame. The centers of the FFT input blocks are aligned with the center of the first and second half of the main data path input block. Model 2 uses the output of the FFT analysis to calculate the masking curves and their associated signal to mask ratios for the coder sub-bands. The higher of the two resulting signal to mask ratios, or the lower of the two masked thresholds, are then selected per each sub-band.

5.3.2 SPL Computation

In Model 2 the frequency-lines are grouped into "threshold calculation partitions" whose widths are roughly 1/3 of a critical band or one FFT line, whichever is wider at that frequency location. For each partition, one single masker SPL is derived from the sum of energy densities in the partition. The total masking energy for the signal frame is computed by first convolving a spreading function with each of the maskers in the signal. This process is equivalent to spreading in frequency the energy of each masker and adding up the relative energies. The spreading function used in Model 2 differs from that used in Model 1 and is a variant of the Schroeder spreading function we discussed in Chapter 7.

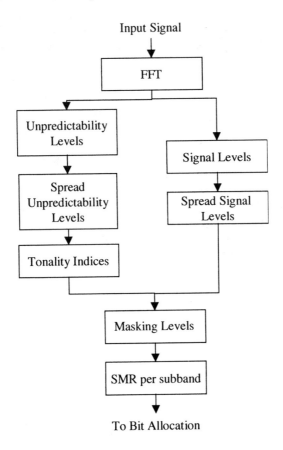

Figure 14. Block diagram of MPEG Psychoacoustic Model 2

5.3.3   Model 2 Spreading Function and Excitation Patterns

The basic spreading function B(dz) measured in dB used by Model 2 is (see also *Figure 15*):

$$B(dz) = 15.8111389 + 7.5*(1.05*dz + 0.474) - 17.5*\sqrt{1.0 + (1.05*dz + 0.474)^2}$$
$$+ 8*MIN(0, (1.05*dz - 0.5)^2 - 2*(1.05*dz - 0.5))$$

where dz is the bark distance between the maskee and the masker. Notice that dz < 0 has the masker at a higher frequency than the maskee. Notice also that, other than a minor factor of 1.05, the first line is exactly equalt to Schroeder's spreading function. The second line lowers the curve for values of dz between 0.5 and 2.5, causing somewhat of a two-sloped shape for positive dz (see *Figure 15*). Furthermore, notice that this spreading function is not level dependent in contrast with the clear level dependence seen in experimental data (see also Chapter 6).

In Model 2 the spreading function is modified by a normalization that preserves power near zero frequency and the upper frequency limit. Namely, a flat spectrum over the positive frequency range of the FFT is convolved with B(dz). The power of the convolved flat spectrum drops at the edges of the positive frequency range due to a lack of spectrum energy outside this range. The spreading function B(dz) is adjusted to remove this drop by dividing the convolved signal power by the power of the convolved flat spectrum. Given the behavior of the threshold in quiet at the boundaries of the human audible range (see Chapter 6), this normalization may not be essential. In order to derive the global masked threshold, this convolved, normalized signal energy is then lowered in each partition by an amount that depends on the tonality of the spectrum in that partition.

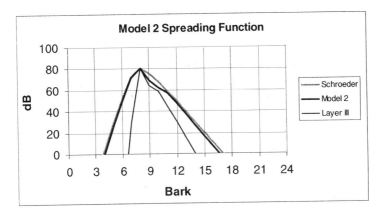

Figure 15. Basic spreading function used by MPEG Psychoacoustic Model 2 compared with the Schroeder spreading function and with the modified version utilized in Layer III.

5.3.4   Tonality Index

The tonality index in each partition is calculated based on how predictable the signal is from the frequency lines in the two prior frames, see also [Brandenburg and Johnston 90]. For each frame m and for each frequency line k, the signal amplitude, $A_m[k]$, and phase, $\phi_m[k]$, are predicted by linear extrapolation from the two prior values as follows:

$$A'_m[k] = A_{m-1}[k] + \{A_{m-1}[k] - A_{m-2}[k]\}$$
$$\phi'_m[k] = \phi_{m-1}[k] + \{\phi_{m-1}[k] - \phi_{m-2}[k]\}$$

where $A'_m[k]$, and $\phi'_m[k]$ represent the predicted values. The predicted values are then mapped into an "unpredictability measure" defined as:

$$c_m[k] = \frac{\sqrt{\{A'_m[k]\cos\phi'_m[k] - A_m[k]\cos\phi_m[k]\}^2 + \{A'_m[k]\sin\phi'_m[k] - A_m[k]\sin\phi_m[k]\}^2}}{A_m[k] + |A'_m[k]|}$$

where $c_m[k]$ is equal to zero when the current value is exactly predicted and equal to 1 when the power of either the predicted or actual signal is dramatically higher than the other.

The unpredictability measure is first weighted with the energy in each partition, deriving a partition unpredictability measure. This partition unpredictability measure is then convolved with the spreading function. The result of this convolution is normalized using the signal energy convolved with the spreading function and then mapped onto a tonality index which is a function of the partition number and whose values vary between zero and one. The tonality index has the property that high unpredictability goes towards zero and low unpredictability goes towards one. Notice that, since the unpredictability measure is convolved using the spreading function that determines the masking energy at a certain frequency location, the resulting tonality index at a frequency location reflects the tonality of the dominant maskers at that frequency location.

5.3.5   Masking Thresholds

The tonality indices for each partition created by this calculation process are utilized to determine the shift $\Delta(z)$ in dB of the re-normalized convolved signal energy that converts it into the global masking level. The values for $\Delta(z)$ are linearly interpolated based on the tonality index from the value of 5.5 dB for zero tonality (a noise masker) to a frequency-dependent value defined in the standard for tonality index equal to one (a tonal masker). The interpolated $\Delta(z)$ is then compared with a frequency-dependent minimum value, also defined in the standard, that controls stereo unmasking effects (see also Section 8 later in this chapter) and the larger value is used for the shift.

5.3.6   SMR Computation

The masking levels in each threshold calculation partition are compared with the threshold in quiet for that partition and the larger is used. Note that this differs from the addition of masking power from the threshold in quiet to that of the maskers in Model 1. This curve represents the masked threshold. For each partition, the masking energy is evenly mapped to each frequency line by dividing the partition masking threshold values by the number of frequency lines in that partition. Finally, the masked threshold and the corresponding power spectral densities are mapped onto the coder scale factors sub-bands (see next section) to calculate the signal to mask ratios (SMRs) for each sub-band. These SMRs are then sent to the allocation routine to determine the number of bits allocated to each sub-band.

5.3.7   An Example

In *Figure 16* and *Figure 17*, an example of Model 2 processing for the signal shown in *Figure 11* is illustrated from [Pan 95]. In *Figure 16*, the signal level and its spread level (top) and the tonality index (bottom) are shown versus the 62, approximately one-third critical-band frequency partitions. The masked threshold is plotted in *Figure 17* (top) as derived from the convolved energy and the tonality index versus the FFT frequency lines. *Figure 17* (center) shows the SMR values for each coder scale factor band. Finally, in *Figure 17* (bottom) the coded audio data at 64 kb/s is shown. Note that the dB values shown differ by roughly 50 dB from those in the Model 1 example due to a different choice of normalization in the SPL calculations. This can be most easily seen by comparing the level of the tonal peak in both cases.

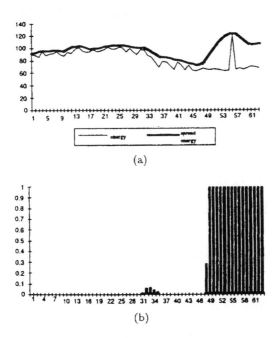

Figure 16. MPEG Psychoacoustic Model 2 processing: the abscissa represents the partition index and the ordinate the signal energy and spread energy in dB (top) and the tonality index (bottom). (From [Pan 95] © 1995 IEEE.)

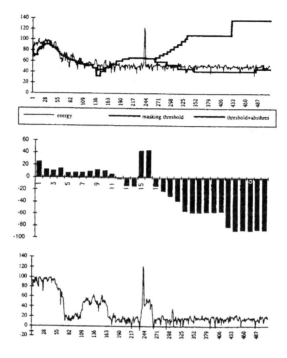

Figure 17. MPEG Psychoacoustic Model 2 processing: original signal and computed masking thresholds (top); SMRs (center); coded signal at 64 kb/s. (From [Pan 95] © 1995 IEEE.)

5.3.8   Model 2 Applied to Layer III

Model 2 is slightly altered when applied to Layer III to take into account the different nature of the Layer III algorithm including the hybrid filter bank and the block-switching capability of Layer III. The main changes are:
- The model is calculated twice in parallel – once for using a long block FFT and once using a short block FFT, where the short block consists of 256 samples.
- The results of both FFTs are combined in calculating the unpredictability measures.
- The spreading function (see *Figure 15*) is changed to drop off faster. Terms of 1.05*dz in the spreading function are replaced with terms of 1.5*dz for negative dz (masker frequency location lower than maskee frequency location) and 3.0*dz for positive dz (masker frequency location higher than maskee frequency location).

- The noise-masking-tone drop has been changed from 5.5 dB to 6.0 dB and the tone-masking-noise drop has been changed from a frequency-dependent table lookup to a constant 29.0 dB.
- Pre-echoes are reduced by comparing the masking threshold with scaled versions of the last 2 frames' thresholds and setting the current masking threshold to the smallest of these before comparing it with the threshold in quiet. (Twice the previous frame's value is used and 16 times the value of the frame before that.) This serves to reduce the masking thresholds following very quiet signals.

Attacks are detected based on a "psychoacoustic entropy" or perceptual entropy, PE (see also Chapter 7), calculation [Johnston 88] equal to the logarithm of a geometric mean of the threshold-weighted energy across the block. The exact formula as described in the informative section of the standard is

$$PE = \sum_{partition\, b} n_b \, \log_2 (1. + \sqrt{energy_b \, / \, threshold_b})$$

where n_b is the number of frequency lines in partition b, $energy_b$ is the aggregate signal energy in partition b and $threshold_b$ is the energy relative to the masked threshold in partition b. The PE as described above measures the minimum number of bits per block required to achieve transparency in a signal given a masking curve. Psychoacoustic entropy above a certain trigger (PE >1800 bits from [ISO/IEC 11172]) signals an attack. The PE is effectively a measure of the perceptual flatness of the frequency spectrum since it effectively measures the geometric mean of the spectral energy (normalized by the masking threshold). Recall from Chapter 8 that a high geometric mean signifies a flat spectrum. A flat spectrum is associated with a sharp attack in this model since sudden changes lead to broad frequency content. Detection of an attack determines which window-type is used according to *Figure 18*. Signal-to-mask ratios are computed for each scale factor partition, rather than for each PQMF sub-band, corresponding to the chosen window-type.

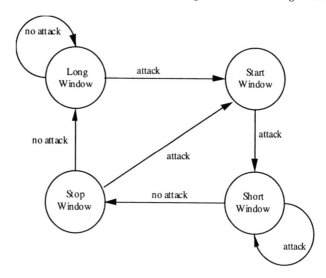

Figure 18. Block switching state diagram for Layer III from [ISO/IEC 11172-3]

6.   MPEG-1 AUDIO SYNTAX

The MPEG Audio bitstream provides periodically spaced frame headers to identify the coded data. In *Figure 19* the audio data frame structure is shown for all Layers. For Layer I the frame size consists of 12 * 32 = 384 samples. For Layers II and III, three granules of 12 samples per sub-band are grouped into a frame for a total of 1152 samples. The 32-bit MPEG Audio header is shown in *Figure 20*. After a 12-bit string of 1 which represents the synch word, there is an MPEG ID bit (=1 implies an MPEG-1 Audio bitstream) and the layer identification (two bits). The error protection bit, the bitrate index (four bits), the sampling frequency bits (two), padding and private bit and mode (stereo, joint stereo, dual channel, single channel), mode extension, etc., follow. The frame format of the three layers is shown in *Figure 21*. In order to understand the underlying structure of the MPEG Audio layers syntax, we now examine the details of different coding options adopted by the layers.

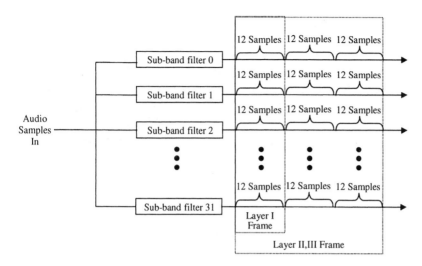

Figure 19. Data frame structure for MPEG-1 Audio Layers I, II, and III

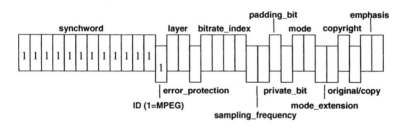

Figure 20. MPEG-1 Audio Header

LAYER I	Header (32)	CRC (0,16)	Bit Allocation (128-256)	Scale Factors (0-384)	Samples	Ancillary Data

LAYER II	Header (32)	CRC (0,16)	Bit Allocation (26-188)	SCFSI (0-120)	Scale Factors (0-1080)	Samples	Ancillary Data

LAYER III	Header (32)	CRC (0,16)	Side Information (130-246)	Main Data (May start at a previous frame)

Figure 21. MPEG-1 Audio frame format

6.1   Layer I

Layer I encodes the audio samples in groups of 384 samples. The audio data in each frame represent 8 ms of audio at a 48 khz sampling rate. The encoded audio data is described in terms of bit allocation information, scale factors, and quantized samples. In addition to the header and the audio data, each Layer I frame contains (see top part of *Figure 21*) a 16-bit optional cyclic redundancy code (CRC) error check word, and optional ancillary data.

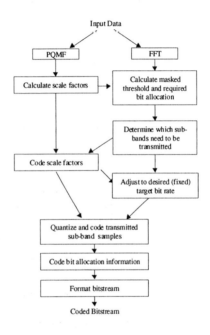

Figure 22. Basic structure of the encoding process for MPEG-1 Audio Layers I and II from [ISO/IEC 11172-3]

6.1.1   Scale Factors Computation

The Layer I (and Layer II) encoding process is shown in *Figure 22*. The analysis PQMF stage is followed by the computation of scale factors for each sub-band.   One scale factor is computed for each 12 sub-band frequency samples (called a "granule") and it is represented using 6 bits. The maximum absolute value of the 12-sample granule is determined and mapped to a scale factor value via a look up table defined in the standard and shown in *Table 1*. The samples in the granule are divided by the scale factor prior to the quantization stage. The dynamic range covered by the scale factors is 120 dB.

The scale factor is transmitted only if the bit allocation for that band is non-zero. The maximum number of bits used in a Layer I frame to transmit scale factors (in the case the where the bit allocation for each sub-band is non-zero) is 6*32*2 = 384 bits for stereo mode and 6*32 = 192 bits for mono mode.

Table 1. MPEG Audio Layers I and II Scale Factors [ISO/IEC 11172-3]

Scale Factor Index	Scale Factor Value	Scale Factor Index	Scale Factor Value
0	2.00000000000000	32	0.00123039165029
1	1.58740105196820	33	0.00097656250000
2	1.25992104989487	34	0.00077509816991
3	1.00000000000000	35	0.00061519582514
4	0.79370052598410	36	0.00048828125000
5	0.62996052494744	37	0.00038754908495
6	0.50000000000000	38	0.00030759791257
7	0.39685026299205	39	0.00024414062500
8	0.31498026247372	40	0.00019377454248
9	0.25000000000000	41	0.00015379895629
10	0.19842513149602	42	0.00012207031250
11	0.15749013123686	43	0.00009688727124
12	0.12500000000000	44	0.00007689947814
13	0.09921256574801	45	0.00006103515625
14	0.07874506561843	46	0.00004844363562
15	0.06250000000000	47	0.00003844973907
16	0.04960628287401	48	0.00003051757813
17	0.03937253280921	49	0.00002422181781
18	0.03125000000000	50	0.00001922486954
19	0.02480314143700	51	0.00001525878906
20	0.01968626640461	52	0.00001211090890
21	0.01562500000000	53	0.00000961243477
22	0.01240157071850	54	0.00000762939453
23	0.00984313320230	55	0.00000605545445
24	0.00781250000000	56	0.00000480621738
25	0.00620078535925	57	0.00000381469727
26	0.00492156660115	58	0.00000302772723
27	0.00390625000000	59	0.00000240310869
28	0.00310039267963	60	0.00000190734863
29	0.00246078330058	61	0.00000151386361
30	0.00195312500000	62	0.00000120155435
31	0.00155019633981		

6.1.2   Bit Allocation and Quantization

The number of bits available to quantize audio samples in each frame is determined by the data rate less the bits needed to transmit the header, CRC, bit allocations, scale factors, and any ancillary data.

The number of bits that can be used to quantize sub-band samples in Layer I ranges between zero and 15, excluding the allocation of 1 bit because of the nature of the midtread quantizer (see also Chapter 2). The bit allocation for each sub-band of each channel is communicated through a four-bit code where, other than a code of zero representing zero bits, the value of the four-bit code is equal to one less than the number of allocated bits. For example, a code of 5 indicates six bits for that sub-band. A code value of 15 is forbidden. The number of bits used to transmit bit allocation information is equal to 4*32 = 128 bits for single channel mode and 4*32*2 = 256 bits for stereo mode.

The bit allocation routine is an iterative process where, after initializing the process by setting all bit allocation codes to zero and assuming no bits are needed to transmit scale factors, in each iteration additional bits are allocated to the sub-band with the highest noise-to-mask ratio (NMR) until there are not enough additional bits left for the next iteration pass.

The NMR for each sub-band is calculated as the difference between the signal-to-mask ratio (calculated in the psychoacoustic model) and the signal-to-noise ratio (estimated from a table lookup based on the number of bits allocated to the sub-band). In each iteration, an additional quantization bit is allocated to the sub-band with the highest NMR that can still accept an additional bit (i.e. has less than 15 bits already). If a sub-band that has not already been given any bits is the one with the highest NMR, that sub-band is given two bits and bits are also set aside to transmit the scale factor for that sub-band.

Once the bit allocations are determined, each sub-band sample is divided by its scale factor and quantized using a midtread uniform quantizer having a number of steps determined by the bit allocation for that sub-band.

6.2   Layer II

The Layer II algorithm builds on the basic structure of Layer I (see *Figure 22*). The frame size is increased for Layer II to 3 granules of 12 sub-band samples corresponding to a total of 12*3*32 = 1152 input samples per frame so that one can take advantage of commonality between consecutive granules. The audio encoded data in each frame represent 24 ms of data at a 48 kHz sampling rate.

6.2.1 Scale Factors and Scale Factors Select Information

The scale factors in Layer II can be shared among the three consecutive granules. The scale factors are computed in the same manner as in Layer I (see *Table 1*). When the values of the scale factors for the consecutive granules are sufficiently close or when temporal post-masking can hide distortion, only one or two scale factors need to be transmitted.

The scale factor select information, SCFSI (see mid portion of *Figure 22*), is coded with two bits and determines whether one, two, or three scale factors will be transmitted for the three consecutive granules in the frame. A SCSFI equal to zero corresponds to three scale factors, a SCSFI equal to one corresponds to transmitting two scale factors (the first for the first two granules and the second for the third granule), a SCSFI equal to two corresponds to transmitting a single scale factor to be used for all granules, and a SCSFI equal to three corresponds to transmitting two scale factors (where the first is used for the first granule and the second is used for the last two granules).

Both SCFSI and scale factors are transmitted only if the bit allocation is different from zero for that sub-band. The bits used for SCFSI vary between zero and $30 * 2 * 2 = 120$ bits for each frame and the bits used for scale factors vary between zero and $6 * 3 * 30 * 2 = 1080$ bits per frame. (Note: Layer II does not allow any bits to be allocated to the two highest sub-bands so there are at most only 30 sub-bands that are allocated bits. See [ISO/IEC 11172-3].)

6.2.2 Bit Allocation and Quantization

The psychoacoustic model and bit allocation process are computed in a similar manner for Layer II as for Layer I, but they are now computed relative to a frame of 36 sub-band samples. Also, bit allocation tables are now specified in the standard that determine the possible quantization levels that can be used to quantize the samples in any sub-band. As in Layer I, quantization is carried out by dividing each sub-band sample by its scale factor and then using a uniform mid-tread quantizer with the prescribed number of steps.

Bit allocations are coded based on tables specified in the standard that depend on the sample rate and the target data rate. Bit allocation codes range from zero to four bits per sub-band, where a greater number of bits are used to code the bit allocations of lower frequency sub-bands. In all cases, no bits are allocated to the two highest frequency sub-bands in Layer II. In other sub-bands, the number of quantization levels can reach that of 16-bit quantization. The total number of bits employed to describe the bit allocation for one frame of the stereo signal varies between 26 and 188 depending on which table is applicable.

An example of a Layer II bit allocation table is shown in *Table 2*. This particular table is valid for sample rates of 44.1 and 48 kHz when they are running at data rates of 32 or 48 kb/s per channel. The first column represents the sub-band number and the second column represents the number of bits used to transmit that sub-band's bit allocation. The successive columns in each row represent the possible number of quantization levels that can be used to quantize the 36 samples in that frame's sub-band. The numbers in the top row indicate the bit allocation code corresponding to that number of quantization steps. Notice that four bits are used to encode the bit allocation for the first two sub-bands, three bits are used to encode the bit allocation for the next six sub-bands, and no bits are allocated to any higher frequency sub-bands in this particular bit allocation table. For this bit allocation table, the total number of bits employed to describe the bit allocation is equal to 26.

Table 2. Example of MPEG Audio Layer II bit allocation table [ISO/IEC 11172-3]

B	N	0	1	2	3	4	5	6	7	8	9	10	11	12	13	14	15
0	4	-	3	5	9	15	31	63	127	255	511	1023	2047	4095	8191	16383	32767
1	4	-	3	5	9	15	31	63	127	255	511	1023	2047	4095	8191	16383	32767
2	3	-	3	5	9	15	31	63	127								
3	3	-	3	5	9	15	31	63	127								
4	3	-	3	5	9	15	31	63	127								
5	3	-	3	5	9	15	31	63	127								
6	3	-	3	5	9	15	31	63	127								
7	3	-	3	5	9	15	31	63	127								
8	0	-															
9	0	-															
10	0	-															
11	0	-															
12	0	-															
13	0	-															
14	0	-															
15	0	-															
16	0	-															
17	0	-															
18	0	-															
19	0	-															
20	0	-															
21	0	-															
22	0	-															
23	0	-															
24	0	-															
25	0	-															
26	0	-															

B	N	0	1	2	3	4	5	6	7	8	9	10	11	12	13	14	15
27	0	-															
28	0	-															
29	0	-															
30	0	-															
31	0	-															

Notice also that the bit allocation table shown in *Table 2* has entries with numbers of levels not usually seen in midtread quantizers. Recall that an N-bit mid-tread quantizer usually has 2^N-1 levels. For example, a 4-bit quantizer would have 15 levels. In contrast, this table has cases with 5 and 9 levels. A 5-step midtread quantizer needs 3 bits to encode its values (2 bits can only count up to 4 levels) while a 9-step quantizer needs 4 bits to encode its values (3 bits can only count up to 8 levels). At first blush, it seems quite wasteful of bits to allow such quantizers, however, grouping of consecutive samples allows for quite efficient packing of the quantized data.

In the cases of 3, 5, and 9-step quantization, three consecutive samples are (vector) coded with one single code word, v, consisting of 5, 7, 10 bits respectively as follows:

$$v_3 = 9z + 3y + x$$
$$v_5 = 25z + 5y + x$$
$$v_9 = 81z + 9y + x$$

where x, y, z, are the quantization levels corresponding to three consecutive sub-band samples (counting from zero up to N_{levels}-1). The result of this vector coding is that quantization can be carried out with few quantizer levels without wasting too many bits.

This vector coding approach saves bits by allocating bits to enumerate possible triplets of coded values rather than allocating bits to each quantized value individually. For example, 3 consecutive samples using a 3-step quantizer would require 3*2 = 6 bits but are here encoded using only 5 bits. This packing can be achieved since there are only 3*3*3 = 27 possible code triplets for 3 samples – a number of possibilities that easily can be enumerated with a 5-bit code (which can enumerate up to 32 items). Similarly, 3 consecutive 5-step quantized values would require 3*3 = 9 bits but the 5*5*5 = 125 possible code triplets can be enumerated with a 7-bit code (which can enumerate up to 128 items). Finally, 3 consecutive 9-step quantized values could require 3*4 = 12 bits while the 9*9*9 = 729 possible triplets can be enumerated with a 10-bit code (which can enumerate up to 1024 items).

In general Layer II represents bit allocation, scale factors, and quantized samples in a more compact way than Layer I, so that more bits are available to represent the coded audio data. In general, Layer II provides higher quality than Layer I at any given data rate.

6.3   Layer III

In addition to the differences in the filter bank and psychoacoustic model computation (see previous sections), Layer III core allocation/quantization routines are more complex. Moreover, a locally variable data rate is employed in Layer III to respond to the demand of difficult signals. The mechanism employed to dynamically provide additional bits is defined in the standard as a bit reservoir mechanism. The frame size for Layer III is 1152 samples as in Layer II, but the coded audio data may expand beyond the current frame, their starting point being identified by a backwards pointer in the bitstream.

6.3.1   Scale Factors

The basic Layer III algorithm (in the single channel mode) feeds the output of the hybrid filter bank to the quantization and coding stage (see also *Figure 4*). The spectrum is subdivided in 21 or 12 scale factor bands (rather than the 32 PQMF sub-bands as in Layers I and II) for long blocks and short block respectively, where the width of these bands loosely follows the critical bandwidth rate and is specified in look-up tables in the standard. In *Table 3* and *Table 4* an example of Layer III scale factors partition at 48 kHz for long and short blocks is shown.

Table 3. MPEG Audio Layer III scale factors partition for long blocks, $F_s = 48$ kHz [ISO/IEC 11172-3]

Scale factor band	Width	Start	End
0	4	0	3
1	4	4	7
2	4	8	11
3	4	12	15
4	4	16	19
5	4	20	23
6	6	24	29
7	6	30	35
8	6	36	41
9	8	42	49
10	10	50	59
11	12	60	71
12	16	72	87
13	18	88	105
14	22	106	127

Scale factor band	Width	Start	End
15	28	128	155
16	34	156	189
17	40	190	229
18	46	230	275
19	54	276	329
20	54	330	383

Table 4. MPEG Audio Layer III scale factors partition for short blocks, Fs = 48 kHz [ISO/IEC 11172-3]

Scale factor band	Width	Start	End
0	4	0	3
1	4	4	7
2	4	8	11
3	4	12	15
4	6	16	21
5	6	22	27
6	10	28	37
7	12	38	49
8	14	50	63
9	16	64	79
10	20	80	99
11	26	100	125

The scale factors in Layer III are employed to modify the quantization step size to ensure that the resulting quantization noise level falls below the computed masked threshold [Brandenburg 87 and Johnston 89], see also next section. Layer III scale factors differ from Layers I and II scale factors in that they are not the result of a normalization process. A set of scale factor select information is coded every two granules, where for Layer III a granule is defined as 576 frequency lines. The SCFSI in Layer III is represented by one bit for each sub-band, which is set to 0 if different scale factors are transmitted for each granule, or to 1 if the scale factor transmitted for the first granule is valid also for the second. If the coder operates in short block mode, then different scale factors are always transmitted for each granule. In addition to the sub-band scale factors, a global gain scale factor for the entire granule determines the overall quantization step size.

6.3.2 Non-Uniform Quantization and Huffman Coding

A non-uniform midtread quantizer is used in Layer III. The quantizer raises its input to the 3/4 power before applying a midtread quantizer (see also Chapter 2). In the decoder before inverse quantizing, the quantized values are re-linearized by raising them to the 4/3 power. In this fashion, bigger values are quantized with less accuracy than smaller values, increasing the signal to noise ratio at low level input.

In addition, in Layer III static Huffman coding is also employed. The encoder subdivides the quantized spectral values into three regions and each region is coded with a different set of Huffman code tables tuned for the statistics of that region. At high frequencies, the encoder identifies a region of "all zeros'. The size of this region can be deduced from the size of the other two regions and does not need to be coded, the only restriction being that it must contain an even number of zeros since the other two regions group their values in even numbered sequences. The second region, called "count 1" region, contains a series of contiguous values consisting only of – 1, 0, +1. In this region, four consecutive values are Huffman encoded at the same time and the size of this region is a multiple of four. Finally the third and last region, the "big values" region, covers the remaining spectral values which are encoded in pairs. This region is further subdivided in three parts each covered by a separate Huffman table. A set of 16 different Huffman code tables is utilized. For each partition, the Huffamn table which best matches the signal statistics is selected during the encoder process. This dynamic search allows for both an increased coding efficiency and a decreased error sensitivity. The largest Huffman table carries 16 by 16 entries. Larger values are accommodated by using an escape mechanism.

The encoder iteratively quantizes the spectral values based on the data rate available, computes the number of Huffman codes needed to represent the quantized spectral values, and derives the distortion noise. If the distortion noise in certain scale factor bands is above the masked threshold values estimated in the psychoacoustic model, it varies the quantizer step size for each scale factor band by amplifying the relative scale factor. The quantization process is repeated until none of the scale factor bands have more than the allowed distortion, or the amplification for any band exceeds the maximum allowed value.

6.3.3   Bit Reservoir

The mechanism put in place in Layer III bitstream formatting routine allows for donating bits to the bit reservoir when the signal analyzed at a certain time requires less than the average number of bits to code a frame or utilizing bits from the reservoir for peak demands. In this fashion, the bitstream formatting routine is designed to support a locally-variable data rate. This mechanism not only allows to respond to local variations in the bit demand, but also is utilized to mitigate possible pre-echo effects. Although the number of bits employed to code a frame of audio data is no longer constant, its long term average is. The deviation from the target data rate is always negative, i.e. the channel capacity is never exceeded. The maximum deviation from the target bit rate is fixed by the size of the maximum allowed delay in the decoder. This value is fixed in the standard by defining a code buffer limited to 7680 bits. For the data rate of 320 kb/s at 48 kHz sampling rate the average number of bits per frame is

Number of bits per frame = 1152* 320/48 = 7680

In this case, no variation is allowed. The first element in the Layer III audio data bitstream is a 9-bit pointer (main_data_begin) which indicates the location of the starting byte of the audio data for that frame. In *Figure 23* an example of the Layer III frame structure is shown.

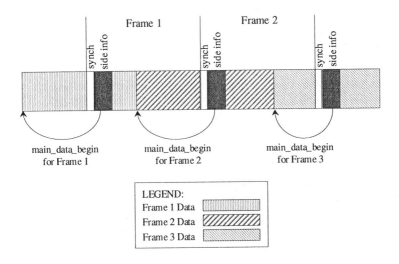

Figure 23. Example of Layer III frame structure from [ISO/IEC 11172-3]

7.   STEREO CODING

Spatial redundancies and irrelevancies are exploited in joint stereo coding in order to further reduce the audio coding system data rates. While for most stereo signals there is little correlation between the time representation of the left and the right channels, typically for the signal power spectra strong correlations exist [Brandenburg 98].   Regarding stereo irrelevancies, we know that at high frequencies the ability of the human auditory system to discriminate the exact source location is decreased [Blauert 83].   Above about 2 kHz, for each critical band the human auditory system bases its perception of stereo imaging from power maxima in space rather than the signal fine temporal structure.

In general, jointly coding the left and right channel provides higher coding gains. There are instances, however, in which stereo coding data rates may exceed twice the rate needed to transparently code one mono signal. In other words, the artifacts masked in single channel coding may become audible when presented as a stereo signal encoded as a dual mono. This effect is related to differences between the masked threshold recorded when the signal is presented as a single channel signal and the masked threshold under binaural conditions. This difference is called the binaural masking level difference (BMLD) [Blauert 83] and it is most pronounced at low frequencies. An example of this effect is related to "the cocktail party effect", where a listener is capable to focus on a conversation in spite of the louder background noise. If the listener plugs one ear, suddenly the same level conversation becomes more difficult to understand. The conversation is less effectively masked when the subject listens to it binaurally than when the subject listens to it monaurally. Applying these findings to stereo coding, one may find that coding artifacts that are masked in single channel mode can become unmasked when presented simultaneously in two channels, i.e. in a dual mono system.

The basic idea in joint stereo coding is to appropriately rotate the stereo plane for each critical band into the main axis direction of the input stereo signal. Applying this idea in audio coding means transmitting side information to convey the direction parameters. This sometimes translates into a large amount of additional side information with no net coding gain. Simplified approaches can be found in literature such as mid/sum, M/S, and intensity stereo coding [Johnston and Ferreira 92 and van der Waal and Veldhuis 91]. While these approaches have some overlapping, their main focus is very different.

M/S stereo coding focuses mainly on redundancies removal and it is a transparent process in absence of quantization and coding. In M/S stereo coding only two directions are considered. Instead of transmitting separately the left and the right signal, the normalized sum and difference signals are transmitted. These signals are also referred to as the middle or mid (sum) and side (difference) signals. The coding gain achieved by utilizing M/S stereo coding is signal dependent. The maximum gain is reached when the left and right signals are equal or phase shifted by π. M/S stereo processing can be applied to the full signal spectrum since it is a lossless process and, in particular, it preserves the spatial attributes of the signal. In some cases, however, stereo irrelevancy effects can also be utilized in this approach.

Intensity stereo coding focuses mainly on irrelevancy removal, where at high frequencies the signal is coded with reduced spatial resolution. This approach is based on the fact that the human auditory system detects direction at high frequency primarily based upon relative intensity in each ear rather than using phase cues. In intensity stereo coding, only one channel resulting from the combination of the right and left channel is transmitted for each critical band. The directional information is conveyed with independent scale factors for the left and the right channels. While the main spatial cues are retained in this method, some details may be missing. Intensity stereo coding preserves the energy of the stereo signal, but some of the signal components may not be properly transmitted, resulting in a potential loss of spatial information. In general, the loss of spatial information is considered less annoying than coding artifacts from bit starvation. For this reason, intensity stereo coding is employed mainly at low data rates. It is important to emphasize that intensity stereo coding is applied for high frequency only. Extending this approach to low frequencies may cause severe distortion such as a major loss in spatial information. Assuming that intensity coding is applied to half of the signal spectrum, a saving of about 20% in the system data rate can be achieved [Brandenburg 98]. Finally it should be mentioned that extensions to multichannel of the general intensity coding technique were developed in more recent years and different flavors are known as coupling channels [Davis 93], dynamic crosstalk [Stoll et al. 94] or generalized intensity coding (see also next chapter for more details).

7.1   MPEG Audio Stereo Coding

In MPEG-1 audio coding mid/sum, MS, stereo coding and intensity stereo coding techniques are applied. M/S stereo coding is employed in Layer III only, while intensity stereo coding is employed in all MPEG-1 Audio layers. In the joint stereo mode (bits 24 and 25 in the MPEG Audio header) the sum of the left and right sub-band samples is transmitted with scale factors for the left and the right channels for selected sub-bands. First an estimate of the required data rate for the left and right channels is performed. If the data rate exceeds the target data rate, a number of sub-bands are set to intensity stereo mode. Depending on the data rate, the allowed set of sub-bands for intensity coding is specified in the standard (bits 26 and 27 in the MPEG Audio header). For the quantization of the combined sub-bands, the higher of the bit allocation for the left and right channel is used. The combined sub-bands are coded in the same fashion as the independent sub-bands. In addition to the common scale factors, directional scale factors are unique to the sub-bands coded in the joint stereo mode and they are transmitted in the bitstream.

7.1.1 M/S Stereo Coding

M/S stereo coding is applied in Layer III when in joint stereo mode when:

$$\sum_{k=0}^{511} (l_k^2 - r_k^2) < 0.8 \sum_{k=0}^{511} (l_k^2 + r_k^2)$$

Where l_k and r_k correspond to the FFT spectral line amplitudes computed in the psychoacoustic model. The values M_i and S_i are transmitted, instead of the left and right channel values L_i and R_i, where L_i and R_i correspond to the hybrid filter bank spectral line amplitudes as follows:

$$M_i = \frac{R_i + L_i}{\sqrt{2}} \quad \text{and} \quad S_i = \frac{L_i - R_i}{\sqrt{2}}$$

In order to code the M/S signals, specially tuned thresholds are computed in the psychoacoustic model and utilized in the allocation routines.

8. SUMMARY

In this chapter we reviewed the main features of MPEG-1 Audio. MPEG-1 is the first international standard of high quality audio and opened the door to a variety of applications from digital audio broadcasting to internet distribution of music. Organized in three layers of increasing sophistication and complexity design, the basic building blocks consist of a time to frequency mapping stage, psychoacoustic model, quantization and coding and bitstream formatting. MPEG-1 allows for audio coding between 32 and 448 kb/s per channel targeting perceptually lossless quality. In the next chapters, we discuss further developments in MPEG Audio including MPEG-2 and MPEG-4.

9. REFERENCES

[Blauert 83]: J. Blauert, *Spatial Hearing*, MIT Press, Cambridge, MA 1983.

[Bosi, Todd and Holman 93]: M. Bosi, C. Todd and T. Holman, "Aspects of Current Standardization Activities for High-Quality, Low Rate Multichannel Audio Coding,"

Proc. of the 1993 IEEE Workshop on Applications of Signal Processing to Audio and Acoustiscs, 2a.3, New Paltz, New York, October 1993.

[Brandenburg 87]: K. Brandenburg, "OCF-A New Coding Algorithm for High Quality Sound Signals", in Proc. ICASSP, pp. 141-144, May 1987.

[Brandenburg 98]: K. Brandenburg, "Perceptual Coding of High Quality Digital Audio", in *Applications of Digital Signal Processing to Audio and Acoustics*, M. Kahrs and K. Brandenburg (ed.), pp. 39-83, 1998.

[Brandenburg and Johnston 90]: K. Brandenburg and J. D. Johnston, "Second Generation Perceptual Audio Coding: The Hybrid Coder", presented at the 88th AES Convention, Preprint 2937, Montreux 1990.

[Brandenburg et al. 91]: K. Brandenburg, J. Herre, J. D. Johnston, Y. Mahieux, and E. F. Schroeder, "ASPEC-Adaptive Spectral Perceptual Entropy Coding of High Quality Music Signals", presented at the 90th AES Convention, Preprint 3011, May 1991.

[Chiariglione 95]: L. Chiariglione, "The development an integrated audio-visual coding standard", Proceedings of the IEEE, Vol. 83 n. 2, pp. 151-157, February 1995.

[Davis 93]: M. Davis, "The AC-3 Multichannel Coder", presented at the 95th AES Convention, Preprint 3774 New York, October 1993.

[Dehéry, Stoll and Kerkhof 91]: Y. F. Dehéry, G. Stoll and L. v.d. Kerkhof, "MUSICAM Source Coding for Digital Sound", In Symp. Rec. Broadcast Sessions of the 17th Int. Television Symp., pp. 612-617, Montreux Switzerland, June 1991.

[Edler 92]: B. Edler, "Aliasing reduction in sub-bands of cascaded filter banks with decimation ", Electronics Letters, Vol. 28, pp. 1104-1105, 1992.

[Edler 89]: B. Edler, "Coding of Audio Signals with Overlapping Transform and Adaptive Window Shape" (in German), Frequenz, Vol. 43, No. 9, pp. 252-256, September 1989.

[ETS 300 401 v2]: The European Telecommunications Standards Institute (ETSI), ETS 300 401 v2, "Radio Broadcasting Systems; Digital Audio Broadcasting (DAB) to Mobile, Portable and Fixed Receivers", 1995-1997.

[ETS 300 421]: The European Telecommunications Standards Institute (ETSI), ETS 300 421, "Digital Video Broadcasting (DVB); Framing Structure, Channel Coding and Modulation for 11/12 GHz Satellite Services", August 1997.

[ETS 300 429]: The European Telecommunications Standards Institute (ETSI), ETS 300 429, "Digital Video Broadcasting (DVB); Framing Structure, Channel Coding and Modulation for Cable Systems", August 1997.

[ETS 300 744]: The European Telecommunications Standards Institute (ETSI), ETS 300 744, "Digital Video Broadcasting (DVB); Framing Structure, Channel Coding and Modulation for Digital Terrestrial Television", August 1997.

[Fielder et al. 96]: L. D. Fielder, M. Bosi, G. A. Davidson, M. Davis, C. Todd, and S. Vernon" AC-2 and AC-3: Low Complexity Transform-Based Audio Coding," in N. Gielchrist and C. Grewin (ed.), *Collected Papers on Digital Audio Bit-Rate Reduction*, pp. 54-72, AES 1996.

[ISO/IEC 11172]: ISO/IEC 11172, Information Technology, "Coding of moving pictures and associated audio for digital storage media at up to about 1.5 Mbit/s", 1993.

[ISO/IEC 11172-3]: ISO/IEC 11172-3, Information Technology, "Coding of moving pictures and associated audio for digital storage media at up to about 1.5 Mbit/s, Part 3: Audio", 1993.

[ISO/IEC 13818]: ISO/IEC 13818, Information Technology, "Generic coding of moving pictures and associated audio", 1994-1997.

[ISO/IEC 13818-7]: ISO/IEC 13818-7, Information Technology, "Generic coding of moving pictures and associated audio, Part 7: Advanced Audio Coding", 1997.

[ISO/IEC 14496]: ISO/IEC 14496, Information Technology, "Coding of audio-visual objects", 1999-2001.

[ISO/IEC 15938]: ISO/IEC 15938, Information Technology, "Multimedia content description interface", 2001.

[ISO/IEC MPEG 91/010]: ISO/IEC JTC 1/SC 29/WG 11 MPEG 91/010, "The MPEG/AUDIO Subjective Listening Test" Stockholm, April/May 1991.

[ISO/IEC MPEG N1229]: ISO/IEC JTC 1/SC 29/WG 11 N1229, "MPEG-2 Backwards Compatible CODECS Layer II and III: RACE dTTb Listening Test Report" Florence, March 1996.

[ISO/IEC MPEG N1420]: ISO/IEC JTC 1/SC 29/WG 11 N1420, "Overview of the Report on the Formal Subjective Listening Tests of MPEG-2 AAC multichannel audio coding" Maceio', November 1996.

[ISO/IEC MPEG N4318]: ISO/IEC JTC 1/SC 29/WG 11 N4318, "MPEG-21 Overview" Sydney, July 2001.

[ISO/IEC MPEG N271]: ISO/IEC JTC 1/SC 29/WG 11 N271, "New Work Item Proposal for Very-Low Bitrates Audiovisual Coding" London, November 1992.

[ITU-R BS.1115]: International Telecommunication Union BS.1115, "Low Bitrate Audio Coding", Geneva, 1994.

[Johnston 88]: J. D. Johnston, "Estimation of Perceptual Entropy Using Noise Masking Criteria", in Proc. ICASSP, pp. 2524-2527, May 1988.

[Johnston 89]: J. D. Johnston, "Transform Coding of Audio Using Perceptual Noise Criteria", in IEEE J. Select. Areas Common., vol. 6, pp. 314-323, 1989.

[Johnston and Ferreira 92], J. D. Johnston and A. J. Ferreira, "Sum-Difference Stereo Transform Coding", in Proc. ICASSP pp. 569-571, May 1992.

[MPEG]: MPEG Home Page, http://mpeg.telecomitalialab.com/.

[Pan 95]: D. Pan, "A Tutorial on MPEG/Audio Compression", in IEEE Multimedia, pp. 60-74, Summer 1995.

[Searing 91]: S. Searing, "Suggested Formulas for Audio Analysis and Synthesis Windows", ISO/IEC JTC 1/SC 29/WG 11 MPEG 91/328, November 1991.

[Sinha, Johnston, Dorward and Quackenbush 98]: D. Sinha, J. D. Johnston, S. Dorward and S. R. Quackenbush, "The perceptual Audio Coder (PAC)", in *The Digital Signal Processing Handbook*, V. Madisetti and D. Williams (ed.), CRC Press, pp. 42.1-42.18, 1998.

[Stoll et al. 94]: G. Stoll, G. Theile, S. Nielsen, A. Silzle, M. Link, R. Sedlmayer and A. Breford, "Extension of ISO/MPEG-Audio Layer II to Multi-Channel Coding: The Future Standard for Broadcasting, Telecommunication, and Multimedia Applications", presented at the 94th AES Convention, Preprint 3550, Berlin 1994.

[Thiede et al. 00]: T. Thiede, W. Treurniet, R. Bitto, C. Schmidmer, T. Sporer, J. Beerends, C. Colomes, M. Keyhl, G. Stoll, K. Brandenburg and B. Feiten,, "PEAQ-The ITU Standard for Objective Measurement of Perceived Audio Quality", J. Audio Eng. Soc., Vol. 48, no. 1/2, pp. 3-29, January/February 2000.

Chapter 12

MPEG-2 Audio

1. INTRODUCTION

This chapter describes the MPEG-2 Audio coding family which was developed to extend the MPEG-1 Audio functionality to lower data rate and multichannel applications. First we will discuss MPEG-2 LSF and "MPEG-2.5" systems in which lower sampling frequencies were used as a means to reduce data rate in MPEG-1 audio coding. (The MP3 audio format as usually implemented as the MPEG-2.5 extension of Layer III.) Next we will discuss the MPEG-2 BC system which extended the MPEG-1 Audio coders to multichannel operation while preserving backwards compatibility so that MPEG-2 BC data streams would be playable on existing MPEG-1 players. In the next chapter, we will discuss the MPEG-2 AAC system which made use of all of the advanced audio coding techniques available at the time to reach higher quality in a multichannel system than was achievable while maintaining compatibility with MPEG-1 players.

2. MPEG-2 LSF, "MPEG-2.5" AND MP3

Motivated by the increase of low data rate applications over the Internet, the goal of MPEG-2 LSF was to achieve MPEG-1 or better audio quality at lower data rates [ISO/IEC 13818-3]. One way to achieve this goal without requiring major modifications in the MPEG-1 system was to decrease the sampling rate of the audio signals which is the approach taken by MPEG-2 LSF. Instead of the 48, 44.1, and 32 kHz sampling rates seen in MPEG-1, the sampling rates for MPEG-2 LSF are 24, 22.05, and 16 kHz. Of course, reducing the sampling rate by a factor of 2 also reduces the audio bandwidth by a factor of 2. This loss in high frequency content was deemed an acceptable compromise for some target applications in order to reduce the data rate by a factor of 2.

The MPEG-2 LSF coder and its bitstream have a very similar structure to that of MPEG-1 Audio. The three audio "layers" of increasing complexity are again defined almost identically to the layers of MPEG-1 Audio. Single, dual channel, stereo, and joint stereo modes are again supported in MPEG-2 LSF. The bitstream header for MPEG-2 LSF differs from that of MPEG-1 Audio solely in the settings of a single bit (see *Figure 1*). The 13th bit (called the "ID bit") is set equal to zero for MPEG-2 LSF while it was equal to one for MPEG-1. The bit setting for MPEG-2 LSF signals the use of different tables for sampling rates and target data rates.

The main difference in the MPEG-2 LSF implementation is the set of allowed sampling rate and data rate pairings in its operation. An additional difference is the adaptation of the psychoacoustic parameters for the lower sampling frequencies. The frame size is reduced from 1152 samples to 576 samples in Layer III to make the audio frame more manageable for packetizing in internet applications. (*Table 1* lists the duration of the audio frame for the different sampling rates and layers of MPEG-2 LSF.)

The resulting data rates in MPEG-2 LSF are lower than the data rates for MPEG-1. The nominal operating rates for MPEG-1 are from 32 kb/s up to 224 kb/s but good quality is usually found around 128 kb/s per channel. In contrast, MPEG-2 LSF typically produces adequate quality for various applications throughout its operating range from 32 to 128 kb/s per channel for Layer I, and from 8 to 80 kb/s per channel for Layers II and III.

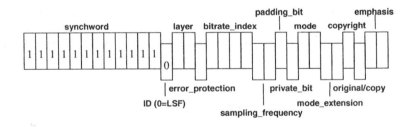

Figure 1. MPEG-2 Audio LSF header

Table 1. MPEG-2 LSF audio frame duration [ISO/IEC 13818-3]

Layer	Sampling Frequency in kHz		
	16	22.05	24
I	24 ms	17.41 ms	16 ms
II	72 ms	52.24 ms	48 ms
III	36 ms	26.12 ms	24 ms

The decrease in data rates, especially for Layer III, made MPEG-2 LSF useful for low bandwidth Internet applications. This led the audio group at the Fraunhofer Institute to create an even lower sampling rate modification of Layer III that they named "MPEG-2.5" (see also [Brandenburg 99]). "MPEG-2.5" reduced the sampling rates by another factor of 2 from MPEG-2 LSF Layer III. In "MPEG-2.5" the allowed sampling rates are 12 kHz, 11.025 kHz, and 8 kHz. The addition of these extensions allow Layer III coders to range from samples rates of 8 kHz ("MPEG-2.5") up to 32 kHz (MPEG-1).

To allow "MPEG-2.5" decoders to work with the same bitstream format as used in MPEG-1 Audio and MPEG-2 LSF, they removed the final bit from header synchword and merged it with the ID bit into a 2-bit ID code. The result was that "MPEG-2.5" decoders work with an 11 bit synchword (rather than 12 bit) but have 2 bits to identify the bitstream format. The "MPEG-2.5" ID codes are [00] for "MPEG-2.5", [11] for MPEG-1, [10] for MPEG-2 LSF, and [01] reserved for future extensions. Notice that using a [1] as the first bit of the two-bit ID code bit for the prior formats leads to compatibility with the MPEG-1 and MPEG-2 formats. Hence, the "MPEG-2.5" bitstream is identical to that of MPEG-1 Audio and MPEG-2 LSF when those formats are being encoded.

The high quality of the Layer III encoder coupled with the wide range of sample rates and data rates that can be encoded using the MPEG-2 LSF and "MPEG-2.5" extensions made Layer III a natural choice for Internet applications. The so-called "MP3" file format is typically implemented as MPEG-1 Layer III with both of these extensions supported. Low bandwidth users typically use the 16 kHz sampling rate (for a bandwidth of roughly 8 kHz) and encode stereo sound at 32 kb/s. Compared with the CD format (44.1 kHz stereo at 16 bits/sample), this represents a data rate reduction of over a factor of 40 and allows for an entire CD's worth of music (about 800 MB) to be stored in under 20 MB. Higher bandwidth users are more likely to operate with the full CD 44.1 kHz sample rate at 128 kb/s for "near CD-quality" sound at a data rate reduced by more than a factor of ten from CD format.

The use of MP3 files for sharing audio over the Internet has spread so widely that it has become the de facto standard for Internet audio. In addition, MP3 players and portable devices are in widespread use for listening to audio and home audio digital components (e.g. CD players, DVD players) increasingly tout MP3 format playback as one of their features.

3.   INTRODUCTION TO MULTICHANNEL AUDIO

Since the main focus of the MPEG-2 Audio work evolved around multichannel audio, a brief introduction on the evolution of spatial representation of sound is presented in this section. Starting with monophonic technology, and partially pushed by the progress in the film industry, the art of multichannel sound developed towards stereophonic, quadraphonic, and more.

3.1   History and Channel Configurations

The cinema industry embraced multichannel formats [Holman 91] because of their flexibility and the greater enveloping experience they provided. In the eighties, with the introduction of the CD format, stereophonic sound became well established, while a few artists were mastering in quadraphonic and a very small audience had access to reproduction systems with more than two channelsat home. During the nineties when the migration from the cinema halls to the living rooms started to take place, the pace was set for standards like that of high definition television in North America [ATSC A/52/10] and the DVD [DVD-Video] standards. In the early nineties the focus of the audio standardization efforts shifted from mono and stereo audio signals to multichannel audio signals. Today, the general evolution of digital technology and a steady growth in transmission bandwidth and storage capacity have made multichannel audio a more realistic option for widespread audio reproduction.

MPEG-2 Audio was one of the first audio coding standards embracing the 5.1-channel audio configuration. This now widely adopted multichannel configuration encompasses five full-bandwidth (20 kHz) channels and a low frequency enhancement channel, LFE, covering frequencies below 200 Hz (hence the .1 denomination since the LFE channel covers less than 10% of a 20-kHz bandwidth signal). The 5.1 configuration was first introduced by SMPTE in 1987 [Holman 91] and later adopted by a number of standardization bodies including ITU-R, MPEG, the North American HDTV, DVD.

Reference loudspeaker arrangement with
loudspeakers L/C/R and LS/RS

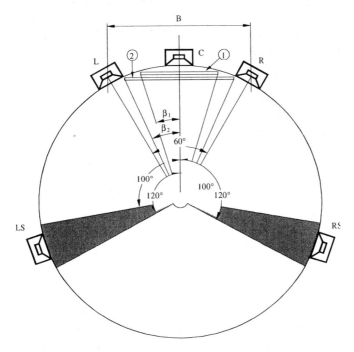

Screen 1 HDTV – Reference distance $= 3\,H\,(2\beta_1 = 33°$
Screen 2 $= 2\,H\,(2\beta_2 = 48°$
H: height of screen
B: loudspeaker base width

Loudspeaker	Horizontal angle from centre (degrees)	Height (m)	Inclination (degrees)
C	0	1.2	0
L, R	30	1.2	0
LS, RS	100 ... 120	≥ 1.2	0 ... 15 down

Figure 2. 3/2 multichannel configuration from [ITU-R BS.775-1]

The 5.1-channel audio configuration is often referred to as the 3/2/.1 configuration, since three loudspeakers are typically placed in front of the listener, and two in the side/rear (see *Figure 2*). This arrangement is described in detail in the ITU-R recommendation BS.775-1 [ITU-R BS.775-1]. According to the ITU-R specifications, the five full-bandwidth loudspeakers are placed on the circumference of a circle centered on the reference listening position(see *Figure 2*). Three front loudspeakers are placed at angles from the listener axis of -30^0 (left channel, L), $+30^0$ (right channel, R), and 0^0 (center channel, C); the two surround loudspeakers are placed at angles between -100^0 and -120^0 (left surround channel, LS) and $+100^0$ and $+120^0$ (right surround channel, RS). The LFE speaker is typically placed in the front, although its exact location is not specified in the ITU-R layout, since the human auditory system does not take strong localization cues from one single low frequency sound channel. The purpose of the LFE channel is to enable high-level sounds at frequencies below 200 Hz without overloading the main channels. In the cinema practice, this channel has 10 dB more headroom than the main channels.

In the following sections of this book, we will be generally referring to the 5.1-channel configuration unless otherwise specified. If more than 5.1 channels are employed, the basic ideas and main principles we present still hold true, but even greater compression requirements for the coding technology should be adopted.

3.2   System Demands

Going from stereophonic to multichannel sound reproduction adds to the demands on storage and delivery media. If we consider the CD format, that is PCM sampled at a frequency of 44.1 kHz and quantized using uniform quantization with of 16 bits per sample, the total data rate for the 5.1 multichannel configuration is 3.598 Mb/s. An hour of multichannel music in the CD format requires 1.62 GB, way above the CD storage capacity of about 800 MB per disk. If we consider current multichannel applications such as digital broadcasting, internet audio and electronic music distribution, in each case bandwidth/capacity are serious challenges.

As in the monophonic and stereo case, the challenge of multichannel audio coding is to minimize the data rate without sacrificing audio quality. While PCM is a well-understood and well-established coding method that offers very low-complexity implementations, it requires very high capacity/bandwidth to provide high-quality audio signals. One should also notice that CD format audio signals may suffer from some degradation. It was shown in [Fielder 87] by comparing the hearing threshold with the CD signal resolution levels and typical 16-bit converters, that audible quantization noise can be introduced in the mid-range frequencies. The implication is that, expensive as it may be, one may need to increase the PCM sample precision, going for example from R=16 to R=24.

In addition to the augmented sample precision, we are witnessing a new trend to adopt higher sampling rates, going from F_s = 44.1 kHz or F_s = 48 kHz to F_s = 96 kHz, and F_s = 192 kHz. While there is no scientific evidence or published experimental results to the authors' knowledge that unequivocally prove the advantages of adopting higher sampling frequencies, many recording engineers and industry "golden ears" feel that adopting sampling frequencies of 96 kHz or higher and the equivalent audio sample word length of 24 bits, and multichannel audio are essential in providing the end user with high quality audio and a truly enveloping experience. This leads to different choices as how to balance the emerging market desire for high resolution audio with current delivery media restrictions. *Table 2* shows examples of different approaches in the marketplace to multichannel high-resolution audio.

At high sampling rates media restrictions become even more binding, prohibiting multichannel audio even for emerging new high-capacity technologies like, for example, DVD-Audio applications [DVD-Audio]. If we consider 5 full-bandwidth channels sampled at 96 kHz with 24-bit precision, the total throughput is 11.52 Mb/s, which exceeds the maximum 9.6 Mb/s throughput of DVD-Audio.

Table 2. Different approaches to multichannel in the marketplace

	MPEG-2 BC	AC-3	MPEG-2 AAC	DVD-Audio
Audio Channels	1-5.1	1-5.1	1-48	1-6
Fs (kHz)	32, 44.1, 48	32, 44.1, 48	8-96	44.1, 48, 88.2, 96, 176.4, 192
R (bits/sample)	16-24	16-24	16-24	16-20-24
I (kb/s)	32-1,130	32-640	Up to 576 per channel	9600
Frame (samples)	384-1152	1536	1024	

4. MPEG-2 MULTICHANNEL BC

MPEG-2 multichannel BC audio coding provides a multichannel extension to MPEG-1 Audio which is backwards compatible with the MPEG-1 bitstream, where backwards compatibility implies that an MPEG-1 decoder is able to decode MPEG-2 encoded bitstreams. The audio sampling rates in MPEG-2 BC for the main channels are the same as in MPEG-1, however, the channel configuration supports up to five full-bandwidth main audio channels plus a low frequency enhancement, LFE, channel whose bandwidth is less than 125 Hz. The sampling frequency of the LFE channel corresponds to the sampling frequency of the main channels divided by 96.

There are different strategies for achieving backward compatibility with monophonic and two-channel systems. One procedure is to provide both monophonic and two-channel information concurrently with the multichannel data stream (simulcast). The advantage of this situation is that no constraints are imposed in the multichannel technology, and the monophonic/two-channel services can simply be discontinued at a given time. The disadvantage of this scenario is an increase in the data capacity required (at least initially). Another approach is to embrace backward compatible matrixing techniques. In this approach, the left and right channels contain the down-mixed multichannel information. The MPEG-2 BC audio coding standard follows this second approach. In general, the advantage of this approach is that little additional data capacity is required for the multichannel service. The disadvantage is that this approach greatly constrains the design of the multichannel coder, which in turns limits quality of the multichannel reproduction for certain classes of signals at a given data rate. In particular, this approach may lead to unmasking of quantization noise in the decoder after the de-matrixing stage [Bosi, Todd and Holman 93]. As we discussed in the introduction to this chapter, the backwards compatibility with the MPEG-1 bitstream requirement plays a very important role in the design of the MPEG-2 BC system [ISO/IEC 13818-3]. In this section we discuss this approach in detail.

Since the MPEG-1 equivalent left and right channels need to carry a complete description of the MPEG-2 main multichannel signal, the five-channel information is down-mixed into two channels as follows, see also [Ten Kate 92]:

$$L_0 = c\,(L + aC + bL_S)$$

$$R_0 = c\,(R + aC + bR_S)$$

Where L represents the left channel, R the right channel, C the center channel, and L_S and R_S the left and right surround channels respectively. L_0 and R_0 represent the MPEG-1 compatible left and right channels (see *Figure 3*). The down-mix coefficients are specified in the standard. For example, commonly used values for the down-mix coefficients are given by:

$$a = b = \frac{1}{\sqrt{2}} \quad \text{and} \quad c = \frac{1}{1 + \sqrt{2}}$$

While the basic MPEG-2 BC bitstream structure is essentially the same as in MPEG-1 (see also *Figure 4* and *Figure 5*), additional channels (see *Figure 3*), T_3, T_4, T_5 (corresponding to C, L_s, and R_s respectively) and LFE channels are stored along with the multichannel header as an optional extension bitstream in the ancillary data field.

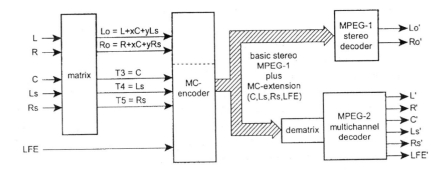

Figure 3. MPEG-2 multichannel BC configuration from [Stoll 96]

For Layers I and II the LFE channel is coded using block floating point. For Layer III the LFE channel is coded following a simplified MPEG-1 Layer III encoding process which includes Huffman coding and operates only in the MDCT short block mode (see also Chapter 11). The audio sample precision adopted in MPEG-2 BC is higher than the precision adopted in MPEG-1 with an equivalent word precision of up to 24 bits per sample. The overall data rate may be increased above that specified by the MPEG-1 standard by using the extension bitstream field so that the MPEG-2 encoded data are still compatible with the MPEG-1 data specifications. The maximum data rates at 48 kHz sampling rate are 1.13 Mb/s for Layer I, 1.066 Mb/s for Layer II, 1.002 Mb/s for Layer III. Notice that for Layers II and III the multichannel extension is encoded as a Layer II or III extension respectively; for Layer I, however, only Layer II extensions are allowed.

In the decoder, the corresponding de-matrixing process takes place as follows:

$$L = L_o/c - aC - bL_s$$

$$R = R_o/c - aC - bR_s$$

During the de-matrixing process quantization noise that was masked in certain channels in the encoder stage may appear as audible quantization noise in different channels in the decoder stage. An example that illustrates this point is given when a signal in a particular channel is derived from two channels with signals out of phase, i.e. canceling each other. If the corresponding quantization noise is not out of phase, then it becomes audible since the masking portion of the signal is canceled out. This is obviously an extreme case, but, in general, partial unmasking of quantization noise may occur frequently after the de-matrixing process. To compensate for this effect an increase in the coding margin (and therefore in the global data rate) is required to maintain perceptual lossless coding.

In addition to the 5.1 or 3/2/.1 configuration, other channel configurations supported in MPEG-2 include the 3/1, 3/0, 2/2, 2/1, 2/0, 1/0 configurations, where the number before the slash symbol indicates the number of channels in the frontal plane, and the number after the slash symbol indicates the number of channels in the rear plane (see also [ITU-R BS.775-1]). Different combination of audio input channels are encoded and transmitted. The first two channels represent always the two basic channels compatible with the MPEG-1 specifications. In general the up to three additional channels or their combination can be dynamically selected to be conveyed in the transmission channels of the ancillary data field.

4.1   Bitstream Format

MPEG-2 BC uses the ancillary data field in the MPEG-1 format to store the additional multichannel information. *Figure 4* shows the structure of an MPEG-2 audio frame. The first section of the frame is identical to the MPEG-1 audio frame. The second section, the multichannel field, is stored in the ancillary data portion of the MPEG-1 bitstream, along with the ancillary data. After the MPEG-1 header, CRC, and audio data, in the MPEG-1 ancillary data field, the multichannel (MC) header (see also *Figure 6*), MC CRC, MC composite status information (see also *Figure 8*), MC audio data, multilingual (see next section) and ancillary data follow. If the multichannel data rate exceeds the MPEG-1 data rates, the extension bit in the MC header is set. In this case, in addition to the MPEG-1 compatible part of the audio frame, an extension part (see *Figure 5*) consisting of the extension (Ext) header, (see also *Figure 7*, namely Ext synchronization word, Ext CRC, Ext length, and Ext ID bit), Ext MC audio data, and multilingual and MC ancillary data are also present.

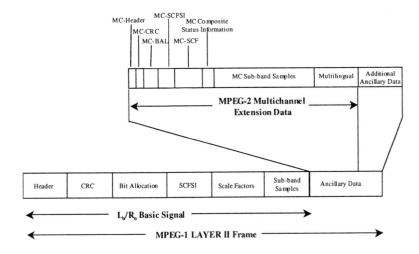

Figure 4. MPEG-2 multichannel bitstream backwards compatible with MPEG-1 Layer II from [ISO/IEC 13818-3]

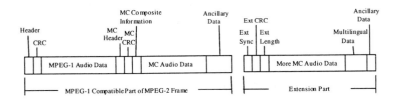

Figure 5. MPEG-2 multichannel extension from [ISO/IEC 13818-3]

As described above, the extension bitstream is utilized when the multichannel and the multilingual information does not fit in one basic MPEG-1 frame. The first bit in the multichannel header indicates whether or not the extension field in the MPEG-2 frame exists. If the extension bitstream exists, then an eight-bit unsigned value describes the length in bytes of the ancillary data field. The other bits in the multichannel header convey information on the multichannel channels, de-matrix procedure, and multilingual channels (see next section), etc. For example, the center bits indicate whether or not the center channel is present and if its bandwidth is limited, i.e. phantom coding of the center channel is applied. The number of multilingual channels, their sampling rate and the multilingual layer is also specified in the multichannel header.

In the composite status information field, the multichannel coding employed such as transmission channel allocation, crosstalk, and prediction are specified. For Layer III the MDCT block size and the beginning of the multichannel data is also indicated in this field via an 11 bits value, since Layer III maintains a locally variable data rate structure.

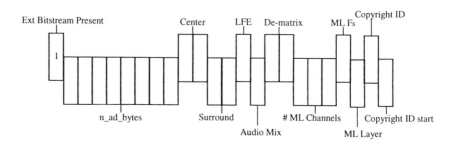

Figure 6. MPEG-2 multichannel header

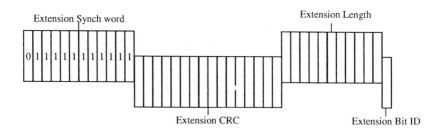

Figure 7. MPEG-2 multichannel extension header

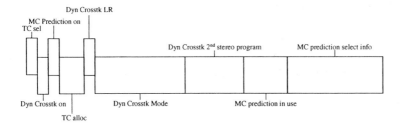

Figure 8. MPEG-2 Layers I and II multichannel composite status information

4.2 Multilingual Channels

In addition to the main audio channels, MPEG-2 BC supports multilingual audio. Up to seven multilingual channels can be transmitted along with the main audio channels (see *Figure 9*). The sampling frequency of the multilingual channels can be the same as or half the sampling frequency of the main channels. When the main channels are coded with Layer I the multilingual extension is always coded with the Layer II algorithm. When the main channels are coded with Layers II and III, the multilingual extension can be coded either with Layer II or Layer III by appropriately setting the multilingual layer bit in the multichannel header.

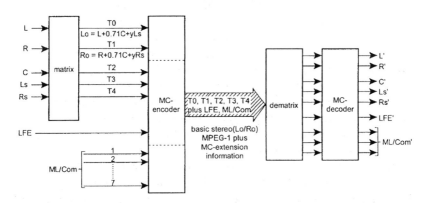

Figure 9. MPEG-2 multichannel BC with multilingual channels from [Stoll 96]

4.3   Encoding Process

In *Figure 10* a block diagram of the MPEG-2 multichannel Layer II coding scheme is shown. If we compare it with the coding scheme of the encoding process for MPEG-1 shown in *Figure 3* and *Figure 22* of Chapter 11, we notice that the basic structure, i.e. the sub-band filter analysis, scale factors computation, psychoacoustic modeling, basic quantization and coding stays the same. The main difference is that instead of a single or stereo channel in MPEG-2 multiple channels are considered and composite channel coding techniques are applied.

Each channel of input signal is first fed to the PQMF sub-band filter and then matrixed. The scale factors calculation, transmission channel allocation, dynamic crosstalk and multichannel prediction stages follow next (see below). In parallel, Psychoacoustic Model 1 (see Chapter 11) is computed. The corresponding SMR values are then modified to take into consideration the prediction values for the sub-band samples. The prediction coefficients are quantized and the prediction error signals are then computed. Finally, the quantized sub-band samples, scale factors, bit allocation values, multichannel control parameters, etc., are multiplexed and formatted into the MPEG-2 bitstream.

MPEG-2 BC adopts composite channel coding techniques such as dynamic crosstalk, adaptive multichannel prediction, and the phantom coding of the center channel, see also [Stoll 96]. The dynamic crosstalk technique exploits the same principles utilized in intensity stereo coding wherein the high frequency content of a stereo signal is transmitted as a single audio signal along with scaling information describing how the intensity is split between the two channels. Similarly, in dynamic crosstalk mode only one audio signal is sent in the high frequency region and scaling coefficients define how to split it between each channel. Dynamic crosstalk encoding can be done independently for several different frequency regions. Adaptive multichannel prediction reduces redundancies by exploiting statistical correlations between channels. Rather than of transmitting the actual signal in each channel, only the prediction error from an adaptive predictor is sent for the center and surround channels. In MPEG-2 BC a predictor of up to the 2nd order with delay compensation is employed [Fuchs 93]. Finally, above a certain frequency, the center channel signal can be split and added to the left and right main channels creating a phantom image of the center signal being anchored in the center by the two side speakers.

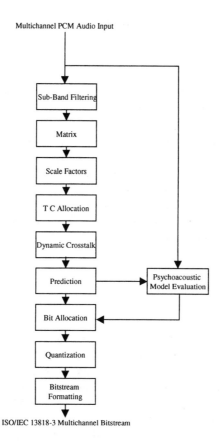

Multichannel PCM Audio Input

Sub-Band Filtering

Matrix

Scale Factors

T C Allocation

Dynamic Crosstalk

Prediction

Psychoacoustic Model Evaluation

Bit Allocation

Quantization

Bitstream Formatting

ISO/IEC 13818-3 Multichannel Bitstream

Figure 10. MPEG-2 Layer II multichannel encoding process block diagram

The multilingual section of the MPEG-2 bitstream is encoded as an MPEG-1 bitstream if the sampling frequency employed is the same as for the main channels. In this case, however, no intensity coding mode is allowed. If the sampling rate is half the sampling rate employed for the main channels, the encoding of the multilingual channel is done according to the MPEG-2 LSF specifications allowing a significant gain in coding efficiency to be achieved at the expense of a reduction in the channel bandwidth. Since in speech signals usually have bandwidth below 10 kHz, this limitation does not typically cause a strong degradation in quality.

4.4 Applications and Quality Evaluation

MPEG-2 multichannel BC applications involve mostly Layer II implementations. MPEG-2 Layer II was selected as a standard recommendation by ITU-R for digital broadcasting applications including digital radio and digital television. It is currently in use in the digital audio broadcasting, DAB, standard [ETS 300 401 v2] and the digital video broadcasting, DVB, standard [ETS 300 421] worldwide. Layer II was also adopted for DVD-Video applications in selected countries.

A number of subjective tests were carried out according to the ITU-R BS.1116 specifications (see also Chapter 13) to evaluate MPEG-2 Layer II performance [ISO/IEC MPEG N1229]. For MPEG-2 Layer II very good quality is achieved for data rate in average of 640 kb/s in the 5.1 multichannel configuration.

5.   SUMMARY

In this chapter we reviewed the main features of the MPEG-2 Audio standard. Initially the goals of MPEG-2 were to address low sampling frequencies and multichannel extensions to MPEG-1. MPEG-2 LSF, motivated by the demands of very low data rates applications, found widespread penetration with one of its derivate, the MP3 format. MPEG-2 multichannel BC was deployed to secure compatibility with an installed base of mono/stereo systems. Based on this requirement, constraints on the multichannel design lead to further investigation on non backwards compatible multichannel systems and the development of MPEG-2 Advanced Audio Coding, AAC. In the next chapter we discuss the main features of MPEG-2 AAC.

6.   REFERENCES

[ATSC A/52/10]: United States Advanced Television Systems Committee Digital Audio Compression (AC-3) Standard, Doc. A/52/10, December 1995.

[Bosi et al. 97]: M Bosi, K. Brandenburg, S. Quackenbush, L. Fielder, K. Akagiri, H. Fuchs, M. Dietz, J. Herre, G. Davidson and Y. Oikawa, "ISO/IEC MPEG-2 Advanced Audio Coding," J. Audio Eng. Soc., vol. 45, pp. 789 – 812, October 1997.

[Bosi, Todd and Holman 93]: M. Bosi, C. Todd, and T. Holman, "Aspects of Current Standardization Activities for High-Quality, Low Rate Multichannel Audio Coding," Proc. of the 1993 IEEE Workshop on Applications of Signal Processing to Audio and Acoustics, 2a.3, New Paltz, New York, October 1993.

[Brandenburg 99]: K. Brandenburg, "MP3 and AAC Explained", from the Proceedings of the AES 17th International Conference, High-Quality Audio Coding, pp. 99-111, Florence, Italy, September 1999.

[Damaske and Ando 72]: P. Damaske and Y. Ando, "Interaural Crosscorrelation for Multichannel Loudspeaker Reproduction," Acustica, vol. 27, pp. 232-238, October 1972.

[Davis 93]: M. Davis, "The AC-3 Multichannel Coder," presented at the 95th AES Convention, New York, pre-print 377, October 1993.

[DVD-Audio]: DVD Specifications for Read-Only Disc, Part 4: AUDIO SPECIFICATIONS Ver. 1.2, Tokyo 1999-2001.

[DVD-Video]: DVD Specifications for Read-Only Disc, Part 3: VIDEO SPECIFICATIONS Ver. 1.1, Tokyo 1997-2001.

[ETS 300 401 v2]: The European Telecommunications Standards Institute (ETSI), ETS 300 401 v2, "Radio Broadcasting Systems; Digital Audio Broadcasting (DAB) to Mobile, Portable and Fixed Receivers", 1995-1997.

[ETS 300 421]: The European Telecommunications Standards Institute (ETSI), ETS 300 421, "Digital Video Broadcasting (DVB); Framing Structure, Channel Coding and Modulation for 11/12 GHz Satellite Services", August 1997.

[Fielder 87]: L. Fielder, "Evaluation of the Audible Distortion and Noise Produced by Digital Audio Converters," J. Audio Eng. Soc., vol. 35, pp. 517-535, July/August 1987.

[Fuchs 93]: H. Fuchs, "Improving Joint Stereo Audio Coding by Adaptive Interchannel Prediction", Proc. of the 1993 IEEE Workshop on Applications of Signal Processing to Audio and Acoustics, New Paltz, New York, October 1993.

[Holman 91]: T. Holman, "New Factors in Sound for Cinema and Television," J. Audio Eng. Soc., vol. 39, pp. 529-539, July/August 1991.

[ISO/IEC 13818-3]: ISO/IEC 13818-3, "Information Technology - Generic Coding of Moving Pictures and Associated Audio, Part 3: Audio," 1994-1997.

[ISO/IEC MPEG N1229]: ISO/IEC JTC 1/SC 29/WG 11 N1229, "MPEG-2 Backwards Compatible CODECS Layer II and III: RACE dTTb Listening Test Report," Florence, March 1996.

[ITU-R BS.775-1]: International Telecommunications Union BS.775-1, "Multichannel Stereophonic Sound System with and without Accompanying Picture", Geneva, Switzerland, 1992-1994.

[Stoll 96]: G. Stoll, "ISO-MPEG-2 Audio: A Generic Standard for the Coding of Two-Channel and Multichannel Sound", in N. Gielchrist and C. Grewin (ed.), *Collected Papers on Digital Audio Bit-Rate Reduction*, pp. 43-53, AES 1996.

[Stoll et al. 94]: G. Stoll, G. Theile, S. Nielsen, A. Silzle, M. Link, R. Sedlmayer and A. Breford, "Extension of ISO/MPEG-Audio Layer II to Multi-Channel Coding: The Future Standard for Broadcasting, Telecommunication, and Multimedia Applications", presented at the 94th AES Convention, Preprint 3550, Berlin 1994.

[Ten Kate 92]: W. ten Kate, P. Boers, A. Maekivirta, J. Kuusama, K. E. Christensen, and E. Soerensen, "Matrixing of Bit-Rate Reduced Signals", in Proc. ICASSP, vol. 2, pp.205-208, May 1992.

[Van der Waal and Veldhuis 91]: R. G. v.d. Waal and R. N. J. Veldhuis, "Subband Coding of Stereophonic Digital Audio Signals", Proc. ICASSP, pp. 3601 – 3604, 1991.

Chapter 13

MPEG-2 AAC

1. INTRODUCTION

This chapter describes the MPEG-2 Advanced Audio Coding, AAC, system.[5] Started in 1994, another effort of the MPEG-2 Audio committee was to define a higher quality multichannel standard than achievable while requiring MPEG-1 backwards compatibility. The so called MPEG-2 non-backwards compatible audio standard, later renamed MPEG-2 Advanced Audio Coding (MPEG-2 AAC) [ISO/IEC 13818-7] was finalized in 1997. AAC made use of all of the advanced audio coding techniques available at the time of its development to provide very high quality multichannel audio.

2. OVERVIEW

The aim of the MPEG-2 AAC development was to reach "indistinguishable" audio quality as specified by the ITU-R TG10-2 [ITU-R TG10-2/3] at data rates of 384 kb/s (or lower) for five full-bandwidth channel audio signals. Tests carried out in the fall of 1996 at BBC, UK, and NHK, Japan, showed that MPEG-2 AAC satisfies the ITU-R quality requirements at 320 kb/s per five full-bandwidth channels (or lower

---

[5] The material in this chapter has significant overlap with that of [Bosi et. al. 1997] and the authors gratefully acknowledge the contribution of MB's co-authors K. Brandenburg, S. Quackenbush, L. Fielder, K. Akagiri, H. Fuchs, M. Dietz, J. Herre, G. Davidson and Y. Oikawa to the material presented here.

according to the NHK data) [ISO/IEC MPEG N1420]. (The MPEG-2 AAC tools also constitute the kernel of the MPEG-4 main, scalable, high quality audio, low delay, natural audio and mobile audio internetworking profiles, see also Chapter 15 [ISO/IEC 14496-3].) The MPEG-2 AAC specifications are the result of a collaborative effort among companies around the world each of which contributed advanced audio coding technology. AAC combines the coding efficiency of a high-resolution filter bank, prediction techniques, and Huffman coding to achieve very good quality audio at low data rates. The AAC specifications have undergone a number of revisions since the first submission of proposals (November 1994). In order to define the AAC system, the audio committee selected a modular approach in which the full system is broken down into a series of self-contained modules or tools, where a tool is defined as a coding module that can be used as a separate component of the overall system. The AAC reference model (RM) described the characteristics of each tool and how they fit together. Each aspect of the RM has been evaluated via core experiments, consisting of informal listening tests that followed the ITU-R BS.1116 (see also Chapter 10) guidelines that were carried out between January '95 and July '96.

The following AAC tools (see *Figure 1* and *Figure 2*) are described in this chapter:

- Gain Control (included in *Figure 1* in the pre-processing stage)
- Filter Bank
- Prediction
- Quantization and Coding
- Noiseless Coding
- Bitstream Multiplexing
- Temporal Noise Shaping (TNS)
- Mid/Side (M/S) Stereo Coding
- Intensity Stereo Coding.

In order to allow a tradeoff between quality and memory/processing power requirements, the AAC system offers three profiles: Main Profile, Low Complexity (LC) Profile, and Scaleable Sampling Rate (SSR) Profile. In the Main Profile configuration, the AAC system provides the best audio quality at any given data rate. With exception of the preprocessing tool, all parts of the AAC tools may be used. Memory and processing power required in this configuration are higher than the memory and processing power required in the LC Profile configuration (see also next sections). It should be noted that a Main Profile AAC decoder can decode an LC Profile encoded bitstream.

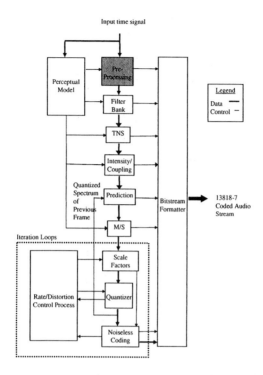

Figure 1. MPEG-2 AAC encoder block diagram from [Bosi et al. 97]

In the LC Profile configuration, the prediction and preprocessing tools are not employed and the TNS order is limited. While quality performance of the LC Profile is very high (see also next sections), the memory and processing power requirements are considerably reduced in this configuration.

In the SSR configuration, the gain control tool is required. The preprocessing performed by the gain control tool consists of a CQF (see also Chapter 4), gain detectors and gain modifiers. The prediction tool is not used in this profile, and TNS order and bandwidth are limited. The SSR Profile has lower complexity than the Main and LC Profiles and it can provide a frequency scaleable signal.

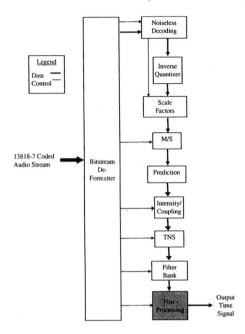

Figure 2. MPEG-2 AAC decoder block diagram from [Bosi et al. 97]

The AAC encoder process can be described as follows (see also *Figure 1*). First, an MDCT-based filter bank (see also Chapter 5) is used to decompose the input signal into sub-sampled spectral components (time-frequency domain). At 48 kHz, the AAC filter bank allows for a frequency resolution of 23 Hz and time resolution of 2.6 ms. Based on the input signal, an estimate of the time dependent masking thresholds are computed (see also Chapter 7). A perceptual model similar to the MPEG Psychoacoustic Model 2 (see also Chapter 11) is used for the AAC system. SMR values are utilized in the quantization stage in order to minimize the audible distortion of the quantized signal at any given data rate.

After the analysis filter bank, the TNS performs an in-place filtering operation on the spectral values, i.e. replaces the spectral coefficients with their prediction residuals. The TNS technique permits the encoder to exercise a control over the temporal fine structure of the quantization noise even within a single filter-bank time-window.

For multichannel signals, intensity stereo coding (see also Chapter 11) can also be applied; in this operation, only the energy envelope is transmitted. Intensity stereo coding allows for a reduction in the spatial information transmitted and it most effective at low data rates.

The time-domain prediction tool can employed in order to take advantage of correlations between sub-sampled spectral components of subsequent frames resulting in an increased redundancy reduction for stationary signals. Instead of transmitting the left and right signal, the normalized sum (M as in Mid) and difference signals (S as in Side) only can be transmitted. Enhanced M/S stereo coding is used in the multichannel AAC encoder at low data rates.

The spectral components are quantized and coded with the aim of keeping the quantization noise below the masked threshold. This step is done by employing an analysis-by-synthesis stage and using additional noiseless compression tools. A mechanism called "bit reservoir" similar to the one adopted in MPEG Layer III (see also Chapter 11) allows for a locally-variable data rate in order to satisfy the signal demands on a frame-by-frame basis. Finally, a bitstream formatter is used to assemble the bitstream, which consists of the quantized and coded spectral coefficients and control parameters.

The MPEG-2 AAC system supports up to 48 audio channels. Default configurations include monophonic, two-channel and five-channel plus LFE channel configurations. In the default five-channel plus LFE configuration, the 3/2-loudspeaker arrangement is adopted as per [ITU-R BS. 775]. In addition to the default configurations, sixteen possible program configurations can be defined in the encoder. Downmix capabilities are also supported [ISO/IEC MPEG N1623].

The sampling rates supported by the AAC system vary from 8 kHz to 96 kHz as shown in *Table 1*. In *Table 1* the maximum data rate per channel, which depends on the sampling rate and the bit reservoir buffer size of 6144 bits per channel, is also shown.

Table 1. MPEG-2 AAC sampling frequencies and data rates [Bosi et al. 97]

Sampling Frequency (Hz)	Maximum Bit Rate per Channel (kb/s)
96000	576
88200	329.2
64000	384
48000	288
44100	264.6
32000	192
24000	144
22050	132.3
16000	96
12000	72
11025	66.25
8000	48

3.   GAIN CONTROL

In the SSR Profile, the gain control block is added in the input stage of the encoder. The gain control module consists of a PQMF filter bank (see also Chapter 4), gain detectors and gain modifiers. The PQMF filter bank splits each audio channel's input signal into four frequency bands of equal width, which are critically sampled. Each filter bank's output has gain modification as necessary and is processed by the MDCT tool to produce 256 spectral coefficients, for a total of 1024 coefficients. Gain control can be applied to each of the four bands independently.

SSR gain control in the decoder has the same components as does in the encoder, but in an inverse arrangement. The distinctive feature of the SSR Profile is that lower bandwidth output signals, and hence lower sampling rate output signals, can be obtained by neglecting the signal from the upper bands of the PQMF. This leads to output bandwidths of 18 kHz, 12 kHz and 6 kHz when one, two or three PQMF outputs are ignored, respectively. The advantage of this signal scalability is that decoder complexity can be reduced as output signal bandwidth is reduced. The gain control module in the encoder receives as input the time-domain signals and produces as outputs the gain control data and a gain modified signal whose length is equal to the length of the MDCT window (see also next section). The block diagram of the gain control tool is shown in *Figure 3*.

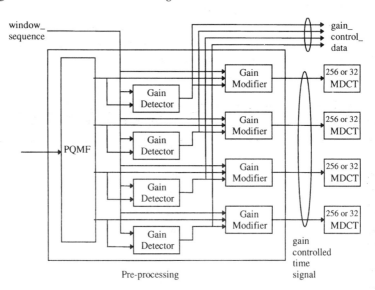

Figure 3. MPEG-2 AAC gain control module from [Bosi et al. 97]

The PQMF analysis and synthesis filters are given by:

$$h_k[n] = \tfrac{1}{4} Q[n] \cos\left((k+\tfrac{1}{2})(n+\tfrac{5}{2}) \frac{\pi}{4} \right)$$
$$g_k[n] = Q[n] \cos\left((k+\tfrac{1}{2})(n-\tfrac{3}{2}) \frac{\pi}{4} \right) \qquad 0 \le n \le 95, \ 0 \le k \le 3$$

where the coefficients of the 96-tap prototype filter $Q[n]$ are specified in [ISO/IEC 13818-7]. The PQMF stage is followed by the gain detector and modifier stages. The gain detector produces gain control data that identify the bands receiving gain modification, the number of modified signal segments, and indices indicating location and level of gain modification for each segment. Note that the output gain control data is for the signal of the previous frame, so that the gain detector has a one-frame delay. The time resolution of the gain control is approximately 0.7 ms at 48 kHz sampling rate. The step size of gain control is 2^n where n is an integer between -4 and 11, allowing the signal to be amplified or attenuated by the gain control tool. The gain modifier applies gain control to the signal in each PQMF band by applying the gain control function to the signal.

In the decoder the gain control module is placed at the end of the decoding process in the SSR Profile. Post-processing performed by the gain control tool consists of applying gain compensation to the sequences produced by each of the four IMDCT stages, overlapping and adding successive sequences with appropriate time alignment and combining these sequences in the inverse PQMF, IPQMF. The block diagram of the decoder gain control is shown in *Figure 4*. Gain compensation in the decoder requires the following three steps for each of the PQMF bands:

    (1) Decoding of gain control data
    (2) Calculation of the gain control function
    (3) Windowing and overlap-adding.

In decoding the gain control data, the gain modification elements are extracted from the bitstream elements. From this information the gain control function is calculated, and is used to multiply the output of the IMDCT. Consecutive sequences are overlapped and added with appropriate time alignment. Finally, the IPQMF combines the separate four frequency bands to synthesize the output signal.

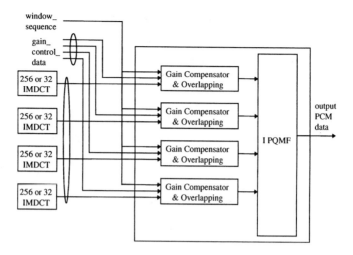

Figure 4. MPEG-2 AAC decoder gain control module from [Bosi et al. 97]

4. FILTER BANK

A fundamental component of the MPEG-2 AAC system is the conversion of the time domain signals at the input of the encoder into an internal time-frequency representation and the reverse process in the decoder. This conversion is done by applying a time-variant MDCT and a IMDCT (see also Chapter 5). The transform block length N can be set to either 2048 or 256 time samples. Since the window function has a significant effect on the filter-bank frequency response, the MPEG-2 AAC filter bank has been designed to allow a change in window shape to best adapt to input signal conditions. The shape of the window is determined in the encoder and transmitted to the decoder.

The use of 2048 time-domain samples transform allows for high coding efficiency for signals with complex spectra, but it may create problems for transient signals. We know from Chapter 7 that quantization errors extending more than a few milliseconds before a transient event are not effectively masked by the transient itself. This leads to a phenomenon called pre-echo in which quantization error from one transform block is spread in time and becomes audible (see also Chapter 7). The MPEG-2 AAC system addresses this problem by allowing the block length of the transform to vary as a function of the signal conditions, with a block switching mechanism based on [Edler 89], see also Chapter 5. Signals that are quasi-stationary are best accommodated by the long transform, while transient signals are generally reproduced more accurately by short transforms. The transition between long and short transforms is seamless in the sense that aliasing is completely cancelled in the absence of transform coefficient quantization.

4.1.1 Filter Bank Resolution and Window Design

As discussed in Chapter 5, the frequency resolution of an MDCT filter bank depends on the window function. A natural choice that satisfies the MDCT perfect reconstruction requirements is the sine window. This window produces a filter bank with good resolution for the signal spectral components, improving coding efficiency for signals with a dense harmonic content. For other types of signals, however, a window with better ultimate rejection may provide better coding efficiency. The KBD window (see [Fielder et al. 96] and also Chapter 5) better satisfies this requirement. In AAC the window shape can be varied dynamically as a function of the signal. The AAC system allows seamless switching between KBD and sine windows while perfect reconstruction and critical sampling are preserved (see also Chapter 5) as shown in *Figure 5*. A single bit per frame is transmitted in the bitstream to indicate the window shape. The window shape is variable for the 2048-length transform blocks only. Window shape decisions made by the encoder are applicable to the second half of the window function only since the first half is constrained by the window shape from the preceding frame.

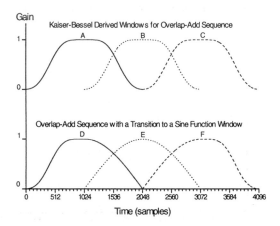

Figure 5. MPEG-2 AAC window shape switching process from [Bosi et al. 97]

The adaptation of the time-frequency resolution of the filter bank to the characteristics of the input signal is done by shifting between transforms whose input lengths are either 2048 or 256 samples. The 256 sample length for transient signal coding was selected as the best compromise between frequency resolution and pre-echo suppression at a data rate of around 64 kb/s per channel. Transform block switching is an effective tool for adapting the time/frequency resolution of the filter bank but potentially creates a problem of block synchrony between the different channels being coded. If one channel uses a 2048 transform length and during the same time interval another channel uses three 256 transforms, the long blocks following the block switch interval will no longer be time aligned. This lack of alignment between channels is undesirable since it creates problems in combining channels during encoding and bitstream formatting/de-formatting. This problem of maintaining block alignment between each channel of the MPEG-2 AAC system has been solved as follows. During transitions between long and short transforms a start and stop bridge window is used that preserves the time domain aliasing cancellation properties of the MDCT and IMDCT transforms and maintains block alignment. These bridge transforms are designated the "Start" and "Stop" sequences, respectively. The conventional long transform with the 2048 sample length is termed a "Long" sequence, while the short transforms occur in groups called the "Short" sequence. The "Short" sequence is composed of eight short block transforms arranged to overlap 50 % with each other and have the half transforms at the sequence boundaries to overlap with the "Start" and "Stop" window shapes. This overlap sequence and grouping of transform blocks into Start, Stop, Long and Short sequences is shown in *Figure 6.*

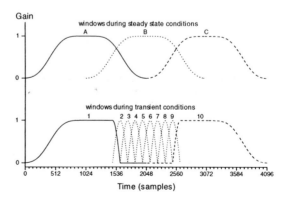

Figure 6. MPEG-2 AAC block switching process from [Bosi et al. 97]

Figure 6 displays the window-overlap process appropriate for both steady state and transient conditions. Curves A, B, and C represent this process when block switching is not employed and all transforms have a 2048 samples and are composed of "Long" sequences only. The windowed transform blocks A, B, C are 50 % overlapped with each other, and assembled in sequential order. The lower part of the figure shows the use of block switching to smoothly transition to and from the shorter N = 256 time sample transforms that are present in the region between sample numbers 1600 to 2496. The figure shows that short length transforms (#2 - #9) are grouped into a sequence of eight 50% overlapped transforms of length 256 samples each and employing a sine function window of the appropriate length. The Start (#1) and Stop (#10) sequences allow a smooth transition between short and long transforms. The first half of the window function for the Start sequence, i.e. time-domain samples numbered 0 - 1023, is the either the first half of the KBD or sine window that matches the previous Long sequence window type. The next section of the window has a value of unity between sample numbers 1024 to 1471, then followed by a sine window. The sine window portion is given by the following formula:

$$w[n] = \sin\left[\frac{\pi}{256} * (n - 1343.5)\right] \qquad \text{where} \quad 1472 \leq n < 1600$$

This region is followed by a final region with zero valued samples to sample number 2047. The "Stop" sequence window is the time-reversed version of the "Start" window and both are designed to ensure a smooth transition between transforms of both lengths and the proper time domain aliasing cancellation properties for the transforms used. For transients which are closely spaced, a single sequence of eight short windows can be extended by adding more consecutive short windows, subject to the restriction that short windows must be added in groups of eight.

5. PREDICTION

Prediction can be used in the AAC coding scheme for improved redundancy reduction. Prediction is especially effective in the case of signals presenting strong stationary components and very demanding in terms of data rate. Because the use of a short window in the filter bank indicates signal changes, i.e., non-stationary signal characteristic, prediction is only used for long windows.

For each channel, prediction is applied to the spectral components resulting from the spectral decomposition of the filter bank. For each spectral component up to 16 kHz, there is one corresponding predictor, resulting in a bank of predictors, where each predictor exploits the auto-correlation between the spectral component values of consecutive frames. If prediction is activated, the quantizer is fed with a prediction error instead of the original spectral component, resulting in a higher coding efficiency. *Figure 7* shows the block diagram of the prediction unit for one single predictor of the predictor bank. The predictor control operates on all predictors of one scale factor band. In *Figure 7* the REC box indicates the reconstruction process of the last quantized value and Q indicates the quantizer. (Note that the complete prediction process is shown only for predictor P_i).

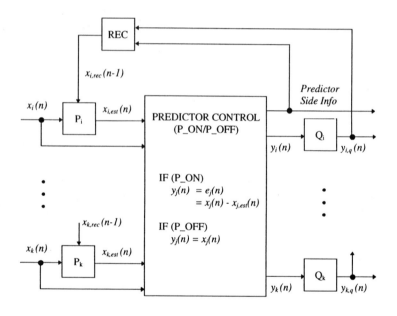

Figure 7. MPEG-2 AAC prediction unit for one scale factor band from [Bosi et al. 97]

In each predictor an estimate $x_{est}[n]$ of the current value of the spectral component $x[n]$ is calculated from preceding reconstructed values $x_{rec}[n-1]$ and $x_{rec}[n-2]$. This estimate is then subtracted from the actual spectral component $x[n]$ resulting in a prediction error $e[n]$ which is quantized and transmitted. In the decoder, $x_{est}[n]$ is recreated and added to the dequantized prediction error that was transmitted data to create the reconstructed value

$x_{rec}[n]$ of the current spectral component $x[n]$. The predictor coefficients are calculated from preceding recovered spectral components in the encoder as well as in the decoder. In this backward-adaptive approach, no additional side information is needed for the transmission of predictor coefficients – as would be required if forward adaptive predictors were to be used.

The predictor is implemented using a lattice structure wherein two so-called basic elements are cascaded. The predictor parameters are adapted to the current signal statistics on a frame-by-frame basis, using an LMS-based (least mean square) adaptation algorithm. A more detailed description of the principles can be found in [Fuchs 95] and the implementation equations can be found in the standard [ISO/IEC 13818-7].

In order to guarantee that prediction is only used if this results in an increase in coding gain, an appropriate predictor control is required and a small amount of predictor control information has to be transmitted to the decoder. For the predictor control, the predictors are grouped into scale factor bands. The predictor control information for each frame is determined in two steps. First, for each scale factor band one determines whether or not prediction gives a coding gain and all predictors belonging to a scale factor band are switched on/off accordingly. Then, one determines whether prediction in the current frame creates enough additional coding gain to justify the additional bits needed for the predictor side information. Only if it does is prediction activated and the side information transmitted. Otherwise, prediction is not used in the current frame and only one bit of side information is transmitted to communicate that decision.

In order to increase the stability of the predictors and to allow defined entry points in the bitstream, a cyclic reset mechanism is applied in the encoder and decoder, in which all predictors are initialized again during a certain time interval in an interleaved way. The whole set of predictors is subdivided into 30 reset groups *(Group 1: P_1, P_{31}, P_{61}, ...; Group 2: P_2, P_{32}, P_{62}, ...; ...; Group 30: P_{30}, P_{60},...)* which are then periodically reset, one after the other with a certain spacing. For example, if one group is reset every eighth frame, then all predictors are reset within an interval of 8 x 30 = 240 frames. The reset mechanism is controlled by a reset on/off bit, which always has to be transmitted as soon as prediction is enabled and a conditional five bit index specifying the group of predictors to be reset. In case of short windows prediction is always disabled and a full reset, i.e. all predictors at once, is carried out.

The various listening tests during the development phase of the standard have shown that significant improvement in sound quality up to 1 grade on the ITU-R five-grade impairment-scale is achieved by prediction for stationary signals, like for example "Pitch Pipe", "Harpsichord".

6.   QUANTIZATION AND CODING

The primary goal of the quantization and coding stage is to quantize the spectral data in such a way that the quantization noise satisfies the demands of the psychoacoustic model. At the same time, the number of bits needed to code the quantized signal must be below a certain limit, normally the average number of bits available for a block of audio data. This value depends on the sampling frequency and, of course, on the desired data rate. In AAC, a bit reservoir gives the possibility of influencing the bit distribution between consecutive audio blocks on a short-time basis. These two constraints, fulfilling the demands of the psychoacoustic model on the one hand and keeping the number of allocated bits below a certain number on the other, are linked to the main challenges of the quantization process. What can be done when the psychoacoustic model demands cannot be fulfilled with the available number of bits? What should be done if not all bits are needed to meet the requirements?

There is no standardized strategy for optimum quantization, the only requirement is that the bitstream produced be AAC-compliant. One possible strategy is using two nested iteration loops as described later in this section. This technique was used for the formal AAC test (see also test description later in this chapter). Other strategies are also possible. One important issue, however, is the fine-tuning between the psychoacoustic model and the quantization process, which may be regarded as one of the 'secrets of audio coding', since it requires a lot of experience and know-how.

The main features of the AAC quantization process are:

- Non-uniform quantization.
- Huffman coding of the spectral values using different tables.
- Noise shaping by amplification of groups of spectral values (so-called scale factor bands). The information about the amplification is stored in the scale factors values.
- Huffman Coding of differential scale factors.

The non-uniform quantizer used in AAC is described as follows (see also MPEG Layer III description in Chapter 11):

$$ix(i) = sign(xr(i)) \cdot nint\left(\left(\frac{|xr(i)|}{\sqrt[4]{2}^{quantizer\_stepsize}}\right)^{0.75} - 0.0946\right)$$

The main advantage of the non-uniform quantizer is the built-in noise shaping depending on coefficient amplitude. The signal to noise ratio remains constant with a wider range of signal energy values when compared to a uniform quantizer. The range of quantized values is limited to +/- 8191. In the above expression, quantizer_stepsize represents the global quantizer step size. Thus the quantizer may be changed in steps of 1.5 dB. The quantized coefficients are then encoded using Huffman coding. A highly flexible coding method allows the use of several Huffman tables for a given set of spectral data. Two and four-dimensional tables (signed or unsigned) are available. The Huffman coding process is described in detail in the next sections. To calculate the number of bits needed to encode the quantized data, the coding process has to be performed and the number of bits needed for the spectral data and the side information has to be computed.

The use of a non-uniform quantizer is, of course, not sufficient to fulfill psychoacoustic demands. In order to fulfill the requirements as efficiently as possible, it is desirable to be able to shape the quantization noise in units similar to the critical bands of the human auditory system. Since the AAC system offers a relatively high frequency resolution for long blocks of 23.43 Hz/line at 48 kHz sampling frequency, it is possible to build groups of spectral values which very closely reflect the bandwidth of the critical bands. *Figure 8* shows the width of the scale factor bands for long blocks (for several reasons the width of the scale factor bands is limited to 32 coefficients except for the last scale factor band). The total number of scale factor bands for long blocks at a sampling frequency of 48 kHz is 49.

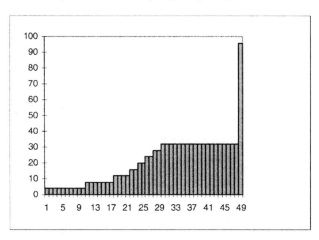

Figure 8. The number of MDCT coefficients in each MPEG-2 AAC scale factor band for long blocks at 48 kHz sampling rate from [Bosi et al. 97]

The inverse scale factor amplification has to be applied in the decoder. For this reason the amplification information, stored in the scale factors in units of 1.5 dB steps is transmitted to the decoder. The first scale factor represents the global quantizer step size and is encoded in a PCM value called global_gain. All following scale factors are differentially encoded using a special Huffman code. This will be described in detail in the next sections.

The decision as to which scale factor band has to be amplified is, within certain limits, left up to the encoder. The thresholds calculated by the psychoacoustic model are the most important criteria, but not the only ones, since only a limited number of bits may be used. As mentioned above, the iteration process described here is only one method for performing the noise shaping. This method is however known to produce very good audio quality. Two nested loops, an inner and an outer iteration loop are used for determining optimum quantization. The description given here is simplified to facilitate understanding of the process. The task of the inner iteration loop is to change the quantizer step size until the given spectral data can be encoded with the number of available bits. For that purpose an initial quantizer step size is chosen, the spectral data are quantized and the number of bits necessary to encode the quantized data is counted. If this number is higher than the number of available bits, the quantizer step size is increased and the whole process is repeated. The inner iteration loop is shown in *Figure 9*.

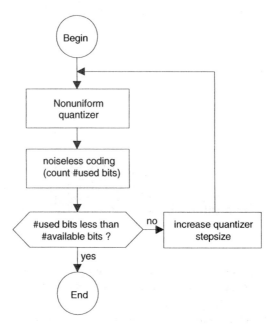

Figure 9. MPEG-2 AAC block diagram of the inner iteration loop from [Bosi et al. 97]

The task of the outer iteration loop is to amplify the scale factor bands (sfbs) in such a way that the demands of the psychoacoustic model are fulfilled as much as possible.

1. At the beginning, no scale factor band is amplified.
2. The inner loop is called.
3. For each scale factor band, the distortion caused by the quantization is calculated (analysis by synthesis).
4. The actual distortion is compared with the permitted distortion calculated via the psychoacoustic model.
5. If this result is the best result so far, it is stored. This is important, since the iteration process does not necessarily converge.
6. Scale factor bands with an actual distortion higher than the permitted distortion are amplified. At this point, different methods for determining the scale factor bands that are to be amplified can be applied.
7. If all scale factor bands were amplified, the iteration process stops. The best result is restored.
8. If there is no scale factor band with an actual distortion above the permitted distortion, the iteration process will stop as well.
9. Otherwise the process will be repeated with the new amplification values.

There are some other conditions not mentioned above which cause a termination of the outer iteration loop. Since the amplified parts of the spectrum need more bits for encoding, but the number of available bits is constant, the quantizer step size has to be changed in the inner iteration loop to decrease the number of used bits. This mechanism shifts bits from spectral regions where they are not required to those where they are required. For the same reason the result after an amplification in the outer loop may be worse than before, so that the best result has to be restored after termination of the iteration process. The outer iteration loop is shown in *Figure 10*. The quantization and encoding process for short blocks is similar to that for long blocks, but grouping and interleaving must be taken into account. Both mechanisms will be described in more detail in the next section.

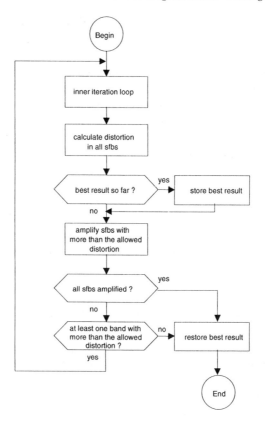

Figure 10. MPEG-2 AAC block diagram of the outer iteration loop from [Bosi et al. 97]

7.   NOISELESS CODING

The input to the noiseless coding module is the set of 1024 quantized spectral coefficients. As a first step a method of noiseless dynamic range compression may be applied to the spectrum. Up to four coefficients can be coded separately as magnitudes in excess of one, with a value of ±1 left in the quantized coefficient array to carry the sign. The "clipped" coefficients are coded as integer magnitudes and an offset from the base of the coefficient array to mark their location. Since the side information for carrying the clipped coefficients costs some bits, this noiseless compression is applied only if it results in a net saving of bits.

Noiseless coding segments the set of 1024 quantized spectral coefficients into sections, such that a single Huffman codebook is used to code each section (the method of Huffman coding is explained in a later section). For reasons of coding efficiency, section boundaries can only be at scale factor band boundaries so that for each section of the spectrum one must transmit the length of the section, in terms of the number of scale factor bands it comprises, and the Huffman codebook number used for the section.

Sectioning is dynamic and typically varies from block to block, so that the number of bits needed to represent the full set of quantized spectral coefficients is minimized. This is done using a greedy merge algorithm starting with the maximum possible number of sections each of which uses the Huffman codebook with the smallest possible index. Sections are merged if the resulting merged section results in a lower total bit count, with merges that yield the greatest bit count reduction done first. If the sections to be merged do not use the same Huffman codebook then the codebook with the higher index must be used.

Sections often contain only coefficients whose value is zero. For example, if the audio input is band limited to 20 kHz or lower, then the highest coefficients are zero. Such sections are coded with Huffman codebook zero, which is an escape mechanism that indicates that all coefficients are zero and it does not require that any Huffman codewords be sent for that section.

If the window sequence is eight short windows then the set of 1024 coefficients is actually a matrix of 8 by 128 frequency coefficients representing the time-frequency evolution of the signal over the duration of the eight short windows. Although the sectioning mechanism is flexible enough to efficiently represent the 8 zero sections, grouping and interleaving provide for greater coding efficiency. As explained earlier, the coefficients associated with contiguous short windows can be grouped so that they share scale factors amongst all scale factor bands within the group. In addition, the coefficients within a group are interleaved by interchanging the order of scale factor bands and windows. To be specific, assume that before interleaving the set of 1024 coefficients c are indexed as

$$c[g][w][b][k]$$

where
　g is the index on groups
　w is the index on windows within a group
　b is the index on scale factor bands within a window
　k is the index on coefficients within a scale factor band
　and the right-most index varies most rapidly.

After interleaving the coefficients are indexed as

c[g][b][w][k]

This has the advantage of combining all zero sections due to band limiting within each group.

The coded spectrum uses one quantizer per scale factor band. The step size of each of these quantizers is specified as a set of scale factors and a global gain that normalizes these scale factors. In order to increase compression, scale factors associated with scale factor bands that have only zero-valued coefficients are not transmitted. Both the global gain and scale factors are quantized in 1.5 dB steps. The global gain is coded as an 8-bit unsigned integer and the scale factors are differentially encoded relative to the previous frequency band scale factor value (or global gain for the first scale factor) and then Huffman coded. The dynamic range of the global gain is sufficient to represent full-scale values from a 24-bit PCM audio source.

Huffman coding is used to represent n-tuples of quantized coefficients, with the Huffman code drawn from one of 12 codebooks. The spectral coefficients within n-tuples are ordered from low to high and the n-tuple size is two or four coefficients. The maximum absolute value of the quantized coefficients that can be represented by each Huffman codebook and the number of coefficients in each n-tuple for each codebook is shown in *Table 2*. There are two codebooks for each maximum absolute value, with each representing a distinct probability distribution function. The best fit is always chosen. In order to save on codebook storage (an important consideration in a mass-produced decoder), most codebooks represent unsigned values. For these codebooks the magnitude of the coefficients is Huffman coded and the sign bit of each non-zero coefficient is appended to the codeword.

Table 2. MPEG-2 AAC Huffman codebooks [Bosi et al. 97]

Codebook Index	n-Tuple size	Maximum Absolute Value	Signed Values
0		0	
1	4	1	yes
2	4	1	yes
3	4	2	no
4	4	2	no
5	2	4	yes
6	2	4	yes
7	2	7	no
8	2	7	no
9	2	12	no
10	2	12	no
11	2	16 (ESC)	no

Two codebooks require special note: codebook 0 and codebook 11. As mentioned previously, codebook 0 indicates that all coefficients within a section are zero. Codebook 11 can represent quantized coefficients that have an absolute value greater than or equal to 16. If the magnitude of one or both coefficients is greater than or equal to 16, a special escape coding mechanism is used to represent those values. The magnitude of the coefficients is limited to no greater than 16 and the corresponding 2-tuple is Huffman coded. The sign bits, as needed, are appended to the codeword. For each coefficient magnitude greater or equal to 16, an escape code is also appended, as follows:

escape code = <escape_prefix><escape_separator><escape_word>

where
<escape_prefix> is a sequence of N binary "1's"
<escape_separator> is a binary "0"
<escape_word> is an N+4 bit unsigned integer, msb first
and N is a count that is just large enough so that the magnitude of the quantized coefficient is equal to

$$2^{(N+4)} + <escape\_word>$$

8. BITSTREAM MULTIPLEXING

The MPEG-2 AAC system has a very flexible bitstream syntax. Two layers are defined: the lower specifies the "raw" audio information while the higher specifies a specific audio transport mechanism. Since any one transport cannot be appropriate for all applications, the raw data layer is designed to be parsable on its own, and in fact is entirely sufficient for applications such as compression for computer storage devices. The composition of a bitstream is shown in *Table 3*.

Table 3. General structure of the MPEG-2 AAC bitstream [Bosi et al. 97]

<stream>	{<transport>}<block>{<transport>}<block> ...
<block>	[<prog_config_ele>]<audio_ele>[<audio_ele>][<coupling_ele>][<data_ele>] [<fill_ele>]<term_ele>

The tokens in the bitstream are indicated by angle brackets (<>). The bitstream is indicated by the token <stream> and is a series of <block> tokens each containing all information necessary to decode 1024 audio frequency samples. Furthermore each <block> token begins on a byte boundary relative to the start of the first <block> in the bitstream. Between <block> tokens there may be transport information, indicated by <transport>, such as would be needed for synchronization on break-in or for error control. Brackets ({ }) indicate an optional token and brackets ([]) indicate that the token may appear zero or more times.

Since the AAC system has a data buffer that permits its instantaneous data rate to vary as required by the audio signal, the length of each <block> is not constant. In this respect the AAC bitstream uses variable-rate headers (header being the <transport> token). These headers are byte-aligned so as to permit editing of bitstreams at any block boundary.

An example of tokens within a <block> is shown in *Table 4*.

Table 4. Example of tokens within an MPEG-2 AAC <block> [Bosi et al. 97]

Token	Meaning
prog_config_ele	program configuration element
audio_ele	audio element, one of:
single_channel_ele	single channel
channel_pair_ele	stereo pair
low_freq_effects_ele	low frequency effects channel
coupling_ele	multichannel coupling
data_ele	data element, segment of data stream
fill_ele	fill element, adjusts data rate for constant rate channels
term_ele	terminator, signals end of block

The prog_config_ele is a configuration element that maps audio channels to an output speaker assignment so that multichannel coding can be as flexible as possible. It can specify the correct voice tracks for multi-lingual programming and specifies the analog sampling rate.

There are three possible audio elements: single_channel_ele is a monophonic audio channel, channel_pair_ele is a stereo pair and low_freq_effects_ele is a sub-woofer channel. Each of the audio elements is named with a 4-bit tag such that up to 16 of any one element can be represented in the bitstream and assigned to a specific output channel. At least one audio element must be present.

The coupling_ele is a mechanism to code signal components common to two or more audio channels (see also next section).

The data_ele is a tagged data stream that can continue over an arbitrary number of blocks. Unlike other elements, the data element contains a length count such that an audio decoder can strip it from the bitstream without knowledge of its meaning. As with the audio elements, up to 16 distinct data streams are supported.

The fill_ele is a bit-stuffing mechanism that enables an encoder to increase the instantaneous rate of the compressed audio stream such that it fills a constant rate channel. Such mechanisms are required as, first, the encoder has a region of convergence for its target bit allocation so that the bits used may be less than the bit budget, and second, the encoder's representation of a digital zero sequence is so much less than the average coding bit budget that it must resort to bit stuffing.

The term_ele signals the end of a block. It is mandatory as this makes the bitstream parsable. Padding bits may follow the term_ele such that the next <block> begins on a byte boundary.

An example of one <block> for a 5.1 channel bitstream, (where the .1 indicates the LFE channel), is

<block> <single_channel_ele>   <channel_pair_ele>   <channel_pair_ele>
<low_freq_effects_ele> <term_ele>

Although discussion of the syntax of each element is beyond the scope of this section, all elements make frequent use of conditional components. This increases flexibility while keeping bitstream overhead to a minimum. For example, a one-bit field indicates whether prediction is used in an audio channel in a given block. If set to one, then the set of bits indicating which scale factor bands use prediction follows. Otherwise the bits are not sent. For additional information see [ISO/IEC 13818-7].

9.   TEMPORAL NOISE SHAPING

A novel concept in perceptual audio coding is represented by the temporal noise shaping, TNS, tool of the AAC system [Herre and Johnston 96]. This tool is motivated by the fact that the handling of transient with long temporal input block filter banks presents a major challenge. In particular, coding of transients is difficult because of the temporal mismatch between masking threshold and quantization noise.

The TNS technique permits the coder to exercise control over the temporal fine structure of the quantization noise even within a filter bank block. The concept of TNS uses the duality between time and frequency domain to extend predictive coding techniques. Signals with an "un-flat" spectrum can be coded efficiently either by directly coding spectral values or by applying predictive coding methods to the time signal. Consequently, the corresponding dual statement relates to the coding of signals with an "un-flat" time structure, i.e. transient signals. Efficient coding of transient signals can thus be achieved by either directly coding time domain values or by applying predictive coding methods to their spectral representation. Such predictive coding of spectral coefficients over frequency constitutes the dual concept to the intra-channel prediction tool described in the previous section. While intra-channel prediction over time increases the coder's spectral resolution, prediction over frequency enhances its temporal resolution.

If forward predictive coding is applied to spectral data over frequency, the temporal shape of the quantization error will appear adapted to the temporal shape of the input signal at the output of the decoder. This effectively localizes the quantization noise in time under the actual signal and avoids problems of temporal masking, either in transient or pitched signals. This type of predictive coding of spectral data is therefore referred to as the TNS method.

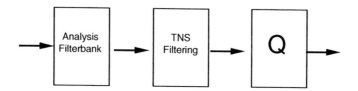

Figure 11. MPEG-2 AAC encoder TNS from [Bosi et al. 97]

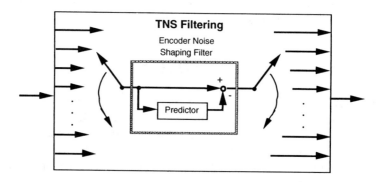

Figure 12. MPEG-2 AAC block diagram of the TNS encoder filtering stage from [Bosi et al. 97]

The TNS processing can be applied either for the entire spectrum, or for only part of the spectrum. In particular, it is possible to use several predictive filters operating on distinct frequency regions.

The predictive encoding/decoding process over frequency can be realized easily by adding one building block to the standard structure of a generic perceptual encoder and decoder. This is shown for the encoder in *Figure 11* and *Figure 12*. Immediately after the analysis filter bank an additional block, "TNS Filtering", is inserted which performs an in-place filtering operation on the spectral values, i.e., replaces the target spectral coefficients (set of spectral coefficients to which TNS should be applied) with the prediction residual. This is symbolized by a rotating switch circuitry in *Figure 12*. Both sliding in the order of increasing and decreasing frequency is possible. Similarly, the TNS decoding process is done by inserting an additional block, inverse TNS filtering, immediately before the synthesis filter bank (see *Figure 13* and *Figure 14*). An inverse in-place filtering operation is performed on the residual spectral values so that the spectral coefficients are replaced with the decoded spectral coefficients by means of the inverse prediction (all-pole) filter. The TNS operation is signaled to the decoder via a TNS on/off flag, the number and the frequency range of the TNS filters applied in each transform window, the order of the prediction filter (max. 12 or 20, depending on the profile) and the filter data itself.

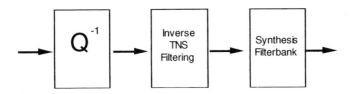

Figure 13. MPEG-2 AAC decoder TNS from [Bosi et al. 97]

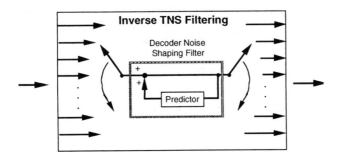

Figure 14. MPEG-2 AAC TNS decoder inverse filtering stage from [Bosi et al. 97]

The properties of the TNS technique can be described as follows. The combination of filter bank and adaptive prediction filter can be interpreted as a continuously signal adaptive filter bank. In fact, this type of adaptive filter bank dynamically provides a continuum in its behavior between a high-frequency resolution filter bank (for stationary signals) and a low- frequency resolution filter bank (for transient signals). Secondly, the TNS approach permits a more efficient use of masking effects by adapting the temporal fine structure of the quantization noise to that of the masker signal. In particular, it enables a better encoding of "pitch-based" signals such as speech, which consist of a pseudo-stationary series of impulse-like events where traditional transform block switching schemes do not offer an efficient solution. Thirdly, the TNS method reduces the peak bit demand of the coder for transient signal segments by exploiting irrelevancy. Finally, the technique can be applied in combination with other methods addressing the temporal noise shaping problem, such as transform block switching and pre-echo control.

During the standardization process of the MPEG-2 AAC system, the TNS tool demonstrated a significant increase in performance for speech stimuli. In particular, an improvement in quality of approximately 0.9 in the five-grade ITU-R impairment scale for the most critical speech item "German Male Speech" was shown during the AAC core experiments. Advantages were also shown for other transient signal (for example in the "Glockenspiel" item).

10.   JOINT STEREO CODING

The MPEG AAC system includes two techniques for stereo coding of signals: M/S stereo coding and intensity stereo coding. Both stereo coding strategies can be combined by selectively applying them to different frequency regions. The concept of joint stereo coding in the MPEG-2 AAC system is discussed in greater detail in [Johnston et al. 96].

10.1   M/S Stereo Coding

In the MPEG-2 AAC system, M/S stereo coding is applied within each channel pair of the multichannel signal, i.e. between a pair of channels that are arranged symmetrically on the left / right listener axis. To a large degree, M/S processing helps to avoid imaging problems due to spatial unmasking.

M/S stereo coding can be used in a flexible way by selectively switching in time on a block-by-block basis, as well as in frequency on a scale factor band by scale factor band basis, see [Johnston and Ferreira 92]. The switching state (M/S stereo coding "on" or "off") is transmitted to the decoder as an array of signaling bits ("ms_used"). This can accommodate short time delays between the L and R channels, and still accomplish both image control and some signal-processing gain. While the amount of time delay that it allows is limited, the time delay is greater than the interaural time delay, and allows for control of the most critical imaging issues [Johnston and Ferreira 92].

10.2   Intensity Stereo Coding

The MPEG-2 AAC system provides two mechanisms for applying intensity stereo coding. The first is based on the "channel pair" concept as used for M/S stereo coding and implements an easy-to-use coding concept that covers most of the normal needs without introducing noticeable overhead into the bitstream. For simplicity, this mechanism is referred to as the AAC "intensity stereo coding" tool. While the intensity stereo coding tool only implements joint coding within each channel pair, it may be used for coding of both 2-channel as well as multichannel signals.

In addition, a second, more sophisticated mechanism is available that is not restricted by the channel pair concept and allows better control over the coding parameters. This mechanism is called the AAC "coupling channel" element and provides two functionalities. First, coupling channels may be used to implement generalized intensity stereo coding where channel spectra can be shared across channel boundaries including sharing among different channel pairs [Davis 93]. The second functionality of the coupling channel element is to perform a down-mix of additional sound objects into the stereo image so that, for example, a commentary channel can be added to an existing multichannel program ("voice-over"). Depending on the profile, certain restrictions apply regarding consistency between coupling channel and target channels in terms of window sequence and window shape parameters, see [ISO/IEC 13818-7]. In general, the MPEG-2 AAC system provides appropriate coding tools for many types of stereophonic program material from traditional two channel recordings to 5.1 or more multichannel material.

11.   TEST RESULTS

Since the first submission of AAC proposals in November 1994, a number of core experiments were planned and carried out to select the best performing tools to be incorporated in the AAC RM. The final MPEG-2 AAC system was tested according to the ITU-R BS.1116 specifications in September 1996 in the five channel, full-bandwidth configuration and compared to the MPEG-2 BC Layer II in the same configuration [ISO/IEC MPEG N1420]. The formal subjective tests were carried at BBC, UK, and NHK, Japan. A total of 23 reliable[6] expert listeners at BBC and 16 reliable expert listeners at NHK participated in the listening tests. As specified by ITU-R BS.1116, the tests were conducted according to the triple-stimulus/hidden-reference/double-blind method using the ITU-R five-grade impairment scale (see also Chapter 10). From the 94 submitted critical excerpts, a selection panel selected the ten most critical items (see *Table 5*).

Table 5. MPEG-2 AAC Subjective Test Critical Items [Bosi et al. 97]

No.	Name	Description
1	Cast	Castanets panned across the front, noise in surround
2	Clarinet	Clarinet in front, theatre foyer ambience, rain in surround
3	Eliot	Female and male speech in a restaurant, chamber music
4	Glock	Glockenspiel and timpani
5	Harp	Harpsichord
6	Manc	Orchestra - strings, cymbals, drums, horns
7	Pipe	Pitch Pipe
8	Station	Male voice with steam-locomotive effects
9	Thal	Piano front left, sax in front right, female voice in center
10	Tria	Triangle

The test results, in terms of non-overlapping 95% confidence intervals for SDG as per ITU- R BS.1116 specifications [ITU- R BS.1116], are shown in *Figure 15* through *Figure 17*. These figures show the test results for the following MPEG-2 AAC configurations:

1. MPEG-2 AAC Main Profile at a data rate of 320 kb/s per five full-bandwidth channels
2. MPEG-2 AAC LC Profile at a data rate of 320 kb/s per five full-bandwidth channels
3. MPEG-2 Layer II BC at a data rate of 640 kb/s per five full- bandwidth channels.

In *Figure 15* and *Figure 16* the vertical axis shows the SDG values. In *Figure 17* the vertical axis shows the MPEG-2 AAC SDGs minus the MPEG-2 Layer II BC SDGs. A positive difference indicates that AAC was awarded a better grade than Layer II and vice versa. The MPEG-2 AAC test results show that the AAC system at a data rate of 320 kb/s per five full-bandwidth channels fulfils the ITU-R requirements for indistinguishable quality [ITU-R TG-2/3] in a BS.1116 fully compliant test.

The AAC multichannel system at a data rate of 320 kb/s overall ranks higher than MPEG-2 Layer II BC at 640 kb/s (see *Figure 17*). In particular,

---

[6] Due to the very rigorous test method adopted, only statistically reliable expert listeners were taken into consideration in the final data analysis. A total of 32 listeners at BBC and 24 listeners at NHK originally participated in the tests. After post-screening of the subjects, nine listeners at BBC and eight listeners at NHK were removed.

the difference between the two systems mean scores for the pitch pipe excerpt is more than 1.5 point in the ITU-R five-grade impairment scale according to the BBC data. It should be noted that the test data for MPEG-2 Layer II BC at 640 kb/s were consistent with data obtained in previously conducted subjective tests [ISO/IEC MPEG N1229].

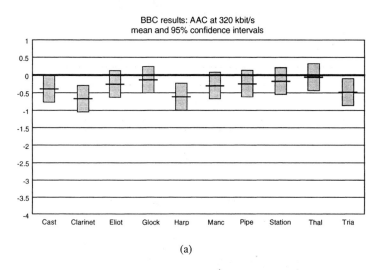

(a)

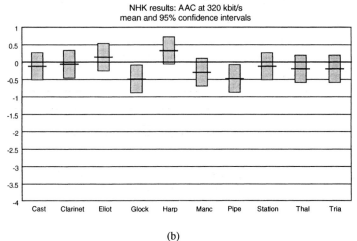

(b)

Figure 15. Results of formal listening tests for MPEG-2 AAC Main Profile at 320 kb/s, five channel configuration from [ISO/IEC MPEG N1420]; a) BBC results, b)NHK results

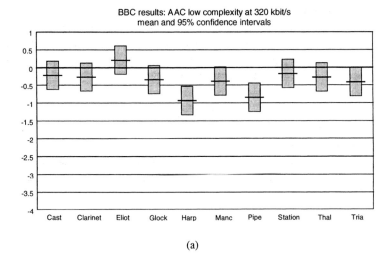

(a)

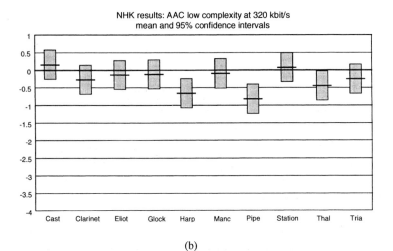

(b)

Figure 16. Results of formal listening tests for MPEG-2 AAC LC Profile at 320 kb/s, five-channel configuration, from [ISO/IEC MPEG N1420].

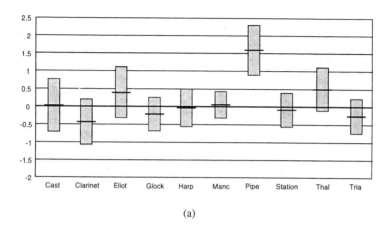

(a)

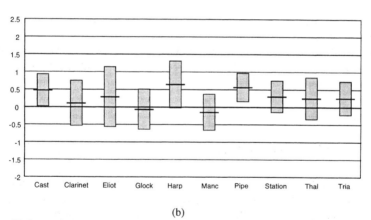

(b)

Figure 17. Comparison between MPEG-2 AAC at 320 kb/s and MPEG-2 BC Layer II at 640 kb/s, five-channel configuration, from [ISO/IEC MPEG N1420]; (a) BBC Results, (b) NHK results

12. DECODER COMPLEXITY EVALUATION

In this section a complexity evaluation of the decoding process in its Main and LC Profile configurations is presented. In order to quantify the complexity of the MPEG-2 AAC decoder, the number of machine instructions, read/write storage locations (RAM), read-only storage locations (ROM) is specified for each module, see also [ISO/IEC MPEG N1712]. For simplicity, the assumption is made that the audio signal is sampled at 48 kHz, 16-bits per sample, the data rate is 64 kb/s per channel, and that there are 1024 frequency values per block.

Two categories of the AAC decoder implementation are considered: software decoders running on general-purpose processors, and hardware decoder running on single-chip ASICs. For these two categories, a summary of the AAC decoder complexity is shown in *Table 6*.

Table 6. MPEG-2 AAC decoder complexity [Bosi et al. 97]

MPEG –2 AAC Configuration	Complexity
2-channel Main Profile software decoder	40 % of 133 MHz Pentium
2-channel LC Profile software decoder	25 % of 133 MHz Pentium
5-channel Main Profile hardware decoder	90 sq. mm die, 0.5 micron CMOS
5-channel LC Profile hardware decoder	60 sq. mm die, 0.5 micron CMOS

12.1   Input/Output Buffers

Considering the bit reservoir encoder structure and the maximum data rate per channel, the minimum decoder input buffer size is 6144 bits. The decoder output, assuming a 16-bit PCM double-buffer system, requires a 1024, 16-bit word buffer. The total number of 16-bit words for the decoder input/output buffer (RAM) is:

384 + 1024 = 1408.

12.2   Huffman Coding

In order to decode a Huffman codeword, the decoder must traverse a Huffman code tree from root node to leaf node. Approximately 10 instructions per bit for the Huffman decoding are required. Given the average of 1365 bits per blocks, the number of instructions per block is 13653. Huffman decoding requires the storage of the tree and the value corresponding to the codeword. The total buffer required is a 995 16-bit word buffer (ROM).

12.3   Inverse Quantization

The inverse quantization can be done by table lookup. Assuming that only 854 spectral coefficients (20 kHz bandwidth) must be inverse quantized and scaled by a scale factor, the 16-bit word ROM buffer is 256 words and the total number of instructions is 1708.

12.4 Prediction

Assuming that only the first 672 spectral coefficients will use prediction and the predictor used is a second order predictor, the number of instructions for predictor is 66 and the total number of instruction per block is 44352. Calculations can be done both in IEEE floating point and/or fixed arithmetic and variables are truncated to 16 bits prior to storage, see also [ISO/IEC MPEG N1628] and [ISO/IEC MPEG N1629]. The required storage buffer is 4032 16-bit word.

12.5 TNS

In the Main Profile configuration, the TNS process employs a filter of order 20 operating on 672 spectral coefficients. The number of instruction per block is 13630. In the LC Profile configuration, the TNS process employs a filter of reduced order 12 with a total number of instructions per block of 8130. TNS requires negligible storage buffers.

12.6 M/S Stereo

This is a very simple module that performs matrixing on two channels of a stereo pair element. Since the computation is done in-place, no additional storage is required. Assuming that only a 20 kHz bandwidth needs the M/S computation, the total number of instruction for stereo pair is 854.

12.7 Intensity Stereo

Intensity stereo coding does not use any additional read-only or read-write storage. The net complexity of intensity stereo coding produces a saving of one inverse quantization per intensity stereo coded coefficient.

12.8 Inverse Filter Bank

The IMDCT of length 1024 requires about 20000 instructions per block while for the 128-point IMDCT, the total number of instructions per 8 short blocks is 24576. The total RAM for the filter bank is 1536 words, the total ROM, including window coefficients, etc., is 2270 words. The storage requirement employs a word length between 16 and 24 bits depending on the stage of the filter bank.

12.9   Complexity Summary

Table 7 through Table 9 summarize the complexity of each decoder module based on number of instructions per block (*Table 7*), amount of read-write storage and amount of read-only (*Table 8* and *Table 9*) in 16-bit words. The tables list complexity on a per-channel basis and for a 5-channel coder. *Table 10* shows a complexity comparison between the MPEG-2 AAC Main Profile and LC Profile.

Table 7. MPEG-2 AAC Main Profile number of instruction per block (decoder) [Bosi et al. 97]

AAC Tool	Single Channel	Five Channels
Huffman Coding	13657	68285
Inverse Quantization	1708	8540
Prediction	44352	221760
TNS	13850	69250
M/S		1708
IMDCT	24576	122880
TOTAL	98143	492423

Table 8. MPEG-2 AAC Main Profile decoder RAM (16-bit words) [Bosi et al. 97]

	Single Channel	Five Channels
Input Buffer	384	1920
Output Buffer	1024	5120
Working buffer	2048	10240
Prediction	4032	20160
IMDCT	1024	5120
TOTAL	8512	42560

Table 9. MPEG-2 AAC Main Profile decoder ROM [Bosi et al. 97]

	Single Channel	Five Channels
Huffman Coding		995
Inverse Quantization		256
TNS		24
Prediction		0
IMDCT		2270
TOTAL		3545

Table 10. MPEG-2 AAC Main Profile and LC Profile (five-channel configuration only) [Bosi et al. 97]

	Main Profile	Low Complexity Profile
Instructions per Block	492423	242063
RAM	42560	22400
ROM	3545	3545

13. SUMMARY

In this chapter we reviewed the features of the MPEG-2 AAC standard. While initially the main goals of MPEG-2 were to address low sampling frequencies and multichannel extensions to MPEG-1 (ISO/IEC 13818-3), an additional multichannel work item, MPEG-2 AAC, was also developed in its framework. The MPEG-2 AAC (ISO/IEC 13818-7) system was designed to provide MPEG-2 with the best multichannel audio quality without any restrictions due to compatibility requirements. The AAC tools provide high coding efficiency through the use of a high-resolution filter bank, prediction techniques, noiseless coding and added functionalities. ITU-R BS.1116 compliant tests have shown that the AAC system achieves indistinguishable audio quality at data rates of 320 kb/s for five full-bandwidth channels and provides similar or better quality than MPEG-2 Layer II BC at 640 kb/s. We anticipate that the MPEG-2 AAC standard will become the audio coding system of choice in applications where high performance at the lowest possible data rate is critical to the success of the application. While MPEG-4 audio addresses speech coding and functionalities in addition to broadband audio coding, AAC plays an important role in this context. Before discussing the main features of MPEG-4 Audio, we complete our review of multichannel audio coding systems with the introduction of Dolby AC-3, the coding standard used in high definition television and DVD applications.

14. REFERENCES

[ATSC A/52/10]: United States Advanced Television Systems Committee Digital Audio Compression (AC-3) Standard, Doc. A/52/10, December 1995.

[Bosi et al. 97]: M Bosi, K. Brandenburg, S. Quackenbush, L. Fielder, K. Akagiri, H. Fuchs, M. Dietz, J. Herre, G. Davidson and Y. Oikawa, "ISO/IEC MPEG-2 Advanced Audio Coding," J. Audio Eng. Soc., vol. 45, pp. 789 – 812, October 1997.

[Davis 93]: M. Davis, "The AC-3 Multichannel Coder," presented at the 95th AES Convention, New York, Preprint 377, October 1993.

[DVD-Video]: DVD Specifications for Read-Only Disc, Part 3: VIDEO SPECIFICATIONS Ver. 1.1, Tokyo 1997-2001.

[Fielder 87]: L. Fielder, "Evaluation of the Audible Distortion and Noise Produced by Digital Audio Converters," J. Audio Eng. Soc., vol. 35, pp. 517-535, July/August 1987.

[Fielder et al. 96]: L. Fielder, M. Bosi, G. Davidson, M. Davis, C. Todd, and S. Vernon, "AC-2 and AC-3: Low-Complexity Transform-Based Audio Coding," *Collected Papers on Digital Audio Bit-Rate Reduction*, Neil Gilchrist and Christer Grewin (ed.), pp. 54-72, AES 1996.

[Fuchs 93]: H. Fuchs, "Improving Joint Stereo Audio Coding by Adaptive Interchannel Prediction", Proc. of the 1993 IEEE Workshop on Applications of Signal Processing to Audio and Acoustics, New Paltz, New York, October 1993.

[Fuchs 95]: H. Fuchs, "Improving MPEG Audio Coding by Backward Adaptive Linear Stereo Prediction", Preprint 4086 (J-1), presented at the 99th AES Convention, New York, October 1995.

[Herre and Johnston 96]: J. Herre, J. D. Johnston, "Enhancing the Performance of Perceptual Audio Coders by Using Temporal Noise Shaping (TNS) ", presented at the 101st AES Convention, Preprint 4384, Los Angeles 1996.

[ISO/IEC 13818-3]: ISO/IEC 13818-3, "Information Technology - Generic Coding of Moving Pictures and Associated Audio, Part 3: Audio," 1994-1997.

[ISO/IEC 13818-7]: ISO/IEC 13818-7, "Information Technology - Generic Coding of Moving Pictures and Associated Audio, Part 7: Advanced Audio Coding", 1997.

[ISO/IEC 14496-3]: ISO/IEC 14496-3, "Information Technology – Coding of Audio Visual Objects, Part 3: Audio ", 1999-2001.

[ISO/IEC MPEG N1229]: ISO/IEC JTC 1/SC 29/WG 11 N1229, "MPEG-2 Backwards Compatible CODECS Layer II and III: RACE dTTb Listening Test Report," Florence, March 1996.

[ISO/IEC MPEG N1420]: ISO/IEC JTC 1/SC 29/WG 11 N1420, "Overview of the Report on the Formal Subjective Listening Tests of MPEG-2 AAC Multichannel Audio Coding" Maceio', November 1996.

[ISO/IEC MPEG N1623]: ISO/IEC JTC 1/SC 29/WG 11 N1623, "Informal Assessment of AAC Downmix Stereo Performance " Bristol, April 1997.

[ISO/IEC MPEG N1628]: ISO/IEC JTC 1/SC 29/WG 11 N1628, "Report on Reduction of Complexity in the AAC Prediction Tool" Bristol, April 1997.

[ISO/IEC MPEG N1629]: ISO/IEC JTC 1/SC 29/WG 11 N1629, " Results of the brief assessments on AAC reduction of prediction complexity " Bristol, April 1997.

[ISO/IEC MPEG N1712]: ISO/IEC JTC 1/SC 29/WG 11 N1712, "Report on Complexity of MPEG-2 AAC Tools" Bristol, April 1997.

[ITU-R BS.775-1]: International Telecommunications Union, Radiocommunication Sector BS.775-1, "Multichannel Stereophonic Sound System with and without Accompanying Picture", Geneva, Switzerland, 1992-1994.

[ITU-R TG10-2/3]: International Telecommunications Union, Radiocommunication Sector Document TG10-2/3- E only, " Basic Audio Quality Requirements for Digital Audio Bit-Rate Reduction Systems for Broadcast Emission and Primary Distribution", 28 October 1991.

[Johnston and Ferreira 92]: J. D. Johnston and A. J. Ferreira, "Sum-Difference Stereo Transform Coding", Proc. ICASSP, pp. 569-571, 1992.

[Johnston et al. 96]: J. D. Johnston, J. Herre, M. Davis and U. Gbur, "MPEG-2 NBC Audio - Stereo and Multichannel Coding Methods," Presented at the 101st AES Convention, Preprint 4383, Los Angeles, November 1996.

[Van der Waal and Veldhuis 91]: R. G. v.d. Waal and R. N. J. Veldhuis, "Subband Coding of Stereophonic Digital Audio Signals", Proc. ICASSP, pp. 3601 – 3604, 1991.

Chapter 14

Dolby AC-3

1. INTRODUCTION

In Chapters 11 through 13 we discussed the goals and the main features of ISO/IEC MPEG-1 and -2 Audio. Other standards bodies addressed the coding of audio based on specific applications. For example, the North American HDTV standard [ATSC A/52/10], the DVD-Video standard [DVD-Video] and the DVB [ETS 300 421] standard all make use of Dolby AC-3, also known as Dolby Digital.

The AC-3 algorithm is based on perceptual coding principles and it is similar in many ways to other perceptual audio coders, such as the MPEG-1 and 2 audio systems described in previous chapters. AC-3, however, was conceived since its very onset as a multichannel system. Originally designed to address the cinema industry needs, its initial goal was the storage of digital multichannel audio on film. First released in 1991 with the film Batman Returns, AC-3 migrated from film applications to consumer products, following the general multichannel audio systems expansion from the cinema halls to the home theatres systems. Born from the design experience of its predecessor AC-2 [Davidson, Fielder and Antill 90.], a TDAC-based, single-channel coding scheme, AC-3 went through many stages of refinement, improvement and fine-tuning. The resulting algorithm is currently in use in a number of standard applications including the North American HDTV standard, DVD-Video, and regional DVB.

This chapter examines the basic functionalities, features, and underlying fundamentals of the AC-3 algorithm, and discusses its resultant ranking within low bit-rate coding standardization efforts.

2.   MAIN FEATURES

The AC-3 algorithm provides a high degree of flexibility with regard to data rate and other operational details [ATSC A/52/10]. One of the main features of AC-3 is that it processes multiple channels as a single ensemble. AC-3 is capable of encoding a number of audio channel configurations into a bitstream ranging between 32 and 640 kb/s. The decoder has the ability to reproduce various playback configurations from one to 5.1 channels from the common bitstream (see also Chapter 12). The AC-3 coding schemes specifically supports 3/2, 3/1, 3/0, 2/2, 2/1, 2/0, and 1/0 channel configurations with an optional LFE channel [ITU-R BS.775-1]. The presence of the LFE channel, although not explicitly included in these configurations, is always an option. The sampling rates supported by AC-3 are 32, 44.1, and 48 kHz. The frame size corresponds to 6 blocks of 512 time-samples, or, equivalently, 1536 frequency-samples. At 48 kHz a frame covers a time interval of 32 ms.

In *Table 1* the different data rates adopted in AC-3 are listed. The six-bit bitstream variable frmsizecod conveys the encoded data rate to the decoder. Although each data rate is applicable to each channel configuration, in practice typical data rates applied for the two-channel and five-channel configurations are 192 kb/s and 384 kb/s respectively.

Table 1. AC-3 data rates [ATSC A/52/10]

frmsizecod	Data Rate (kb/s)	frmsizecod	Data Rate (kb/s)
2	32	20	160
4	40	22	192
6	48	24	224
8	56	26	256
10	64	28	320
12	80	30	384
14	96	32	448
16	112	34	512
18	128	36	640

Table 2. AC-3 channel configurations [ATSC A/52/10]

acmod	Configuration	Number of Channels	Channel Order
000	1+1	2	Channel 1, Channel 2
001	1/0	1	C
010	2/0	2	L, R
011	3/0	3	L, C, R
100	2/1	3	L, R, S
101	3/1	4	L, C, R, S
110	2/2	4	L, R, S_L, S_R
111	3/2	5	L, C, R, S_L, S_R

In *Table 2* the different channel configurations and how they are identified by the three-bit variable acmod in the AC-3 bitstream are described. For example, acmod equal to 000 implies two independent channels (dual mono). For values of acmod ≥ 100, the channel configuration includes one or more surround channels. The optional LFE channel is enabled by a separate variable called lfeon (lfeon = 1 implies the presence of the LFE channel).

AC-3 includes provision for sending of multiple auxiliary data streams, language identifiers, copyright protection, time stamps and control information. Listener features include down-mixing to fewer channels than present in the bitstream, dialog normalization, Dolby Surround compatibility, visually and hearing impaired bitstreams, dynamic range control.

One of the most interesting user features, the dynamic range control for example, allows the program provider to store control parameters that specify the dynamic range reduction appropriate for a particular excerpt in the AC-3 bitstream. In this fashion, the AC-3 encoded audio bitstream always retains full dynamic range, allowing the end user to choose between the compressed or the full dynamic range audio excerpt. The dynamic range control parameters are used by the decoder to alter the level of the decoded audio on a block-by-block basis (every 5.3 ms at a 48 kHz sample rate)[7]. The control parameters may indicate that the decoder gain be raised or lowered and are generated by a dynamic level compression algorithm which may be resident in the AC-3 encoder, or in a subsequent bitstream processor. The dynamic range control parameters have an amplitude resolution of less than 0.25 dB and the block to block gain variations are further smoothed by the gentle overlap add process lasting 5.3 ms (see also the filter bank and overlap and add process description in the next sections), preventing the audibility of gain stepping artifacts. The exact nature of the dynamic range control is determined by the algorithm that generates the control parameters. In general, however, the sound level headroom is reduced, that is loud sounds are brought down towards dialogue level, and quiet sounds are made more audible, that is brought up towards dialogue level. If the dynamic range compression is turned off, the original signal dynamic range will be reproduced. The default for the listener is to reproduce the program with compression as specified by the program originator. Full description of the AC-3 listener features are beyond the scope of this book and can be found in [ATSC A/52/10].

---

[7] In addition to the fine resolution dynamic range control described above, a coarser dynamic range control based on parameters passed on a frame-by-frame basis is also available.

3. OVERVIEW OF THE ENCODING PROCESS

Similarly to the MPEG-1 and -2 Audio coding schemes, in AC-3 achieves high coding gains with respect to PCM by encoding the audio signal in the frequency domain. A block diagram of the encoding process is shown in *Figure 1*.

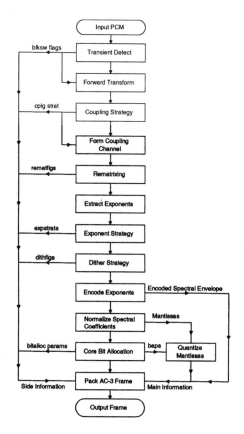

Figure 1. Block diagram of the AC-3 encoder process from [ATSC A/52/10]

The audio input signal is grouped in blocks of 512 PCM time-samples. The internal dynamic range of AC-3 allows for up to 24-bit input sample precision. The first step in the encoder is to assess whether or not the signal under exam presents a transient. Depending on the nature of the input signal, an appropriate filter bank resolution is selected by dynamically adapting the filter bank block size. In steady state conditions, that is when no transient is detected, 512 time-sample blocks are windowed and then mapped into the frequency domain via an MDCT ensuring a frequency resolution of 93.75 Hz at 48 kHz sampling rate. In the presence of a transient, the MDCT block size is reduced to 256 time-samples in order to increase the time resolution of the signal frequency representation. The time resolution in this case is 2.67 ms at 48 kHz sampling rate.

Next, multichannel coding takes place by computing the channel coupling strategy, composing the coupling channel, and rematrixing the frequency coefficient of the signal (see next sections).

The individual MDCT frequency coefficients are converted into a floating-point representation where each coefficient is described by an exponent and a mantissa. Based on the exponent coding strategy, each exponent is then encoded. The set of the coded exponents represent the spectral envelope of the input signal. After the exponents are encoded, the mantissas are quantized according to the bit allocation output.

The bit allocation routine processes the signal spectral envelope to compute hearing masking curves. It applies a parametric model to define the appropriate amplitude precision for the mantissas. Based on control parameters passed in the encoded bitstream, the mantissas can be decoded with dither when they are allocated zero bits for quantization.

The idea behind the dither strategy is that the reproduced signal energy should be maintained even if no bits are allocated. The dither strategy is applied on a per-channel basis and can be bypassed depending on the exponents accuracy and other considerations. In practice, applying dither can result in added audible distortion, therefore a careful monitoring of the resulting bitstream is required.

Finally, the control parameters such as block switching (blksw) flags, coupling strategy (cplg strat), rematrixing (remt) flags, exponent strategy (exps strat), dither flags (dith flags), bit allocation parameters (bitalloc params) are multiplexed with the encoded spectral envelope and mantissas to compose the AC-3 bitstream, where each AC-3 frame corresponds to 6 blocks of 256 frequency-samples.

The block diagram of the decoder is shown in *Figure 2*. Typically the AC-3 bitstream is byte or 16-bit word aligned. After synchronizing the encoded bitstream, the decoder checks for errors and de-formats various type of data in the bitstream such as control parameters and encoded spectral envelope and mantissas. The spectral envelope is decoded to reproduce the exponents.

Based on the exponent values and the bit allocation parameters transmitted, the bit allocation re-computes the precision of the mantissas representation and these data are employed to unpack and de-quantize the mantissas values.

The multichannel inverse allocation takes place (de-coupling and rematrixing), and the dynamic range control parameters are applied.

The final values for the frequency coefficients of the signal are then inverse transformed via an IMDCT applied to 256 blocks of frequency samples. The decoded PCM time-samples are obtained by windowing and overlapping and adding the resulting samples. If required, the down-mix process takes place before delivering the output PCM samples.

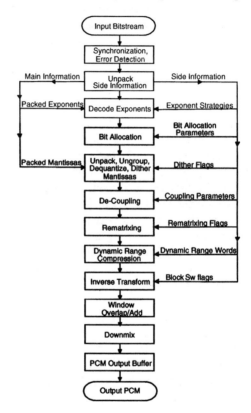

Figure 2. **Block diagram of the AC-3 decoder from [ATSC A/52/10]**

4. FILTER BANK

The filter bank used in AC-3 is based on a 512 time-sample block MDCT [Princen, Johnson and Bradley 87]. The input buffer is build by taking 256 new time-samples and concatenating them to the 256 time-samples from the previous block, with the typical TDAC overlapping of 50% block size between adjacent blocks for steady state signals as we saw in Chapter 5. Next, the PCM samples are windowed with a normalized window derived from the Kaiser-Bessel window [Fielder et al. 96]. As a reminder, the KBD expression follows:

$$W_{KBD-Left}[n] = \sqrt{\frac{\sum_{p=0}^{n} w[p,\alpha]}{\sum_{p=0}^{N/2} w[p,\alpha]}} \qquad 0 \le n < \frac{N}{2}$$

where w (p,α) represents the Kaiser-Bessel window kernel of length N/2 + 1 and N equals 512. Since the KBD window is symmetrical, the remaining N/2 window coefficient can be derived from the first N/2 coefficients by starting at time N/2 and time-reversing the first N/2 coefficients. As a reminder, the Kaiser-Bessel kernel window, also a symmetrical window, is given by:

$$w[n,\alpha] = \frac{I_0\left[\pi\alpha\sqrt{1.0-\left(\frac{n-N/4}{N/4}\right)^2}\right]}{I_0[\pi\alpha]} \qquad 0 \le n \le \frac{N}{2}$$

where

$$I_0[x] = \sum_{k=0}^{\infty}\left[\frac{(x/2)^k}{k!}\right]^2$$

and α is the kernel Kaiser-Bessel window alpha factor that allows for a trade off between close window resolution and ultimate rejection. As we saw in Chapter 5, small values for α imply a small main lobe width for the window (increased close resolution), large values for α imply low side lobe levels (increased ultimate rejection). The KBD window is also employed in MPEG AAC [Bosi et al. 97] with $\alpha = 4$ for steady state conditions and $\alpha = 6$ for transients. In AC-3 α is set equal to five (5). The selection of the window characteristics was based on the study of the shape of masking template curves [Fielder 87]. In particular, the frequency response of the window should be below the worst-case combination of all masking templates. If the filter bank response is below (or coincides) with the worst-case combination of the masking templates, the number of bits necessary to represent the signal can be reduced by exploiting the masking curve information. In *Figure 3* a comparison between a 512-point sine window and the 512-point KBD window employed in AC-3 is shown. The target curve represents the combination of the worst-case masking curve derived from masking data for 20 Hz, 50 Hz, and 1 kHz maskers masking narrow-band noise [Fielder et al. 96]. While both windows fall short of satisfying the frequency selectivity imposed by the masking templates, the AC-3 window is a considerably better match.

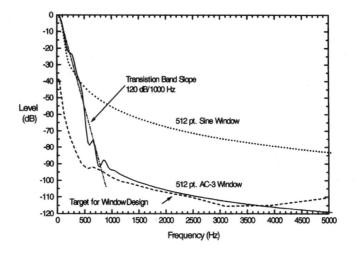

Figure 3. Filter bank resolution comparison from [Fielder et al. 96]

The MDCT operates on overlapping blocks of 512 time samples (long blocks) when the signal is in steady state conditions, or 256 time samples when the signal is transient-like (short blocks). The expressions for the MDCT and IMDCT, as we saw in Chapter 5, are given by:

$$X[k] = \sum_{n=0}^{N-1} x[n]w[n]\cos(\tfrac{2\pi}{N}(n+n_0)(k+\tfrac{1}{2})) \qquad \text{for } k=0,\dots,N/2-1$$

and

$$X[k] = -X[N-k-1] \qquad \text{for } k=N/2,\dots,N-1$$

$$x'[n] = \tfrac{2}{N} \sum_{k=0}^{N-1} X[k]\cos(\tfrac{2\pi}{N}(n+n_0)(k+\tfrac{1}{2})) \qquad \text{for } n=0,\dots,N-1$$

where N= 512 in steady-state conditions and n_0 is the phase term that ensures time-domain aliasing cancellation.

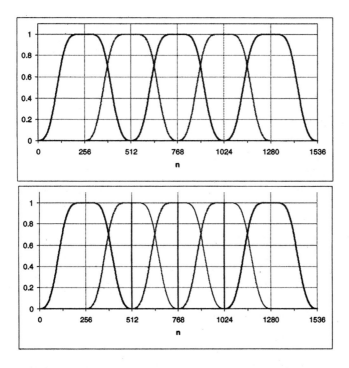

Figure 4. Typical window sequence for AC-3: top for a steady state signal and bottom for a transient-like signal occurring in the region between n = 256 and n = 1024

Time versus frequency resolution trade-offs are achieved by applying block switching as in other coding schemes such as for example MPEG

Layer III [ISO/IEC 11172-3 and ISO/IEC 13818-3] or AAC [ISO/IEC 13818-7] (see also Chapters 11 and 13). The AC-3 block-switching scheme, however, is closer to the AC-2A block switching-scheme [Bosi e Davidson 92] rather than the Edler's [Edler 89] block-switching scheme (see also Chapter 5). In AC-3 the phase term n_0 is not always equal to $(1 + N/2)/2$ where N is the current block length, but depends on the overlapping region between the current block m and the successive block $(m + 1)$. In the case of steady state signals where long blocks are utilized, the overlap region equals 50% of the block length. In the case of transient-like signals where short block are utilized, the overlap region equals 50% of the long block or zero, that is two consecutive short windows are butted together. The phase term n_0 is equal to half the amount of overlapping plus ½.

A typical AC-3 window sequence is shown in *Figure 4*. For steady state conditions and for the second of the short windows, $n_0 = 257/2$; for the first of the short windows $n_0 = ½$. In the presence of transients, the time resolution of the filter bank is increased by selecting the shorter block size. The transition form long to short block is achieved by simply adopting an asymmetrical window that has the first half equal to the long window and then drops to zero. The short window sequence alternates asymmetrical windows that have zero overlapping and 256-sample overlapping. While the frequency response of these windows is sub-optimal, they achieve increased time resolution while keeping the structure of the AC-3 coding algorithm very simple. In the case of short windows, exactly two blocks fit in a long window, keeping the filter bank, packing routines and the overall coding scheme simple.

In AC-2A, a similar concept of varying the overlapping region between adjacent blocks is adopted to dynamically trade-off frequency versus time resolution in the signal representation (see also Chapter 5). In this case, however, asymmetrical windows always overlapping with each other are employed. The frequency response of these window is better than the frequency response of the windows used in AC-3, however, the complexity of the filter bank, packing and other routines is increased. In *Figure 5* a window sequence similar to that adopted in AC-2A is shown. The long window is 512-point long with $\alpha = 4$; the short window is 128-point long with $\alpha = 6$. Time resolution for AC-2A is sharper than in AC-3 (about 1.33 ms versus 2.67 ms at 48 kHz sampling rate). The AC-2A method also allows for greater flexibility in transient representation. The transition windows from long to short (start) and from short to long (stop) are asymmetrical and of size equal to 320 time-samples. In AC-2A the evenly-stacked TDAC [Princen and Bradley 86] transform is utilized, which alternates series of MDCT and modified discrete sine transforms, MDST. The phase factor n_0 equals 257/2 for the long and stop window, and 65/2 for the short and start window. Both the transition and short AC-2A windows have a better frequency response than the AC-3 short windows. The non-power-of-two transition window, however, causes an increase in complexity in the filter bank as well as in other aspects of the AC-2A algorithm. Furthermore, for non-independent channel coding, the multichannel data alignment may cause problems.

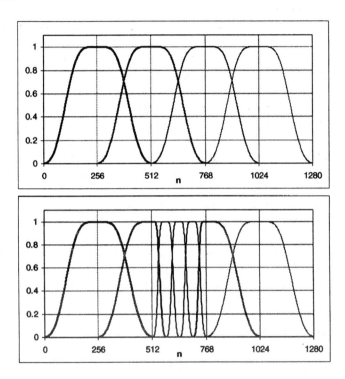

Figure 5. AC-2A window sequence example; top for a steady state signal and bottom for a transient-like signal occurring in the region between n = 512 and n = 768

4.1   Transient Detection

In the AC-3 encoder a transient detection mechanism is utilized to assess the nature of the input signal and, based on this information, whether the filter bank and the AC-3 algorithm operate in long or short block. High-pass filtered version of the full-bandwidth channels are examined to detect a rapid surge in energy.

Typically, abrupt increases in the signal energy at high frequencies are associated with the presence of an attack. If the onset of a transient is detected in the second half of a long block in a certain channel, then that channel switches to a short block. The transient detector input is a block of 512 time samples for each audio block; it processes the time-samples blocks in two steps, each operating on 256 samples. The transient detector output is a one-bit flag, blksw[n] for each full-bandwidth channel, which, when set to one, indicates the presence of a transient in the second half of the 512-point block for the corresponding channel.

The transient detector presents four stages: the high-pass filter, the segmentation of the time samples, a peak amplitude detection for each segment, and the comparison of the peak values with a threshold set to trigger only significant changes in the amplitude values. The high-pass filter is implemented as an infinite impulse response filter with cut-off frequency equal to 8 kHz. The block of the high-passed 256 samples is then decomposed into a hierarchical tree whose shorter segment is 64 samples long. The sample with the largest magnitude is then identified for each segment and then compared to the threshold if there is significant change in level for the current block. First, the overall peak is compared to a silence threshold; if the overall peak is below the threshold, then the signal is in steady state condition and a long block is used for the AC-3 algorithm. If the ratio of peak values for adjacent segments exceeds a pre-defined threshold, then the flag is set to indicate the presence of a transient in the current 256-point segment. The second step follows exactly the above-mentioned stages for the second 256-point input segment and determines the presence of a transient in the second half of the input block.

5.   SPECTRAL ENVELOPE CODING

Once the audio signal is represented in the frequency domain after being processed by the filter bank, the frequency samples or frequency coefficients are coded in floating point form, where each coefficient consists of a scale factor or exponent and a mantissa. As described in Chapter 2, the exponents indicate the number of leading zeros in the binary representation of a coefficient. The set of exponents for each audio block conveys an estimate of the overall spectral content of the signal. This information is often referred to as the spectral envelope of the signal. In coding the spectral envelope of the signal, AC-3 allows exponents to be represented by 5-bit numbers and their values vary between 0 (for the largest value coefficients with no leading zeroes) and 24. The AC-3 bitstream contains coded exponents for all independent channels, all coupled channels, and for the coupling (see also next section) and LFE channels (when they are enabled).

AC-3 spectral envelope coding allows for variable resolution in time and frequency [Todd et al. 94]. In the time domain, two or more consecutive block within one frame can share a common set of exponents. Since audio information is not shared across frames in AC-3, block 0 of every frame always includes new exponents for every channel. In the frequency domain, either one (D15 mode), two (D25 mode) or four (D45 mode) mantissas can share the same exponent. The strategy for sharing exponents in the time or frequency domain is embedded in the encoder process and is based on the signal conditions. For steady state signals, while the maximum frequency resolution allowed is preferred (D15 mode), the signal spectrum is expected not to vary significantly from block to block. In this case, the set of exponents is only transmitted for block zero and kept constant for the other blocks in a frame. This operation results in a substantial saving in the exponent representation therefore freeing up bits for mantissas quantization. On the other hand, for transient-like signals the signal spectrum is expected to vary significantly from block to block. For transient-like signals AC-3 typically transmits the set of exponents for all blocks in a frame, allowing the coded spectral envelope to follow as closely as possible the variations in time of the input signal. In this case, however, fine grain frequency resolution (for example the D15 mode) for the exponent representation is superfluous, therefore a saving in the bit budget can be realized by representing the set of exponents in a coarser frequency scale (for example the D25 or the D45 mode). For each frame and for any given input signal the exponent strategy is based on the minimization of audibility of the quantization noise at the target data rate.

The exponents, with the exception of the first frequency term, are differentially coded across frequency. The first exponent of a full-bandwidth or LFE channel, representing the DC term of that channel, is always coded as a 4-bit absolute number, with values ranging between 0 and 15. The differential exponents are combined into groups in the audio block. The grouping is done employing one of three modes, D15, D25, or D45, where the D15 mode provides the finest frequency resolution, and the D45 mode requires the least amount of data. The number of grouped differential exponents placed in an audio block representation for a particular channel depends on the exponent strategy and on the frequency bandwidth information for that channel. The number of exponents in each group depends only on the exponent strategy.

An AC-3 audio block contains two types of fields with exponent information. The first type defines the exponent coding strategy for each channel, and the second type contains the actual coded exponents for channels requiring new exponent information. For independent channels, frequency bandwidth information is included along with the exponent strategy fields. For coupled channels and the coupling channel (see also next section), the frequency information is found in the coupling strategy fields.

Each differential exponent can take on one of five values: -2, -1, 0, +1, +2, allowing differences of up to ± 12 dB between exponents. These values are then mapped to new values by adding an offset equal to +2 and further combined into a 7-bit unsigned integer as follows:

$$7\text{-bit Value} = (25 \times M_1) + (5 \times M_2) + M_3$$

where M_1, M_2, and M_3 are three differential, adjacent in frequency, exponents mapped values. In this fashion, each exponent requires an average of $7/3 = 2.33$ bits.

Table 3. Data rate for different exponent strategies in terms of bits needed per frequency coefficient

Exponent Strategy	Shared Time Interval in Terms of Number of Audio Blocks					
	1	2	3	4	5	6
D15	2.33	1.17	0.78	0.58	0.47	0.39
D25	1.17	0.58	0.39	0.29	0.23	0.19
D45	0.58	0.29	0.19	0.15	0.12	0.10

In *Table 3* the data rate relative to different exponent strategies is shown. Since in the D25 or the D45 mode a single exponent is effectively shared by 2 or 4 different mantissas, encoders must ensure that the exponent chosen for the pair or quad is the minimum absolute value (corresponding to the largest exponent) needed to represent all the mantissas. In general, the exponent field for a given channel in an AC-3 audio block consists of a single absolute exponent followed by a number of these grouped values.

The coding of the spectral envelope in AC-3 is very different from coding of the spectral envelope in its predecessor, AC-2. In AC-2 the floating-point representation of the signal spectrum is very close to a group linear A-law (see also Chapter 2). First, the spectral coefficients are grouped into bands that simulate the auditory critical bandwidths. For each critical band the maximum is selected, and the exponent for this coefficient is selected as the exponent for the coefficients in that critical band. As a result of this operation the signal quantization floor level may be raised depending on the value of the local spectral maxima. The difference between the AC-2 and AC-3 spectral envelope representation is shown in *Figure 6* [Fielder et al. 96].

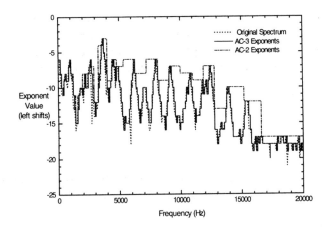

Figure 6. Comparison between AC-2 and AC-3 spectral envelope representation from [Fielder et al. 96]

6. MULTICHANNEL CODING

As we saw in Chapter 12 and 13, the main goal of multichannel audio coding is to reduce the data rate of a multichannel audio signal by exploiting redundancy between channels and irrelevancy in the spatial representation of the multichannel signal while preserving the basic audio quality and the spatial attributes of the original signal. In perceptual audio coding, this goal is achieved by preserving the listener cues that influence the directionality of hearing [Blauert 83]. In AC-3, two techniques are adopted for multichannel coding. One exploits redundancies among pairs of highly correlated channel and it is called rematrixing. Rematrixing is based on a similar principle as M/S stereo coding (see also Chapter 11): sum and differences of correlated channel spectra are coded rather than the original channels [Johnston and Ferreira 92]. The other multichannel technique adopted in AC-3 is channel coupling (see also Chapters 12 and 13), in which two or more correlated channel spectra are combined together and the combined or coupled channel is coded and transmitted with additional side information [Davis 93].

6.1   Rematrixing

In AC-3 rematrixing is applicable only in the 2/0 mode, acmod = 010. In this mode, when rematrixing is applied, rather than separately coding two highly correlated channels, the left, L, and right, R, channel are combined into two new channels , L', and R', which are defined as follows:

$$L' = (L + R)/2$$

$$R' = (L - R)/2$$

Quantization and packing are then applied to the L' and R' channels. In the decoder, the original L and R channel are derived as follows:

$$L = L' + R'$$

$$R = L' - R'$$

In the case of complete correlation between the two channels, like when, for example, the two channels are identical, then L' is the same as L or R, and R' is zero. In this case, no bits are allocated to R', allowing for an increased accuracy in the L and R = L' representation.

Rematrixing is performed independently for different frequency regions. There are up to four frequency regions with boundaries dependent on coupling information. Rematrixing is never used in the coupling channels. If coupling and rematrixing are simultaneously in use, the highest rematrixing region ends at the starting of the coupling region. In *Table 4* the frequency boundaries when coupling is not in use are shown at different sampling rates.

Table 4. AC-3 rematrixing frequency regions boundaries in kHz [ATSC A/52/10]

Frequency Region	Lower Bound Fs = 48 kHz	Upper Bound Fs = 48 kHz	Lower Bound Fs = 44.1 kHz	Upper Bound Fs = 44.1 kHz
1	1.17	2.3	1.08	2.11
2	2.3	3.42	2.11	3.14
3	3.42	5.67	3.14	5.21
4	5.67	23.67	5.21	21.75

6.2   Coupling

Channel coupling exploits the experimental findings that sound sources localization cues depend mostly on the energy envelope of the signal and not its fine temporal structure. Channel coupling can be seen as an extension of intensity stereo coding [Van der Waal and Veldhuis 91] as we described in Chapter 11, although the two technologies were derived independently. In AC-3 two or more correlated channel spectra (coupling channels) are combined together in a single channel (coupled channel) above a certain frequency (coupling frequency) [Davis 93]. The coupled channel is the result of the vector summation of the spectra of all the channels in coupling. In addition to the coupled channel, side information is also conveyed to the decoder in order to enable the reconstruction of the original channels. The set of side information is called coupling coordinates and consists of the quantized version of the power spectra ratios between the original signal and the coupled channel for each input channel and spectral band. The coupling coordinates, floating-point quantized and represented with a set of exponents and mantissas, are computed in such manner that they allow for the preservation of the original signal short-term energy envelope.

In *Figure 7* an example of channel coupling with three input channels is shown. For each input channel an optional phase adjustment is first applied to avoid phase cancellation during the summation. Next, the coupled channel is computed by summing all frequency coefficients above the coupling frequency. The power of the original channels and the coupled channel is then derived. In the simplified case of *Figure 7* only two frequency bands are considered. In general the number of frequency bands vary between 1 and 18: typically 14 bands are considered. In *Table 5* the allowed coupling bands are shown for a sampling rate of 48 kHz. Finally, the power ratios are computed to derive the coupling coordinates. As mentioned before coupling is only active above the coupling frequency, where this frequency may vary from block to block.

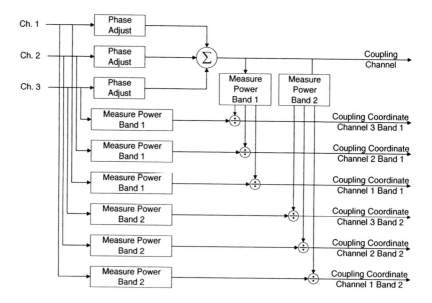

Figure 7. Example of AC-3 coupling: block diagram for three input channels from [Fielder et al. 96]

Table 5. AC-3 coupling bands at a sampling rate of 48 kHz [ATSC A/52/10]

Coupling Band	Lower Bound (kHz)	Upper Bound (kHz)
0	3.42	4.55
1	4.55	5.67
2	5.67	6.80
3	6.80	7.92
4	7.92	9.05
5	9.05	10.17
6	10.17	11.30
7	11.30	12.42
8	12.42	13.55
9	13.55	14.67
10	14.67	15.80
11	15.80	16.92
12	16.92	18.05
13	18.05	19.17
14	19.17	20.30
15	20.30	21.42
16	21.42	22.55
17	22.55	23.67

Coupling parameters such as the coupling frequency and which channels are in coupling are always transmitted in block 0 of a frame; they may also be part of the control information for blocks 1 through 5. The coupling coordinates dynamic range covers a range between – 132 and +18 dB with a resolution varying between 0.28 and 0.53 dB. In the decoder the spectral coefficients corresponding to the coupling channels are derived by multiplying the coupling coordinates by the received coupled channel coefficients as shown in *Figure 8*.

It should be noted that coupling is intended for use only when audio coding at a certain data rate and desired audio bandwidth would introduce audible artifacts due to bit starvation. In these cases, coupling allows for maintaining the coding constraints without significantly altering the original signal.

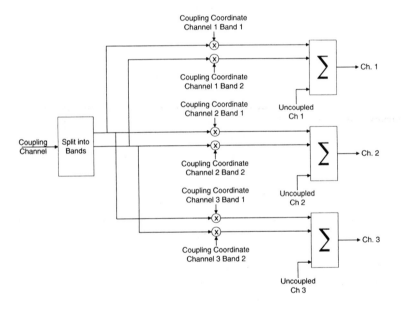

Figure 8. Example of the AC-3 de-coupling process for three input channels from [Fielder et al. 96]

7.   BIT ALLOCATION

In AC-3 a parametric bit allocation is employed in order to distribute the number of bits available per block to the frequency coefficients mantissas given a certain data rate. The AC-3 parametric bit allocation combines forward and backwards adaptive strategies [Davidson, Fielder and Link 94]. In a forward adaptive bit allocation strategy, as adopted in the MPEG audio coders, the allocation is computed in the encoder and then transmitted to the decoder. Advantages of this approach include high flexibility in the allocation without modifying the decoder structure. The backward adaptive strategy calls for a computation of the allocati₋ in both the encoder and the decoder. This method was applied in the AC-3 predecessor, AC-2, bit allocation strategy. While loosing some flexibility, this method has the advantage of saving bits in the representation of the control parameters and therefore it frees resources that become available to encode the frequency mantissas.

In AC-3 both encoder and decoder bit allocation include the core psychoacoustics model upon which the bit allocation is built and the allocation itself, therefore eliminating the need to explicitly transmit the bit allocation in its entirety. Only essential psychoacoustics parameters and a delta bit allocation, in terms of a parametric adjustment to the masking curves, are conveyed to the decoder. This strategy allows for an improvement path since these parameters are computed in the encoder only and don't affect the decoder structure but also minimizes the amount of control data to be transmitted to the decoder.

Bit allocation parameters are always sent in block 0 and are optional in blocks 1 through 5. The main input to the bit allocation routine in the encoder and decoder is the set of the fine grain exponents that represent spectral envelope of the signal for the current block. Another input in the decoder bit allocation is represented by the optional delta bit allocation. The main output in the encoder and decoder bit allocation routine is a bit allocation array; in the encoder control parameters to be conveyed to the decoder are additional outputs.

In order to compute the bit allocation, the excitation patterns are first derived. For each block, the exponent set is mapped to a logarithmic power-spectral density. A logarithmic addition of the power-spectral density over frequency bands that follow the critical band rate as defined by Flecther [Fletcher 40] and Zwicker [Zwicker 61] is computed (see also Chapter 6). In *Figure 9* the band sub-division adopted in AC-3 is shown. A comparison with the AC-2 banding structure is also shown in *Figure 9*. While AC-2 banding structure approximates the critical bandwidths, AC-3 offers an increased resolution, its banding being closer to half critical bandwidths.

The excitation patterns are computed by applying a spreading function to the signal energy levels on a critical band by critical band basis. The spreading function adopted in AC-3 is derived from masking data of 500 Hz, 1 kHz, 2kHz, and 4kHz maskers masking narrow-band noise as shown in *Figure 10* [Fielder et al. 96]. The masking curve towards lower frequencies can be approximated by a single linear curve with a slope of 10 dB per band. Towards higher frequencies, the masking curve can be approximated by a two-piece linear segment curve. The slope and the vertical offset of these segments can be varied based on the frequency of the masking components to better follow the corresponding masking data. Four parameters are transmitted to the decoder to characterize the spreading function shape (namely the offset and the slope of the two segments of the spreading function).

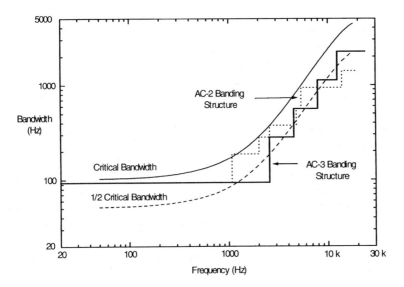

Figure 9. AC-3 and AC-2 bit allocation spectral bandwidths versus critical bandwidths from [Fielder et al. 96]

In order to capture the contribution of all masking components in the block of data under examination, the masking components are weighted by the spreading function and then combined together. This step is sometimes implemented as a convolution (see for example [Schroeder, Atal and Hall

79] and MPEG Psychoacoustic Model 2 [ISO/IEC 11172-3]). The convolution between the masking components of the signal and the spreading function can be computed via a linear recursive filter (or IIR filter), since its output is the result of weighed summation of the input samples. In this case the filter order and coefficients are determined from the spreading function. In AC-3 the linear recursive filter is replaced with an equivalent filter that processes logarithmic spectral samples. To implement the convolution with the two-slope spreading function, two filters are connected in parallel. The computation of the excitation patterns utilizing IIR filters in place of a convolution results in a very efficient implementation, drastically reducing the complexity of the algorithm.

Once the excitation pattern is computed, it is then offset downward by an appropriate amount (about 25 dB). The signal masking curve is then combined with the threshold in quiet by selecting the greater of the two masking levels for each frequency point in order to derive the corresponding global masked curve.

The masking curve computation is present in both encoder and decoder. A number of parameters describing the masking models, however, are conveyed to the decoder. The shape of the spreading function, for example is described in the AC-3 bitstream by four parameters. In addition, optional improvements to the masking models can be transmitted to the decoder via the delta bit allocation. The delta bit allocation is derived from the difference between two masking curves calculated in parallel in the encoder, where one masking curve represent the core model and is recomputed in the decoder and the other represents an improved version of it.

The last step in the bit allocation routine is the derivation of the number of bits to be assigned to each frequency mantissa. The masking curve is subtracted to the fine-grain logarithmic spectral envelope. This difference is right shifted by 5 and then mapped to a vector of values, baptab, to obtain the final bit allocation. In *Table 6* the mapping between the shifted difference values and the final allocation is shown.

It should be noted that, in general, bit allocation strategies are based on the assumption that the quantization noise in a particular band is independent of the number of bits allocated in neighboring bands. While this assumption is reasonably well satisfied when the time-to-frequency mapping of the signal is performed with a high frequency-resolution, aliasing-free filter bank, this is not always the case. This effect is especially pronounced at low frequencies, where the slope of the masking curves can exceed the selectivity of the filter bank. For example, in the downward frequency-masking regions for tonal components with frequencies between 500 Hz - 2.5 kHz the computation of the bit allocation based solely on the differences between the signal spectrum levels and the masking levels may lead to audible quantization noise. In AC-3, a method sometimes nicknamed "Low Comp" is applied in order to compensate for potential audible quantization noise at low frequencies due to the limited frequency resolution of the signal representation. In this scheme, an iterative process is applied, in which the noise contributions from each transform coefficient are examined and an appropriate word length adjustment is adopted in order to ensure that the quantization noise level lie below the computed masking curve. The adoption of the Low Comp scheme often results in the addition of one to three bits per frequency sample in the region of high positive slopes of the masking curves for low frequency or mid-range tonal maskers. The reader interested in a more detailed discussion of the Low Comp method should consult [Davidson, Fielder and Link 94].

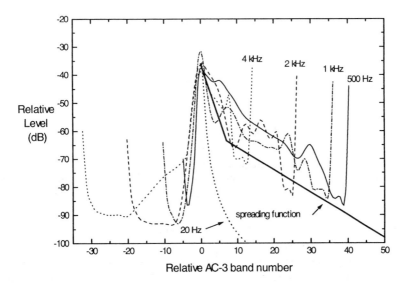

Figure 10. Comparison between the AC-3 spreading function and masking curves for 500 Hz, 1 kHz, 2 kHz, 4kHz sinusoidal maskers from [Fielder et al. 96]

Table 6 . AC-3 bit allocation from shifted SMR values [ATSC A/52/10]

Shifted SMR	Baptab	Shifted SMR	Baptab	Shifted SMR	Baptab
0	0	22	7	44	13
1	1	23	8	45	13
2	1	24	8	46	13
3	1	25	8	47	14
4	1	26	8	48	14
5	1	27	9	49	14
6	2	28	9	50	14
7	2	29	9	51	14
8	3	30	9	52	14
9	3	31	10	53	14
10	3	32	10	54	14
11	4	33	10	55	15
12	4	34	10	56	15
13	5	35	11	57	15
14	5	36	11	58	15
15	6	37	11	59	15
16	6	38	11	60	15
17	6	39	12	61	15
18	6	40	12	62	15
19	7	41	12	63	15
20	7	42	12		
21	7	43	13		

8.   QUANTIZATION

The mantissas are quantized according to the number of bits allocated as indicated in *Table 7*. The baptab value corresponds to a bit allocation pointer, bap, which describes the number of quantizer levels. Depending on the number of levels, the quantizer utilized in AC-3 may be symmetrical or asymmetrical. For levels up to 15 the quantizer is a midtread quantizer (see also Chapter 2). For levels above 15, i.e. 32, 64,..., 65536, the quantizer is a two's complement quantizer. In addition, some quantized mantissas are grouped into a single codeword (see also the MPEG-1 Layer II quantization description in Chapter 11). In the case of a three and five-level quantizer, bap = 1 and bap = 2 respectively, three quantized mantissas are grouped into a five and seven-bit codeword respectively as follows:

bap = 1   codeword = 9 mantissa[a] + 3 mantissa[b] + mantissa[c]

bap = 2   codeword = 25 mantissa[a] + 5 mantissa[b] + mantissa[c]

In the case of an eleven-level quantizer, two quantized values are grouped and represented by a seven-bit codeword as follows:

bap = 4   codeword = 11 mantissa[a] + mantissa[b]

Table 7 shows the correspondence between the bap value and the number of quantization levels and bits used to represent a single mantissa.

Table 7. AC-3 quantizer levels [ATSC A/52/10]

bap	Quantizer levels	Mantissa bits
0	0	0
1	3	1.67 (5/3)
2	5	2.33 (7/3)
3	7	3
4	11	3.5 (7/2)
5	15	4
6	32	5
7	64	6
8	128	7
9	256	8
10	512	9
11	1024	10
12	2048	11
13	4096	12
14	16,384	14
15	65,536	16

The AC-3 decoder may employ optionally a dither function when bap = 0, i.e. the number of mantissa bits is zero. Based on the values of a one-bit control parameter transmitted in the AC-3 bitstream, dithflag, the decoder may substitute random values for mantissas with bap equal to zero. For dithflag equal to zero, true zero values are utilized.

9.   BITSTREAM SYNTAX

The AC-3 bitstream consists of a sequence of frames (see *Figure 11*). Each frame contains six coded audio blocks, each of which represent 256 new audio samples for a total of 1536 samples. A synchronization information header at the beginning of each frame contains information needed to acquire and maintain synchronization. First a synchronization word equal to 0000 1011 0111 0111 is transmitted. An optional cyclic redundancy code, CRC, word follows. This 16-bit CRC applies to the first 5/8 of the frame. An 8-bit field synchronization information (SI) conveys the sample rate code (2 bits) and the frame size code (6 bits). The SI is used to determine the number of two-byte words before the next synchronization word. The length of the above mentioned part of the bitstream (Synch word, CRC and SI information) is fixed and it is always transmitted for each frame.

A bitstream information (BSI) header follows the SI, and contains parameters describing the coded audio service. The coded audio blocks may be followed by an auxiliary data (Aux) field. At the end of each frame is an error check field that includes a CRC word for error detection. With the exception of the CRC, these fields may vary from frame to frame depending on programming parameters such as the number of encoded channels, the audio coding mode, and the number of listener features.

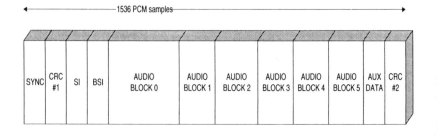

Figure 11. AC-3 frame structure

The BSI field is a variable field containing parameters describing the coded audio services including bitstream identification and mode, audio coding modes, mix levels, dynamic range compression control word, language code, time code, etc.

Within one frame the relative size of each audio block can be adapted to the signal bit demands. Audio blocks with higher bit demand can be weighted more heavily than other in the distribution of the bit pool available per frame. In addition, the rate of the AC-3 frame can be adjusted based on the signal demands, by changing the frame size code parameter in the SI field. In this fashion, variable bit rate on a short and long-term basis can be implemented in AC-3. This feature may prove to be very useful in storage applications.

10.   PERFORMANCE

A number of tests were carried out to measure the performance of AC-3. One of the most recent tests and possibly one of the most interesting because of its assessment in conjunction with the subjective evaluation of other state-of-the-art two-channel audio coders took place at the Communication Research Centre, CRC, Ottawa, Canada [Soulodre, Grusec, Lavoie and Thibault 98] (see *Figure 12*). Other codecs included in the tests were MPEG-2 AAC at 128 kb/s (Main Profile), MPEG Layer II at 192 kb/s and MPEG Layer III at 128 kb/s, and Lucent PAC [Sinha, Johnston, Dorward and Quackenbush 98] at 160 kb/s. At 192 kb/s AC-3 scored in average 4.5 in the five-grade ITU-R impairment scale, i.e. the differences between the AC-3 coded an the original excerpts was deemed by expert listeners in the region of perceptible but not annoying. AC-3 at 192 kb/s together with MPEG-2 AAC at 128 kb/s s ranked the best among the codecs tested.

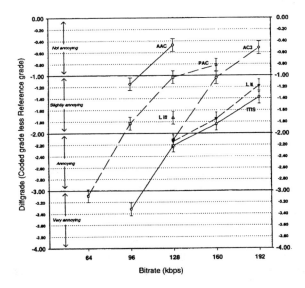

Figure 12. Comparison of AC-3 overall quality with MPEG-2 AAC, MPEG-2 Layer II and MPEG-2 Layer III from [Soulodre, Grusec, Lavoie and Thibault 98]

11. SUMMARY

In this chapter we reviewed the main features of the Dolby AC-3 Audio system. AC-3 was developed for encoding multichannel audio on film and later migrated to consumer applications. Dolby AC-3 is currently in use in the North American HDTV, DVD-Video, and regional DVB standards.

AC-3 is a perceptual audio coding system that allows the encoding of diverse audio channels format. The AC-3 algorithm presents similarities with its predecessor, AC-2, and other perceptual audio coding schemes such as MPEG-1 and -2 Audio, as well as unique, distinctive approaches to audio coding. AC-3 data rates range from 32 kb/s to 640 kb/s, with preferred operational data rates at 192 kb/s in the two-channel configuration and 384 kb/s in the five-channel configuration. User's features include downmixing capability, dynamic range control, multilingual services, and hearing and visual impaired services.

AC-3 was tested in the stereo configuration by the CRC, Canada, during the subjective evaluation tests of state-of-the-art two-channel audio codecs, scoring in average 4.5 in the five-grade ITU-R impairment scale at 192 kb/s in the stereo configuration.

12. REFERENCES

[ATSC A/52/10]: United States Advanced Television Systems Committee Digital Audio Compression (AC-3) Standard, Doc. A/52/10, December 1995.

[Blauert 83]: J. Blauert, *Spatial Hearing*, MIT Press, Cambridge, MA 1983.

[Bosi and Davidson 92]: M. Bosi and G. A. Davidson, "High-Quality, Low-Rate Audio Transform Coding for Transmission and Multimedia Applications", presented at the 93rd AES Convention, J. Audio Eng. Soc. (Abstracts), vol. 40, P. 1041, Preprint 3365, December 1992.

[Bosi et al. 97]: M Bosi, K. Brandenburg, S. Quackenbush, L. Fielder, K. Akagiri, H. Fuchs, M. Dietz, J. Herre, G. Davidson, and Y. Oikawa, "ISO/IEC MPEG-2 Advanced Audio Coding," J. Audio Eng. Soc., vol. 45, pp. 789 – 812, October 1997.

[Davidson, Fielder and Link 94]: G. A. Davidson, L. D. Fielder, and B. D. Link, "Parametric Bit Allocation in a Perceptual Audio Coder ", presented at the 97th AES convention, San Francisco, Preprint 3921, November 1994.

[Davidson, Fielder, and Antill 90]: G. A. Davidson, L. D. Fielder, and M. Antill, "Low-Complexity Transform Coder for Satellite Link Applications," presented at the 89th Convention of the Audio Engineering Society, pre-print 2966, New York, September 1990.

[Davis 93]: M. Davis, "The AC-3 Multichannel Coder," presented at the 95th AES Convention, New York, pre-print 3774, October 1993.

[DVD-Video]: DVD Specifications for Read-Only Disc, Part 3: VIDEO SPECIFICATIONS Ver. 1.1, Tokyo 1997-2001.

[Edler 89]: B. Edler, "Coding of Audio Signals with Overlapping Transform and Adaptive Window Shape" (in German), Frequenz, Vol. 43, No. 9, pp. 252-256, September 1989.

[ETS 300 421]: The European Telecommunications Standards Institute (ETSI), ETS 300 421, "Digital Video Broadcasting (DVB); Framing Structure, Channel Coding and Modulation for 11/12 GHz Satellite Services", August 1997.

[Fielder 87]: L. Fielder, "Evaluation of the Audible Distortion and Noise Produced by Digital Audio Converters," J. Audio Eng. Soc., vol. 35, pp. 517-535, July/August 1987.

[Fielder et al. 96]: L. Fielder, M. Bosi, G. Davidson, M. Davis, C. Todd, and S. Vernon, "AC-2 and AC-3: Low-Complexity Transform-Based Audio Coding," *Collected Papers on Digital Audio Bit-Rate Reduction*, Neil Gilchrist and Christer Grewin, Eds., pp. 54-72, AES 1996,.

[Fletcher 40]: H. Fletcher, "Auditory Patterns," Reviews of Modern Physics, Vol. 12, pp. 47-65, January 1940.

[ISO/IEC 11172-3]: ISO/IEC 11172, Information Technology, "Coding of moving pictures and associated audio for digital storage media at up to about 1.5 Mbit/s, Part 3: Audio", 1993.

[ISO/IEC 13818-3]: ISO/IEC 13818-3, "Information Technology - Generic Coding of Moving Pictures and Associated Audio, Part 3: Audio," 1994-1997.

[ISO/IEC 13818-7]: ISO/IEC 13818-7, "Information Technology - Generic Coding of Moving Pictures and Associated Audio, Part 7: Advanced Audio Coding", 1997.

[ITU-R BS.775-1]: International Telecommunications Union BS.775-1, ""Multichannel Stereophonic Sound System with and without Accompanying Picture ", Geneva, Switzerland, 1992-1994.

[Johnston and Ferreira 92]: J. D. Johnston, A. J. Ferreira, "Sum-Difference Stereo Transform Coding", Proc. ICASSP pp. 569-571, 1992.

[Princen and Bradley 86]: J. P. Princen and A. B. Bradley, "Analysis/Synthesis Filter Bank Design Based on Time Domain Aliasing Cancellation," IEEE Transactions on Acoustics, Speech, and Signal Processing, vol. ASSP-34, no. 5, pp. 1153 – 1161, October 1986.

[Princen, Johnson and Bradley 87]: J. P. Princen, A. Johnson and A. B. Bradley, "Subband/Transform Coding Using Filter Bank Designs Based on Time Domain Aliasing Cancellation", Proc. of the ICASSP 1987, pp. 2161-2164, 1987.

[Schroeder, Atal and Hall 79]: M. R. Schroeder, B. S. Atal and J. L. Hall, "Optimizing Digital Speech Coders by Exploiting Masking Properties of the Human Ear", J. Acoust. Soc. Am., Vol. 66 no. 6, pp. 1647-1652, December 1979.

[Sinha, Johnston, Dorward and Quackenbush 98]: D. Sinha, J. D. Johnston, S. Dorward and S. R. Quackenbush, "The Perceptual Audio Coder (PAC)", in *The Digital Signal Processing Handbook*, V. Madisetti and D. Williams (ed.), CRC Press, pp. 42.1-42.18, 1998.

[Soulodre, Grusec, Lavoie and Thibault 98]: G. A. Soulodre, T. Grusec, M. Lavoie, and L. Thibault, "Subjective Evaluation of State-of-the-Art Two-Channel Audio Codecs", J. Audio Eng. Soc., Vol. 46, no. 3, pp. 164-177, March 1998.

[Todd et al. 94]: C. Todd, G. A. Davidson, M. F. Davis, L. D. Fielder, B. D. Link and S. Vernon, "AC-3: Flexible Perceptual Coding for Audio Transmission and Storage," presented at the 96th Convention of the Audio Engineering Society, Preprint 3796, February 1994.

[Van der Waal and Veldhuis 91]: R. G. v.d. Waal and R. N. J. Veldhuis, "Subband Coding of Stereophonic Digital Audio Signals", Proc. ICASSP, pp. 3601 – 3604, 1991.

[Zwicker 61]: E. Zwicker, "Subdivision of the Audible Frequency Range into Critical Bands (Frequenzgruppen)," J. Acoust. Soc. of Am., Vol. 33, p. 248, February 1961.

Chapter 15

MPEG-4 Audio

1. INTRODUCTION

In Chapters 11, 12 and 13 we discussed the goals of the first two phases of the MPEG Audio standard, MPEG-1 and MPEG-2, and we reviewed the main features of the specifications. MPEG-4 is another ISO/IEC standard that was proposed as a work item in 1992 [ISO/IEC MPEG N271]. In addition to audiovisual coding at very low bit rates, the MPEG-4 standard addresses different functionalities, such as, for example, scalability, 3-D, synthetic/natural hybrid coding, etc. MPEG-4 became an ISO/IEC final draft international standard, FDIS, in October 1998 (ISO/IEC 14496 version 1), see for example [ISO/IEC MPEG N2501, N2506, N2502 and N2503]. The second version of ISO/IEC 14496 was finalized in December 1999 [ISO/IEC 14996]. In order to address the needs of emerging applications, the scope of the standard was expanded in later amendments and, even currently, a number of new features are under development. These features will be incorporated in new extensions to the standard, where the newer versions of the standard are compatible with the older ones.

The MPEG-4 standard targets a wide number of applications including wired, wireless, streaming, digital broadcasting, interactive multimedia and high quality audio/video. Rather than standardize a full algorithm and a bitstream as was done in MPEG-1 and 2, MPEG-4 specifies a set of tools, where a tool is defined as a coding module that can be used as a component in different coding algorithms. Different profiles, that represent a collection of tools and refer to a particular application, are defined in the standard.

MPEG-4 Audio includes, in addition to technology for coding general audio as in MPEG-1 and 2, speech, synthetic audio and text to speech interface technology. Features like scalability, special effects, sound manipulations, and 3-D composition are also included in the standard. While MPEG-1 and 2 Audio typically specify the data rate at the time of the encoding process, the scalability feature in MPEG-4 allows for a system data rate, which is, with some boundaries, dynamically adaptable to the channel capacity. This feature provides significant benefits when dealing with transmission channels with variable capacity, such as internet and mobile channels.

In this chapter, a high level description of MPEG-4, its goals and functionalities are discussed. The development of MPEG-4 Audio is then presented followed by a description of the basic tools and profiles of MPEG-4 Audio. Finally an evaluation of the audio coding tools performance is discussed and intellectual property management issues are introduced.

2.   MPEG-4: WHAT IS IT?

The MPEG-4 standard specifies the coding parameters of elements of audio, visual, or audiovisual information, referred to as "media objects". These objects can be multidimensional, natural or synthetic, i.e. they can be recorded from natural scenes with a microphone and a video recorder or they can be computer-generated [Chiariglione 98].

For example (see *Figure 1*), a talking person can be represented as the ensemble of basic media objects such as the background image (still image object), the talking person without the background (video object) and that person's voice plus background noise (audio object).

In addition, the MPEG-4 standard describes the composition of these objects to create groups of media objects that describe an audiovisual scene. For example, the audio object representing the person's voice can be combined with video object representing the talking person to form a new media object containing both the audio and visual components of the talking person and then further combined into more complex audiovisual scenes.

MPEG-4 defines also the multiplexing and synchronization of the data associated with media objects, so that they can be transported over media channels, and it provides means for interaction with the audiovisual scene generated at the receiver's end. It incorporates identification of intellectual property and supports controlled access to intellectual property through the requirements specified in the "Management and Protection of Intellectual Property", IPMP, part of the standard [ISO/IEC 14496-1, ISO/IEC MPEG N2614].

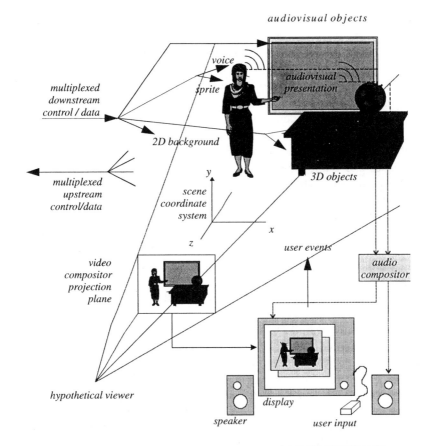

Figure 1. An example of an MPEG-4 scene from [ISO/IEC MPEG N4668]

2.1 MPEG-4 Standard Document Organization

Similarly to the other MPEG standard documents, the ISO/IEC 14496 document, "Coding of audio-visual objects", specifies the decoder process and bistream format. The main sections, listed below, describe the different parts of the standard as follows:

ISO/IEC 14496-1: Systems
ISO/IEC 14496-2: Visual
ISO/IEC 14496-3: Audio
ISO/IEC 14496-6: Delivery Multimedia Integration Framework
ISO/IEC 14496-10 (Proposed) Advanced Video Coding

Part 1 of the standard describes the MPEG-4 Systems specifications. The Systems part of the standard specifies the tools for describing the spatial-temporal relation between the audio-visual objects in a scene. The actual scene is created using the binary format for scenes (BIFS) and, at a lower level, by defining the relations between audio-visual elementary streams using object descriptors. In addition, the Systems part of the standard specifies the MPEG-4 file format, MP4 and interfaces to different aspects of terminals and networks through Java application engines, MPEG-J.

Parts 2 and 3 of the standard define a set of advanced compression tools for visual and audio objects, respectively. The elementary data streams resulting from the coding procedures described in these parts of the standard can be transmitted or stored separately, but eventually need to be composed to create the final multimedia scene.

Part 2, Visual, supports the coding of natural images and video together with synthetic scenes at data rates ranging from 5 kb/s to more than 1Gb/s. In addition, complexity, quality and spatio-temporal scalability is supported together with robustness in error prone environments. Face and body animation and coding of 2-D and 3-D polygonal meshes are also addressed.

Part 3, Audio, addresses the coding of general audio, speech, synthetic audio, text to speech, TTS, interface, as well as additional functionalities such as scalability, time/pitch shift, 3-D, and error robustness. The audio tools will be discussed further later in this chapter.

Part 6 of the standard presents the Delivery Multimedia Integration Framework (DMIF) tools. The FlexMux tool is used to interleave multiple elementary streams into a single stream. The TransMux tool is used to map elementary streams onto transport streams such as the real time transport protocol (RTP) or MPEG-2 transport streams.

Recently, a new video format, proposed as Part 10 of the MPEG-4 standard, reached the final committee draft (FCD) status [ISO/IEC MPEG N4920]. Part 10, or Advanced Video Coding, AVC, jointly developed with ITU-T SG16 by the joint video team, JVT, [ISO/IEC MPEG N4400] addresses a new set of visual compression tools. Part 10 FCD holds the promise of extremely high quality video at increased complexity with respect to Part 2. AVC is expected to be finalized as ISO/IEC 14496-10 by the end of 2002.

New extensions are currently under consideration by the MPEG-4 Audio standard committee [ISO/IEC MPEG N4764]:

- Bandwidth extension – a technology that allows for the reconstruction of the high frequency components of an audio signal at the receiver side. This method significantly improves the compression efficiency of general audio coders [Dietz, Liljeryd, Kjoerling and Kunz 02].

- Parametric coding at higher data rates – a tool that extends the capability of the harmonic individual lines and noise, HILN, parametric coding scheme of the standard [den Brinker, Schuijers and Oomen 02].

In addition, a call for proposals for lossless audio coding was recently issued by the MPEG Committee [ISO/IEC MPEG N5040]. The idea behind this call is to extend the general audio coding capability of MPEG-4 Audio to lossless coding.

3. MPEG-4 AUDIO GOALS AND FUNCTIONALITIES

The scope of MPEG-4 Audio is broader than the scope of MPEG-1 and 2 Audio. The different types of applications that MPEG-4 is addressing, such as telephony and mobile communication, digital broadcasting, internet networks, interactive multimedia, etc., require a high degree of coding efficiency together with flexible access to coded data, including access to subsets of coded data (i.e. scalability of the coded bitstream), and protection against transmission errors. Reflecting the needs of these requirements, the MPEG-4 Audio goals and functionalities include, in addition to highly efficient audio coding, the provision of speech coding to address telephony applications, universal access through scalability of the coded data to address different transmission channel requirements and robustness in error prone environments. Furthermore content-based interactivity through flexible access and manipulation of the coded data and support to synthetic audio and speech through the structured audio, SA, and TTS interface are addressed by the standard functionalities.

Different technologies described in different parts of the audio standard refer to this diverse set of requirements/goals. *Figure 2* shows the typical data rate requirements for different applications versus the bandwidth of the coded signals and which part of the MPEG-4 Audio standard is applicable. Namely, MPEG-4 addresses two basic types of audio, synthetic (TTS and SA) [Vercoe, Gardner and Scheirer 98, Scheirer, Lee and Yang 00] and natural (parametric, code excited linear predictive or CELP, general audio or G/A, and scalable coders) [Edler, Purnhagen and Ferekidis 96, Purnhagen and Meine 00, Johnston, Quackenbush, Herre and Grill 00]. The synchronization and mix of natural with synthetic audio is called Synthetic/Natural hybrid coding, SNHC. In addition, the AudioBIFS [Scheirer, Väänänen and Huopaniemi 99] part of the Systems BIFS framework allows for receiver's mixing and postproduction and 3-D sound presentation.

The TTS interface part of MPEG-4 Audio standardizes a transmission protocol for synthesized speech, where TTS systems translate text information into speech so it can be transferred through speech lines such as telephone lines. In addition, TTS systems can be used for services for the visually impaired, automatic voice response systems, etc. The data rates covered by the TTS systems vary between 200 b/s and 1.2 kb/s.

In the SA part of the audio standard, the delivery of synthetic audio is described. This capability allows for ultra-low data rates (200 b/s as shown in *Figure 2*) and interactivity at the receiver end. The SA bitstream format specifies a set of synthesis algorithms that describe how to create the sound, and a set of synthesis control parameters that describe which sounds to create. The set of synthesis algorithms, which can generate "instruments", (such as real-life instruments like the flute, violin, etc., or instruments that reflect the sound of ocean waves, or synthetic-hybrid "instruments", etc.) is specified in the SA orchestra language, SAOL. The control parameters that govern the creation of specific sounds are specified in the SA score language, SASL. A format designed to represent banks of wave-tables, the SA audio sample bank format, SASBF, is included in the standard and was developed in collaboration with the musical instrument digital interface, MIDI, manufactures association [MIDI]. Wave-table synthesis is ideal for applications that don't need interaction and require low complexity structure, such as, for example, karaoke applications. This technology allows for the synthesis of a desired sound from look-up tables where particular waveform types are stored. In this case, extremely low data rates can be achieved.

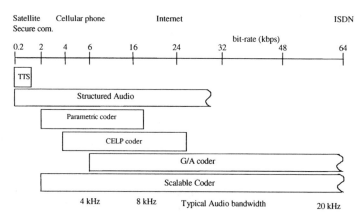

Figure 2. MPEG-4 Audio data rates and target applications from [Edler 97]

Speech signals can be coded with the MPEG-4 Audio standard by utilizing the parametric speech and CELP tools [Nishiguchi and Edler 02].

The parametric speech coder, called also harmonic vector excitation coding, HVXC, achieves good quality at data rates between 2 and 4 kb/s. Lower data rates, such as 1.2 kb/s in average, can be achieved when variable rate coding is enabled. For data rates between 4 and 24 kb/s the CELP coder is utilized. Two sampling rates are supported, 8 and 16 kHz, where the first sampling rate is employed for narrow-band coding of speech and the second for wide-band coding of speech.

General audio covers data rates between 6 kb/s for audio signals with bandwidth of 4 kHz, and 300 kb/s (or above) per channel for signals with bandwidths above 20 kHz for mono to multichannel audio. The work of MPEG-4 Audio in this area represents a continuation of the MPEG-1 and MPEG-2 Audio work with additional tools for addressing natural audio source material.

Data rate scalability allows for the data to be parsed into bitstreams of lower data rates that can be still decoded into a meaningful signal. Encoder/decoder complexity scalability allows for encoder/decoders of lower complexity to generate valid and meaningful bitstreams. Scalability works within a single MPEG-4 Audio tool, such as for example HVXC, and can also be applied to combinations of tools, such as, for example, with CELP as the base coder and other general audio coders such AAC or TwinVQ for the enhancement layers.

In *Figure 3* different audio objects and the combination of the elementary streams through a compositor are shown. The different elementary audio streams can be multiplexed together, and, in addition, encoded video streams can also be multiplexed into the same MPEG-4 bitstream. In the decoder stage, the bitstream is demultiplexed and each elementary stream is decoded. The resulting audio can be played directly or made available for scene composition through the AudioBIFS information. The audio composition tools are specified in the Systems part of the standard. In the compositor multiple audio streams are "mixed" to create a single track. As an example, let's assume we have a speaking voice over background music. One way of coding this signal with high quality would be to use general audio coding tools at 64 kb/s per channel. An alternative approach would be to code the speaking voice using the CELP tools at 16 kb/s per channel, and the music using SA tools at 2 kb/s per channel and then synchronize these two audio objects through the compositor. In the first case a data rate of 64 kb/s per channel was adopted, in the second the data rate was reduced to 18 kb/s per channel. For this approach to be effective for a large class of signals, the encoder should be able to properly separate the different components of the input signal. Encoder specifications and, in particular, the mechanism for source separation are beyond the scope of the MPEG-4 standard.

Theoretical work [Bregman 90] has the potential to carry the foundation to practical guidelines in this field.

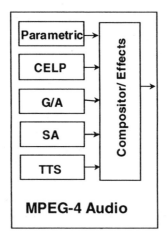

Figure 3. MPEG-4 Audio structure from [Grill 97a]

4.   MPEG-4 AUDIO TOOLS AND PROFILES

MPEG-4 Audio profiles define subsets of the MPEG-4 Audio functionalities appropriate for specific applications. The audio profiles are specified in terms of audio "object types" and audio "levels". An audio object type is a collection of specific coding tools that can be used together for a determined application. An audio level specifies how the coding tools can be used in a determined application in terms of the number of supported audio channels, the number of simultaneous audio objects in use, maximum allowed sampling rates, implementation complexity, and use of error protection. In contrast with MPEG-2 AAC where the profiles are organized in a hierarchical structure (with the MPEG-2 AAC main profile being a superset of MPEG-2 AAC SSR and LC profiles), MPEG-4 is not hierarchical and the profiles are defined based on useful groupings of tools rather than in any a priori "logical" structure.

4.1   MPEG-4 Audio Coding Tools

The tools utilized in MPEG-4 Audio can be grouped into the following categories:

- Speech Tools
- General Audio Tools
- Scalability Tools
- Synthesis Tools
- Composition Tools
- Streaming Tools
- Error Protection Tools

The speech tools include coding tools designed specifically for speech coding. For example, MPEG-4 supports linear predictive (CELP) and parametric speech (HVCX) coding.

The general audio tools include the basic audio coders such as MPEG-2 AAC and transform-domain weighted interleave vector quantization, TwinVQ. Tools that enhance MPEG-2 AAC efficiency, such as perceptual noise substitution (PNS) and long term prediction (LTP) are also included. In addition, parametric coding of general audio signal defined in the harmonic and individual lines plus noise (HILN) tools are enclosed.

The scalability tools allow for the creation and manipulation of data streams that can be decoded at varying data rates. Such tools let a single data stream to be successfully decoded over widely different bandwidths. For example the same bitstream can be utilized to convey music to a slow-dialup computer as well as to listeners with broadband access.

The synthesis tools such as the SA, SASBF and MIDI tools are used in the creation of synthetic sounds (as opposed to the reproduction of natural sounds as carried out using the general audio tools).

The composition tools are used in the creation of audio-visual scenes and are typically used to control the merging of one or more audio and video signals into a single scene. For example, the composition tools could be used to layer a vocal track over a background instrumental mix. Streaming tools allow a remote user to control the way audio signals are streamed to them. Although composition and streaming tools are actually defined in part 1 of the standard (Systems), they are often referred to in the audio specifications.

Finally, error protection tools permit the addition of higher error protection for very susceptible signals or for signals facing particularly noisy transmission channels. The error protection EP tool provides unequal error protection by applying forward error protection codes, FEC, and/or cyclic redundancy codes, CRC, to audio tools [ISO/IEC 14496-3].

The tools relevant to defining audio profiles are listed by category in *Table 1*.

Table 1. Key MPEG-4 Audio coding tools

Category	Audio Tools
Speech	Code Excited Linear Prediction (CELP)
	Harmonic Vector Excitation Coding (HVXC)
	Text to Speech Interface (TTS)
	Variable Bitrate HVXC
	Silence Compression
General Audio	MPEG-2 AAC Main
	MPEG-2 AAC Low Complexity (LC)
	MPEG-2 AAC Scalable Sampling Rate (SSR)
	Low Delay (LD) AAC
	Perceptual Noise Substitution (PNS)
	Long Term Prediction (LTP)
	Harmonic and Individual Lines plus Noise (HILN)
	TwinVQ
Scalability	Bit-Sliced Arithmetic Coding (BSAC)
Synthesis	Tools for Large Step Scalability (TLSS)
	Synthetic Audio (SA) Tools
	Structured Audio Sample Bank Format (SASBF)
	MIDI
Error Protection	Error Robustness Tools

4.1.1   Speech

The CELP-based speech coder (see *Figure 4*) exploits a source model tailored on the vocal mechanism. It distinguishes between voiced and unvoiced excitation and employs linear prediction filters to simulate the vocal tract. The perceptual model in the CELP coder is very simple. The quantization noise is spectrally shaped via spectral weighing filters so that it has a similar shape to the input signal. MPEG-4 CELP can operate in the wideband or narrowband mode. The sampling rates are comprised between 8 and 16 kHz and the data rates between 4 and 24 kb/s per channel. The coding delay depends on the data rate; the maximum delay for 4 kb/s per channel is less than 50 ms; the minimum delay for higher data rates is less than 15 ms. The CELP core allows for layered scalable coding and it performs best for speech only applications and speech with low background noise at low data rates. For speech only applications, the CELP tools showed very good results. For example, for data rates above 6 kb/s the CELP tools scored above 3.0 in the ITU-R five-grade impairment scale (see also Chapter 10) [ISO/IEC MPEG N2424]. In general, MPEG-4 speech coding offers great flexibility and added functionalities when compared to other speech coding standards such as [ITU-T G.722, G.723.1, and G.729]. In version 2 of the standard the CELP silence compression tool was added. This tool reduces the average data rate by transmitting a silence insertion description, SID, when a region with silence or background noise only is detected.

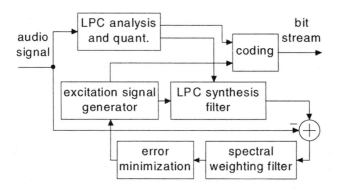

Figure 4. MPEG-4 CELP speech coder structure from [Edler 97]

The parametric coding core exploits a source model that is based upon the decomposition of the audio signals into individual sinusoidal components, harmonic sounds and noise (see also *Figure 7*). HVXC consists of parametric coding technology applied to speech signals only. It operates at data rates of 1.2-4 kb/s per channel at 8 kHz sampling rate. The minimum coding delay is less than 40 ms, and it features a layered scalable coding option. As for HILN (see also next section), the speed/pitch change functionality is an inherent functionality of this coder. The HVXC coder performance at 2 kb/s per channel is good for speech only and speech with low background noise [ISO/IEC MPEG N2424].

The TTS interface is standardized in MPEG-4. The TTS interface component supports features such as speaker age, gender, etc. and can be interfaced with the face animation technology.

4.1.2 General Audio

The general audio coding tools support data rates ranging from 6 kb/s per channel to several hundred of kb/s per channel [Herre and Purnhagen 02]. Monophonic and multichannel configurations (similarly defined as in MPEG-2 AAC) are supported. High quality general audio coding is provided in MPEG-4 Audio by including MPEG-2 AAC as a coding core tool and by adding tools such as PNS and LTP which are capable of improving the basic AAC performance.

PNS works in conjunction with MPEG-2 AAC by identifying scale factor bands that consist primarily of noise and transmitting the total noise power rather than the individual spectral coefficients [Schulz 96, Herre and Schulz 98]. PNS allows for a parametric description of noise-like signal components. At decoding time the original noise-like spectrum in that scale factor band is replaced ("substituted") by pseudo-random noise with the appropriate signal power. The PNS tool improves the basic quality of the MPEG-2 AAC coder for signal containing noise-like spectral components at data rates below 48 kb/s per stereo channel.

The LTP tool is a lower complexity replacement for the prediction tool defined in the MPEG-2 AAC Main Profile that provides comparable performance. Typically the LTP tool provides a 50% saving in terms of memory and processing utilization with respect to MPEG-2 AAC [Ojanperä and Väänänen 99].

In addition, the MPEG-4 Audio specifications allow the quantization/noiseless coding scheme in the MPEG-2 AAC coder to be replaced by either the TwinVQ [Iwakami, Moriya and Miki 95] or the bit-sliced arithmetic coding, BSAC, scheme [Park and Kim 97]. TwinVQ is a quantization scheme based on vector quantization that uses the same spectral representation as MPEG-2 AAC, yet is more effective at remaining intelligible at the lowest supported data rates (below 16 kb/s per channel). TwinVQ performs a spectral flattening and then uses two vector quantization codebooks to quantize the flattened spectrum based on a perceptual distortion measure. AAC with the TwinVQ quantization represents a possible core for the MPEG-4 scalable coding. BSAC represents an alternative for the noiseless coding stage of MPEG-2 AAC. BSAC makes use of an arithmetic coding scheme rather than being based on Huffman coding. As discussed below, BSAC allows for fine-step scalability in the general audio coder.

Figure 5 and *Figure 6* show block diagrams of the MPEG-4 general audio encoder and decoder structures. In comparison with similar block diagrams for MPEG-2 AAC (see Chapter 13 *Figure 1* and *Figure 2*), notice the inclusion of PNS and LTP in the main coding chain and also the choices of using TwinVQ or BSAC instead of the standard MPEG-2 AAC quantization and noiseless coding in the final stage of the encode chain (and in the first stage of the decode chain).

In addition to the tools shown in the figures, error robustness tools and a low delay (LD) version of AAC [ISO/IEC 14496-3] were included in version 2 of the standard. The AAC LD goal is to achieve speech quality at low data rates and low delay (equal or less than 30 ms). AAC LD is derived from the AAC basic structure by reducing the frame length and filter bank delay to 480 samples (instead of 1024), by eliminating the block switching structure so that there is no need for a look-ahead buffer (576 samples for AAC) and by reducing the use of the bit reservoir to a minimum so that no delay is added (instead of 74.7 ms at 48 kHz). The total algorithmic delay introduced by AAC LD at 48 kHz sampling rate is 20 ms (versus 129.4 of AAC).

Based on MPEG-4 verification tests AAC LD at data rates of 32 and 64 kb/s has a comparable performance to MPEG-2 AAC Main Profile at 24 and 56 kb/s with a delay of 30 ms (versus over 300 ms) and 20 ms (versus over 140 ms) respectively [ISO/IEC MPEG N3075].

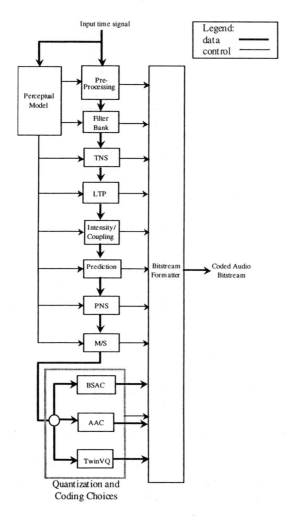

Figure 5. Block diagram of the GA non-scalable encoder from [ISO/IEC 14996-3]

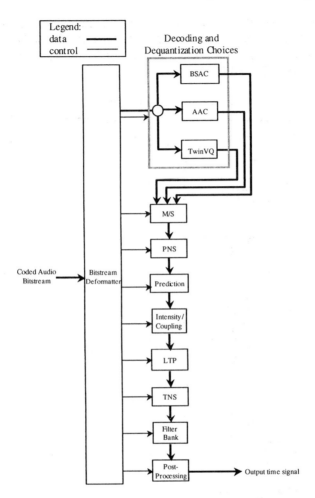

Figure 6. Block diagram of the GA non-scalable decoder from [ISO/IEC 14996-3]

The HILN tool was also added in the second version of the standard. This tool is based on parametric audio coding technology, i.e. the audio signals are described by utilizing some chosen model parameters rather than approximating the audio waveform representation as previously done in MPEG-1 and 2. The source model is based upon the assumption of quasi-stationarity and the coexistence of pure tones, transients and noise in the signal under exam. The parameters selected to describe the source include spectral samples, frequency and amplitude of sinusoids, amplitude envelope and noise spectrum. In *Figure 7* the basic structure of the parametric audio coding tool is shown. The audio signal is first decomposed into individual sinusoidal components, harmonic sounds and noise. Perceptual models are applied in the quantization of the spectral and harmonic components parameters. The parameters quantization is such that the step size covers just noticeable differences. In addition, entropy coding is also employed. HILN operates at data rates of 4-16 kb/s per channel and it has a layered scalable coding option. The speed/pitch change functionality is an inherent functionality of this coder. The HILN tools extend natural audio coding at very low data rates. The optimal area of application for the HILN tools includes monophonic music signals with low content complexity for data rates ranging between 4 kb/s and 8 kb/s. An integrated parametric decoder, which includes both HILN and HVXC, allows for coding of speech and music at very low data rates. For example, speech plus background music can be encoded by using a total of 6 kb/s where 2 kb/s are utilized to code the speech part of the signal with the HVXC tools and 4 kb/s are utilized to code the music with the HILN tools.

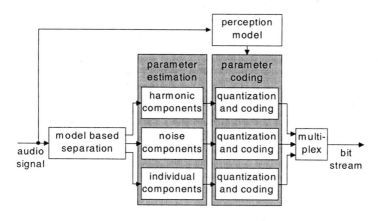

Figure 7. MPEG-4 Audio parametric audio coding (HILN) tool structure from [Edler 97]

4.1.3 Scalability

An important new functionality included in MPEG-4 Audio is that of scalability [Grill 97]. The scalability tools allow a decoder to parse out a subset of the bitstream and decode that into an intelligible audio signal. There are 2 types of scalability implemented in MPEG-4: large step scalability and fine grain scalability. Large step scalability was implemented in version 1 of MPEG-4 Audio and allows for creating an audio bitstream that can be grouped into a small number of subsets of differing data rates.

Large step scalability is implemented in a cascaded encoding approach wherein the audio signal is first encoded at the lowest desired data rate and then the differences between the coded signal and the original signal are then encoded in subsequent stages using additional bits (see *Figure 8*). The decoder can then decide how many stages it can handle and parse out the appropriate portion of the bitstream to decode. The scalable coder structure may include AAC only, TwinVQ only or a combination of AAC and TwinVQ. Typically the large step scalability size is 8 kb/s or larger. Scalable coder combinations such as 6 kb/s per channel CELP or TwinVQ combined with 18 kb/s per channel AAC were successfully tested and scored slightly below 4.0 versus AAC at 24 kb/s per channel, which scored slightly above 4.0 (see also *Table 10*) [ISO/IEC MPEG N2276].

Notice how the scalability is predefined in "large steps" with each step corresponding to the use of an additional coding stage – this is in contrast to the fine grain scalability approach included with the addition of the BSAC tool in version 2 of MPEG-4 Audio. The BSAC tool layers the quantized frequency-domain audio samples in order of the significance of the bits in their representation, allowing for only subsets of the spectrum sample bits to be used in the decode stage for a lower precision copy of the original audio signal. In this manner, the BSAC tool allows for scalability changes in steps of 1 kb/s per channel as it decides to include or exclude particular layers of the quantized spectrum. BSAC performance varies between 4.4 at 96 kb/s per stereo channel and 3.0 at 64 kb/s per stereo channel (see also *Table 10*) [ISO/IEC MPEG N3075].

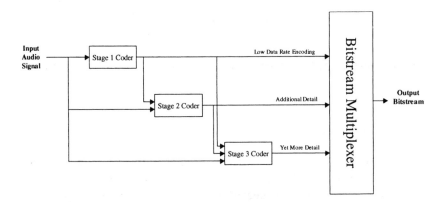

Figure 8. MPEG-4 Audio approach to large step scalable encoding

4.2 MPEG-4 Audio Object Types

MPEG-4 Audio is based on groups of pre-defined "object types" that define sets of functionality that can be used together. *Table 2* shows the object types included in the GA coder structure and the tools available to those types.

Table 2. MPEG-4 Audio tools and object types [ISO/IEC 14496-3, ISO/IEC MPEG N4979]

Object Type / Tools	MPEG-2 AAC Main	MPEG-2 AAC LC	MPEG-2 AAC SSR	PNS	LTP	TLSS	Twin VO	CELP	HVXC	TTSI	SA Tools	SASBF	MIDI	BSAC	HILN	AAC LD	HVXC 4 kb/s VR	Silence Compression	Error Robust	SBR
Null																				
AAC main	X			X																X
AAC LC		X		X																X
AAC SSR			X	X																X
AAC LTP		X		X	X															X
SBR																				X
AAC Scalable		X		X	X	X														
TwinVQ				X			X													
CELP								X												
HVXC									X											
TTSI										X										
Main Synthetic											X	X	X							
Wavetable Synth.												X	X							
General MIDI													X							
AlgSynth/AudFX											X									
ER AAC LC		X		X															X	X
ER AAC LTP		X		X	X														X	X
ER AAC Scale.		X		X		X													X	
ER TwinVQ							X												X	
ER BSAC				X										X					X	
ER AAC LD				X	X											X			X	
ER CELP								X										X	X	
ER HVXC									X								X		X	
ER HILN															X				X	
ER Parametric								X									X	X	X	

The **MPEG-4 AAC Main, MPEG-4 AAC Low Complexity (LC),** and **MPEG-4 AAC Scalable Sampling Rate (SSR)** object types all include the same tools contained in the corresponding MPEG-2 AAC Main, LC and SSR profiles with the addition of the PNS tool. The **MPEG-4 AAC LTP** object type is equivalent to the MPEG-4 AAC LC object type with the addition of the LTP tool. The **TwinVQ** object type contains the TwinVQ and LTP tools. In conjunction with AAC, it operates at lower data rates with respect to AAC, supporting mono and stereo sound.

Error resilient bitstream reordering allows for the use of unequal error protection. In addition to object types described above, the following error resilient, ER, object types are included in the GA description: **ER AAC LC, ER AAC LTP, ER BSAC, ER TwinVQ, ER AAC LD.**

The **AAC Scalable** object type allows a large number of scalable combinations including combinations with TwinVQ and CELP coder tools as the core coders. It supports only mono or 2-channel stereo sound. It contains the AAC LTP object plus TLSS. The **ER AAC Scalable** object type includes error resilient tools.

The **CELP** object type supports 8 kHz and 16 kHz sampling rates at bit rates from 4 to 24 kb/s. CELP bitstreams can be coded in a scalable way using bit rate scalability and bandwidth scalability. **ER CELP** also includes error resilient tools and silence compression tools.

The **HVXC** object type provides a parametric representation of 8 kHz, mono speech at fixed data rates between 2 and 4 kb/s and below 2 kb/s using a variable data rate mode, supporting pitch and speed changes. **ER HVXC** also contains error resilient tools.

In addition to the HVXC technology for the parametric speech coding, the HILN parametric coding tools were added in version 2 of the standard. The **ER HILN** object type includes error resilience tools. The **ER Parametric** object type combines the functionalities of the ER HILN and ER HVXC objects. Only monophonic channels and sampling rates of 8 kHz are supported in this configuration.

The **TTS** interface object type gives an extremely low data rate phonemic representation of speech. While the specific TTS technology is not specified, the interface is fully defined. Data rates range from 0.2 to 1.2 kb/s.

Additional object types are specified for synthetic sound. The **Main Synthetic** object type includes all MPEG-4 SA tools, namely SAOL, SASBF, etc. Sound can be described without input until it is stopped by an explicit command and up to 3-4 kb/s. The **Wavetable Synthesis** object type is a subset of the Main Synthetic object type, making use of the SASBF format and MIDI tools. The **General MIDI** object type provides interoperability with existing content. The **Algorithmic Synthesis and AudioFX** object type provides SAOL-based synthesis capabilities for very low data rate terminals.

Finally, the **NULL** object type provides the possibility to feed raw PCM data directly to the MPEG-4 audio compositor in order to allow mixing in of local sound at the decoder. This means that support for this object type is in the compositor.

Although not yet officially included in the standard specifications, the spectral band replication, *SBR*, tool and object type are also shown in *Table 2* [ISO/IEC MPEG N4764]. SBR is based on bandwidth extension technology, currently under consideration by the MPEG Audio Committee. The bandwidth extension tool, SBR, replicates sequences of harmonics contained in the bandwidth-limited encoded signal representation and is based on control data obtained from the encoder [Dietz, Liljeryd, Kjoerling, Kunz 02]. The ratio between tonal and noise-like components is maintained by adaptive inverse filtering as well as addition of noise and sinusoidal components. Once formally approved by the standard bodies, the SBR tool will be included in AAC Main, LC, SSR, LTP and in ER AAC LC and LTP. SBR allows for compatibility with earlier versions of these tools.

4.3   Profiles

The following eight MPEG-4 Audio profiles are specified by the standard (see *Table 3*):

- **Main**– It encompasses all MPEG-4 Audio natural and synthetic objects, with the exception of the error correction objects.
- **Scalable**– It includes all the audio objects contained in the main profile with the exception of MPEG-2 AAC Main and SSR and SA. It allows for scalable coding of speech and music and it addresses transmission methods such as internet and digital audio broadcasting.
- **Speech**– It includes the CELP, HVXC and TTS interface objects.
- **Synthesis**– It contains all SA and TTS interface objects and provides the capability to generate audio and speech ay very low data rates.
- **Natural**– It encompasses all the natural audio coding objects and includes TTS interface and error correction tools.
- **High Quality**– It includes the AAC LC object plus LTP, the AAC scalable and CELP objects; in this profile, there is the option of employing the error resilient tools for the above-mentioned objects.
- **Low Delay**– It includes AAC LD plus CELP, HVXC, with the option of using the ER tools, and TTS interface objects.
- **Mobile Audio Internetworking (MAUI)**– It includes ER AACLC, ER AAC scalable, ER Twin VQ, ER BSAC and ER AAC LD. This profile is intended to address communication applications using speech coding algorithms and high quality audio coding.

Two additional audio profiles, the *Simple* Audio Profile, which contains the MPEG-4 AAC LC tools but works at sampling rates up to 96 kHz, and the

Simple SBR Audio Profile are currently under consideration [ISO/IEC MPEG N4764 and N4979]. The Simple SBR Profile is equivalent to the Simple Profile with the addition of the SBR object. The conformance specifications of the MPEG-4 standard are tailored around the different profiles.

Table 3. MPEG-4 Audio profiles [ISO/IEC 14496-3, ISO/IEC MPEG N4979, ISO/IEC MPEG N4764]

Object Type	Main	Scalable	Speech	Synthetic	High Quality	Low Delay	Natural	MAUI	Simple	Simple SBR
Null										
AAC main	X						X			
AAC LC	X	X			X		X		X	X
AAC SSR	X						X			
AAC LTP	X	X			X		X			
SBR										X
AAC Scalable	X	X			X		X			
TwinVQ	X	X					X			
CELP	X	X	X		X	X	X			
HVXC	X	X	X			X	X			
TTSI	X	X	X	X		X	X			
Main Synthetic	X			X						
Wavetable Synth.										
General MIDI										
AlgSynth/AudFX										
ER AAC LC					X		X	X		
ER AAC LTP					X		X			
ER AAC Scale.					X		X	X		
ER TwinVQ							X	X		
ER BSAC							X	X		
ER AAC LD						X	X	X		
ER CELP					X	X	X			
ER HVXC						X	X			
ER HILN							X			
ER Parametric							X			

4.3.1   Levels

Profiles may specify different levels that differ with respect to the number of channels, sampling rates, and simultaneous audio objects supported; their implementation complexity; and whether or not they make use of the error protection (EP) tool. *Table 4* through *Table 7* show the main characteristics of the level descriptions for the some of the relevant profiles associated with general audio coding. In these tables, complexity limits are shown both in terms of processing required approximated in PCU or "processor complexity units", which specify an integer number of MOPS or "Millions of Operations per Second" and in memory usage approximated in RCU or "RAM complexity units" which specify an integer number of kWords.

Table 4. High Quality Audio Profile Levels [ISO/IEC 14496-3]

Level	Maximum number of channels/object	Maximum Sampling Rate (kHz)	Maximum PCU	Maximum RCU	EP Tool Present
1	2	22.05	5	8	No
2	2	48	10	8	No
3	5.1	48	25	12	No
4	5.1	48	100	42	No
5	2	22.05	5	8	Yes
6	2	48	10	8	Yes
7	5.1	48	25	12	Yes
8	5.1	48	100	42	Yes

Table 5. Low Delay Audio Profile Levels [ISO/IEC 14496-3]

Level	Maximum number of channels/object	Maximum Sampling Rate (kHz)	Maximum PCU	Maximum RCU	EP Tool Present
1	1	8	2	1	No
2	1	16	3	1	No
3	1	48	3	2	No
4	2	48	24	12	No
5	1	8	2	1	Yes
6	1	16	3	1	Yes
7	1	48	3	2	Yes
8	2	48	24	12	Yes

Table 6. Natural Audio Profile Levels [ISO/IEC 14496-3]

Level	Maximum Sampling Rate (kHz)	Maximum PCU	EP Tool Present
1	48	20	No
2	96	100	No
3	48	20	Yes
4	96	100	Yes

Table 7. MAUI Audio Profile Levels [ISO/IEC 14496-3]

Level	Maximum number of channels	Maximum number of objects	Maximum Sampling Rate (kHz)	Maximum PCU	Maximum RCU	EP Tool Present
1	1	1	24	2.5	4	No
2	2	2	48	10	8	No
3	5.1	-	48	25	12	No
4	1	1	24	2.5	4	Yes
5	2	2	48	10	8	Yes
6	5.1	-	48	25	12	Yes

In addition, the Simple and Simple SBR profiles levels are shown in *Table 8* and *Table 9*.

Table 8. Simple Audio Profile Levels [ISO/IEC MPEG N4764]

Level	Maximum number of channels/objects	Maximum Sampling Rate (kHz)	Maximum PCU	Maximum RCU
1	2	24	3	5
2	2	48	6	5
3	5	48	19	15
4	5	96	38	15

Table 9. Simple SBR Audio Profile Levels [ISO/IEC MPEG N4979]

Level	Maximum number of channels/objects	Maximum AAC Sampling Rate (kHz)	Maximum SBR Sampling Rate (kHz)
1	2	24	24
2	2	48	48
3	5	48	48
4	5	96	48

5. MPEG-1 AND 2 VERSUS MPEG-4 AUDIO

We saw in Chapters 11-13 how MPEG-1 and -2 approach audio coding based on the removal of redundancies and irrelevancies in the original audio signal. The removal of redundancies is based on the frequency representation of the signal, which is in general more efficient than its PCM representation given the quasi-stationary nature of audio signals. In addition, the removal of redundancies is based on models of human perception like, for example, psychoacoustic masking models. In this approach, by additionally removing irrelevant parts of the signal, high quality audio at low data rates is typically achieved. General-purpose audio codecs such as MPEG-1 and 2 audio codecs provide very high quality output for a large class of audio signals at data rates of 128 kb/s or below.

Before perceptual audio coding reached maturity, a number of coding schemes based on removal of redundancies only, such as prediction technologies, were developed. These codecs try to model the source as precisely as possible in order to extract the largest possible amount redundancies. For speech signals, CELP codecs model the vocal tract and work well at data rates of 32 kb/s or below. However, they show serious problems with signals that don't precisely fit the source models, for example music signals. While MPEG-1 and 2 Audio is sub optimal for speech signals, CELP coders are unable to properly code music signals. One possible solution to this problem is to restrict the class of signals in input to a certain type of codec. Another possible solution is to define a useful combination of different codec types. Given the wide scope of its applications, MPEG-4 adopted the second approach.

The MPEG-4 Audio encoder structure is shown in *Figure 9*. As we saw in the previous sections, three types of algorithms can be found:

- Coding based on time/frequency mapping (T/F), like MPEG-1, MPEG-2 audio, which represents the basic structure of the GA tools. The foundation of this type of coding is MPEG-2 AAC. As we saw in previous sections, additional tools that enhance the codec performance and efficiency at very low data rates are also included.

- Coding based on CELP, like for example in the ITU-T G.722, G.723.1 and G. 729 coders. The MPEG-4 CELP codec exploits a source model based on the vocal tract mechanism like the ITU-T speech codecs, but it also applies a simple perceptual model where the quantization noise spectral envelope follows the input signal spectral envelope.

- Coding based on parametric representation (PARA). This coding technique in addition to allow for added functionalities such as pitch/time changes and volume modifications, tends to perform better than CELP (HVXC) for very low data rates speech signals and the T/F scheme (HILN) for very low data rates music signals containing single instruments with a large number of harmonics.

Separate coding depending on the characteristics of the input signal can improve the performance of the overall codec if in the encoder stage an appropriate algorithm selection, manual or automatic, takes place. Unfortunately the MPEG-4 standard does not specify the encoder operations other than in an informative part of the standard. Automatic signal analysis and separation possibly allows for future optimization of the encoder stage.

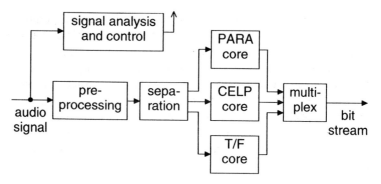

Figure 9. MPEG-4 Audio encoder structure from [Edler 97]

The MPEG-4 audio bitstream represents also a departure from the MPEG-1 or 2 fashion of representing the compressed signal, i.e. there is no multiplex, no synch word, etc. MPEG-4 audio only defines setup information packets and payload for each coder. MPEG-4 Systems specifies "Flex-Mux" to cover multiplex aspects of MPEG-4 functionalities, such as for example scalability. An MPEG-4 file (.MP4) format is also described in the Systems specifications.

6.   THE PERFORMANCE OF THE MPEG-4 AUDIO CODING TOOLS

The primary goal of the MPEG-4 verification tests was to evaluate the subjective performance of specific coding tools operating at a certain data rate. To better enable the evaluation of MPEG-4, several coders from MPEG-2 and ITU-T were included in the tests. The subjective performance of some the MPEG-4 tools is summarized in terms of the ITU-R five-grade impairment scale in *Table 10* [ISO/IEC MPEG N4668] along with the performance of comparable technology such as MPEG-2, ITU-T G.722 and G.723. The reader interested in knowing the details of the audio tests conditions and results should consult [Contin 02].

Table 10. MPEG-4 Audio coding tools subjective performance [ISO/IEC MPEG N4668]

Coding Tool	Number of Channels	Data Rate	Grading Scale	Typical Quality
AAC	5	320 kb/s	Impairment	4.6
95 MPEG-2 LII BC	5	640 kb/s	Impairment	4.6
AAC	2	128 kb/s	Impairment	4.8
AAC	2	96 kb/s	Impairment	4.4
MPEG-1 LII	2	192 kb/s	Impairment	4.3
MPEG-1 LIII	2	128 kb/s	Impairment	4.1
AAC	1	24 kb/s	Quality	4.2
CELP/AAC scal.	1	6 kb/s+18 kb/s	Quality	3.7
TwinVQ/AAC scal.	1	6 kb/s+18 kb/s	Quality	3.6
AAC	1	18 kb/s	Quality	3.2
G.723	1	6.3 kb/s	Quality	2.8
Wideband CELP[8]	1	18.2 kb/s	Quality	2.3
BSAC	2	96 kb/s	Quality	4.4
BSAC	2	80 kb/s	Quality	3.7
BSAC	2	64 kb/s	Quality	3.0
AAC-LD (20ms)	1	64 kb/s	Quality	4.4
G.722	1	32 kb/s	Quality	4.2
AAC-LD (30ms)	1	32 kb/s	Quality	3.4
Narroband CELP	1	6 kb/s	Quality	2.5
Twin VQ	1	6 kb/s	Quality	1.8
HILN	1	16 kb/s	Quality	2.8
HILN	1	6 kb/s	Quality	1.8

7. INTELLECTUAL PROPERTY AND MPEG-4

Recognizing at an early stage of the development of MPEG-4 that one of the biggest potential impediment for a wide adoption of a standard is the clearance of the intellectual property implicated, part of the MPEG-4 Systems specifications are devoted to the identification of intellectual property involved in its implementation. In order to identify intellectual property in the MPEG-4 media objects, MPEG-4 developed the intellectual property management and protection (IPMP) [ISO/IEC MPEG N2614]. MPEG-4 target applications range from low data rate internet telephony to high fidelity video and audio. Anyone can develop applications based on any needed subset of MPEG-4 profiles. The level and type of protection may vary dramatically depending on the content, complexity, and associated business models. In addition, the traditional business model of paying once for hardware devices and then having the associated royalties managed by the device manufacturer is less attractive for software implementations of MPEG-4 clients. While MPEG-4 does not standardize IPMP systems, it

---

[8] The data shown reflect test results for both speech and music signals.

does standardize the IPMP interface as a simple extension to the MPEG-4 systems architecture via a set of descriptors and elementary streams (IPMP-D and IMPM-ES).

In addition to the work of ISO/IEC WG 11 on MPEG-4, the MPEG-4 Industry Forum, M4IF, was established in order to "further the adoption of the MPEG-4 standard, by establishing MPEG-4 as an accepted and widely used standard among application developers, service providers, content creators and end users" [M4IF]. Currently licensing schemes for MPEG-4 AAC are available through Dolby Laboratories [AAC Audio] and for MPEG-4 Visual and Systems through MPEG LA, LLC [M4 Visual and Systems].

8. SUMMARY

In this chapter we reviewed the main features of the MPEG-4 Audio standard. MPEG-4 represents the last phase of work within MPEG that deals directly with the coding of audiovisual signals.

The main goals of MPEG-4 Audio are broader than the goals set for MPEG-1 and -2. In addition to audio coding, coding of speech, synthetic audio, text to speech interfaces, scalability, 3D, and added functionalities were also addressed by MPEG-4. To this date the MPEG-4 Audio, first finalized at the end 1998, went through two revision stages during which added schemes such as HILN for very low data rate audio and additional functionalities for MPEG-4 AAC such as low delay and error robustness versions, were included in the specifications. MPEG-4 targets wireless, digital broadcasting, and interactive multimedia (streaming, internet, distribution and access to content, etc.) applications.

This chapter concludes this book's overview of major audio coding standards. Hopefully, the review of the major coding standards has both provided further insight into how the principles of audio coding have been applied in state-of-the-art coders and also given enough coding details to assist the reader in effectively using the standard documentation to implement compliant coders. The true goal of the book, however, is to have taught some readers enough "coding secrets" to facilitate their personal journeys to create the next generation of audio coders.

9. REFERENCES

[AAC Audio]: http://www.aac-audio.com/, "Dolby Laboratories Announces MPEG-4 AAC Licensing Program," March 2002.

[Bregman 90]: A. S. Bregman, *Auditory Scene Analysis: The Perceptual Organization of Sound,* Cambridge, Mass.: Bradford Books, MIT Press 1990.

[Chiariglione 98]: L. Chiariglione, "The MPEG-4 standard, " Journal of the China Institute of Communications, September 1998.

[Contin 02]: L. Contin, "Audio Testing for Validation," in *The MPEG-4 Book,* pp. 709 - 751, F. Pereira and T. Ebrahimi (ed.), Prentice Hall 2002.

[den Brinker, Schuijers and Oomen 02]: A. C. den Brinker, E. G. P. Schuijers and A. W. J. Oomen, "Parametric Coding for High-Quality Audio," presented at the 112th AES Convention, Munich, pre-print 5553, May 2002.

[Dietz, Liljeryd, Kjoerling and Kunz 02]: M. Dietz, L. Liljeryd, K. Kjoerling and O. Kunz, "Spectral Band Replication, a novel approach in audio coding," presented at the 112th AES Convention, Munich, Preprint 5553, May 2002.

[Edler 97] B. Edler, Powerpoint slides shared with the authors, 1997. Used with permission.

[Edler and Purnhagen 98]: B. Edler and H. Purnhagen, "Concepts for Hybrid Audio Coding Schemes Based on Parametric Techniques," presented at the 105th AES Convention, San Francisco, Preprint 4808, October 1998.

[Grill 97]: B. Grill, "A Bit Rate Scalable Perceptual Coder for MPEG-4 Audio," presented at the 103rd AES Convention, New York, Preprint 4620, October 1997.

[Grill 97a]: B. Grill, Powerpoint slides shared with the authors, 1997. Used with permission.

[Herre and Purnhagen 02]: J. Herre and H. Purnhagen, "General Audio Coding," in *The MPEG-4 Book,* pp. 487 - 544, F. Pereira and T. Ebrahimi (ed.), Prentice Hall 2002.

[Herre and Schulz 98]: J. Herre and D. Schulz, "Extending the MPEG-4 AAC Codec by Perceptual Noise Substitution," presented at the 112th AES Convention, Amsterdam, Preprint 4720, May 1998.

[ISO/IEC 14496-1]: ISO/IEC 14496-1, "Information Technology – Coding of Audio Visual Objects, Part 1: Systems ", 1999-2001.

[ISO/IEC 14496-2]: ISO/IEC 14496-2, "Information Technology – Coding of Audio Visual Objects, Part 2: Visual ", 1999-2001.

[ISO/IEC 14496-3]: ISO/IEC 14496-3, "Information Technology – Coding of Audio Visual Objects, Part 3: Audio ", 1999-2001.

[ISO/IEC 14496-6]: ISO/IEC 14496-6, "Information Technology – Coding of Audio Visual Objects, Part 6: Delivery Multimedia Integration Framework (DMIF)", 1999-2000.

[ISO/IEC MPEG N2276]: ISO/IEC JTC 1/SC 29/WG 11 N2276, "Report on the MPEG-4 Audio NADIB Verification Tests" Dublin, July 1998.

[ISO/IEC MPEG N2424]: ISO/IEC JTC 1/SC 29/WG 11 N2424, "Report on the MPEG-4 speech codec verification tests " Atlantic City, October 1998.

[ISO/IEC MPEG N2501]: ISO/IEC JTC 1/SC 29/WG 11 N2501, "FDIS of ISO/IEC 144496-1" Atlantic City, October 1998.

[ISO/IEC MPEG N2502]: ISO/IEC JTC 1/SC 29/WG 11 N2502, "FDIS of ISO/IEC 144496-2" Atlantic City, October 1998.

[ISO/IEC MPEG N2503]: ISO/IEC JTC 1/SC 29/WG 11 N2503, "FDIS of ISO/IEC 144496-3" Atlantic City, October 1998.

[ISO/IEC MPEG N2614]: ISO/IEC JTC 1/SC 29/WG 11 N2614, "MPEG-4 Intellectual Property Management and Protection (IPMP) Overview and Applications Document" Rome, December 1998.

[ISO/IEC MPEG N271]: ISO/IEC JTC 1/SC 29/WG 11 N271, "New Work Item Proposal for Very-Low Bitrates Audiovisual Coding" London, November 1992.

[ISO/IEC MPEG N3075]: ISO/IEC JTC 1/SC 29/WG 11 N3075, "Report on the MPEG-4 Audio Version 2 Verification Tests" Maui, December 1999.

[ISO/IEC MPEG N4400]: ISO/IEC JTC 1/SC 29/WG 11 N4400, "JVT Terms of Reference " Pattaya, December 2001.

[ISO/IEC MPEG N4668]: ISO/IEC JTC 1/SC 29/WG 11 N4668, "MPEG-4 Overview " Jeju, March 2002.

[ISO/IEC MPEG N4764]: ISO/IEC JTC 1/SC 29/WG 11 N4764, "Text of ISO/IEC 14496-3:2001 PDAM 1" Fairfax, May 2002.

[ISO/IEC MPEG N4920]: ISO/IEC JTC 1/SC 29/WG 11 N4920, "Text if ISO/IEC 14496-10 FCD Advanced Video Coding " Klagenfurt, July 2002.

[ISO/IEC MPEG N4979]: ISO/IEC JTC 1/SC 29/WG 11 N4979, "MPEG-4 Profiles Under Consideration" Klagenfurt, July 2002.

[ISO/IEC MPEG N5040]: ISO/IEC JTC 1/SC 29/WG 11 N5040, "Call for Proposals on MPEG-4 Lossless Audio Coding" Klagenfurt, July 2002.

[ITU-T G.722]: International Telecommunications Union Telecommunications Sector G.722, "7 kHz Audio Coding Within 64 kb/s", Geneva 1998.

[ITU-T G.723.1]: International Telecommunications Union Telecommunications Sector G.723.1, "Dual Rate Speech Coder for Multimedia Communications Transmitting at 5.3 and 6.3 kb/s ", Geneva 1996.

[ITU-T G.729]: International Telecommunications Union Telecommunications Sector G.729, "Coding of Speech at 8 kb/s Using Conjugate Structure Algebraic Code Exited Linear Prediction", Geneva 1996.

[Iwakami, Moriya and Miki 95]: N. Iwakami, T. Moriya, and S. Miki, "High-Quality Audio Coding at Less Than 64 kb/s by Using Transform-Domain Weighted Interleaved Vector Quantization (TwinVQ)," Proc. IEEE ICASSP, pp. 3095-3098, Detroit, May 1995.

[Johnston, Quackenbush, Herre and Grill 00]: J. D. Johnston, S. R. Quackenbush, J. Herre and B. Grill, "Review of MPEG-4 General Audio Coding" in *Multimedia Systems, Standards, and Networks,* pp. 131-155, A. Puri and T. Chen (ed.), Marcel Dekker, Inc. 2000.

[M4 Visual and Systems]: http://www.mpegla.com/, "Final Terms of MPEG-4 Visual and Systems Patent Portfolio Licenses Decided, License Agreements to Issue in September," July 2002.

[M4IF]: MPEG-4 Industry Forum Home Page, www.m4if.org/index.html

[MIDI]: MIDI Manufactures Association Home Page http://www.midi.org/.

[Nishiguchi and Edler 02]: M. Nishiguchi and B. Edler, "Speech Coding," in *The MPEG-4 Book,* pp. 451 - 485, F. Pereira and T. Ebrahimi (ed.), Prentice Hall 2002.

[Ojanperä and Väänänen 99]: J. Ojanperä and M. Väänänen, "Long Term Predictor for Transform Domain Perceptual Audio Coding," presented at the 107th AES Convention, New York, pre-print 5036, September 1999.

[Park and Kim 97]: S. H. Park and Y. B. Kim 97, "Multi-Layered Bit-Sliced Bit-Rate Scalable Audio Coding," presented at the 103rd AES Convention, New York, pre-print 4520, October 1997.

[Purnhagen and Meine 00]: H. Purnhagen and N. Meine, "HILN: The MPEG-4 Parametric Audio Coding Tools," Proc. Intl. Symposium On Circuit and Systems, Geneva, 2000.

[Rubinstein and Kahn 01]: K. Rubinstein and E. Kahn, Powerpoint slides shared with the authors, 2001. Used with permission.

[Scheirer, Lee and Yang 00]: E. D. Scheirer, Y. Lee and Y. J. W. Yang, "Synthetic Audio and SNHC Audio in MPEG-4" in *Multimedia Systems, Standards, and Networks,* pp. 157 - 177, A. Puri and T. Chen (ed.), Marcel Dekker, Inc. 2000.

[Scheirer, Väänänen and Huopaniemi 99]: E. D. Scheirer and R. Väänänen, J. Huopaniemi, "Describing Audio Scenes with the MPEG-4 Multimedia Standard" IEEE Trans. On Multimedia, Vol. 1 no. 3 pp. 237-250, September 1999.

[Schulz 96]: D. Schulz, "Improving Audio Codecs by Noise Substitution," J. Audio Eng. Soc., vol. 44, pp. 593 – 598, July/August 1996.

[Vercoe, Garnder and Scheirer 98]: B. L. Vercoe, W. G. Garnder and E. D. Scheirer, "Structured Audio: The Creation, Transmission, and Rendering of Parametric Sound Representations," Proc. IEEE, Vol. 85 No. 5, pp. 922-940, May 1998.

Index

A

AAC. *See* MPEG-2 AAC

B

Bark Scale, 182–83
Basilar Membrane, 170–74
Binary Numbers, 14–20
Bit Allocation
    Basic Methods, 204–18
    Dolby AC-3, 390–95
    MPEG Layer I, 300
    MPEG Layer II, 301–4
    MPEG Layer III, 305–7
    MPEG-2 AAC, 346–50
Bits, Manipulating, 16–20
Bitstream Format
    Basic Ideas, 230–33
    Dolby AC-3, 395–96
    MPEG-1 Audio, 296–97
    MPEG-2 AAC, 353–55
    MPEG-2 BC, 324–27
Block Switching
    Basics for DFT, 122–23
    Basics for MDCT, 131–36
    Dolby AC-3, 379–82
    MPEG-1 Coders, 275–77
    MPEG-2 AAC, 341–43

C

Cochlea, 170–74
Coding Artifacts, 255–57
Coding Errors, 9–10
Coding Goals, 5–7
Coding Standards
    Background on MPEG Audio, 266–68
    Dolby AC-3, 371–98
    MPEG-1 Audio, 265–310
    MPEG-2 AAC, 333–67
    MPEG-2 BC, 321–30
    MPEG-2 LSF and MP3 Files, 315–18
    MPEG-4 Audio, 401–26
    Pulse Code Modulation (PCM), 7–9
Convolution Theorem, 58–59, 79–80
Critical Bands, 164–68

D

Data Rate, 202–4

Dirac Delta Function, 49–51
Discrete Fourier Transform (DFT), 110–13
Dolby AC-3, 371–98
    Bit Allocation, 390–95
    Bitstream Format, 395–96
    Block Switching, 379–82
    Filter Bank, 377–82
    Scale Factors, 382–85
Down Sampling, 80–83

E

Entropy Coding
    Basic Ideas, 38–43
    MPEG Layer III, 305–6
    MPEG-2 AAC, 350–53
Euler Identity, 49

F

Fast Fourier Transform (FFT), 111–13
Fourier Series, 59–61
Fourier Transform
    Continuous, 51–53
    Discrete, 110–13
    Fast Fourier Transform (FFT), 111–13
    Fourier Series, 59–61
    Z Transform, 77–84

G

Gain Control, 338–40

H

Haar Filter, 87–89
Hearing Threshold, 153–56
Hearing, How It Works, 168–74
Huffman Coding. *See* Entropy Coding

I

Impairment Scales, 240–41
Inner Ear, 170–74

ITU-R 5-Grade Impairment Scale, 431

L

Listening Tests, 240–50, 359–63, 396–97, 424–25
Loudness, 150–51

M

Masking, 149–98
    Addition of, 192–95
    Basic Ideas, 156–60
    Experiments, 160–64
    Hearing Threshold, 153–56
    Heuristic Model of, 180–82
    Masking Curves, 183–92, 223–29
    MPEG Psychoacoustic Model 1, 280–88
    MPEG Psychoacoustic Model 2, 288–96
MDCT, 124–43
    Fast Implementation, 141–43
    Relationship to PQMF, 136–41
    Theory of, 125–31
MP3 Files, 315–18
MPEG-1 Audio, 265–310
    Bit Allocation
        Layer I, 300
        Layer II, 301–4
        Layer III, 305–7
    Bitstream Format, 296–97
    Filter Bank, 273–78
    Layer III Hybrid Filter Bank, 273–75
    Layers, 271–73
    Psychoacoustic Models, 278–96
    Scale Factors
        Layer I, 298–99
        Layer II, 301
        Layer III, 304–5
    Stereo Coding, 309–10
MPEG-2 AAC, 333–67
    Bitstream Format, 353–55

Block Switching, 341–43
Complexity Evaluation, 363–67
Entropy Coding, 350–53
Filter Bank, 340–43
Gain Control, 338–40
Prediction, 343–45
Scale Factors, 346–50
Stereo Coding, 358–59
Temporal Noise Shaping (TNS), 355–58
MPEG-2 BC, 321–30
Bitstream Format, 324–27
Multilingual Channels, 327
MPEG-2 LSF, 315–18
MPEG-4 Audio, 401–26
Coding Tools, 408–16
Goals and Functionalities, 405–8
Object Types, 417–19
Profiles and Levels, 419–22

N

Notation, 48–49

O

Overlap and Add Technique
Basic Approach, 113–19
Window Constraints for DFT, 117–19
Window Constraints in MDCT, 127–28

P

Perceptual Entropy, 196
Perfect Reconstruction Filter Banks
CQF Solution, 89–90
MDCT, 124–43
PQMF, 90–99, 92–99
QMF Solution, 86–87
Two Channel Case, 84–90
PQMF Filter Bank, 90–99
Prediction, 63–67, 343–45
MPEG-2 AAC, 343–45

Pulse Code Modulation (PCM), 7–9

Q

QMF Filter Bank, 86–87
Quality Measurement
Impairment Scales, 431
Listening Tests, 240–50, 359–63, 396–97, 424–25
Objective Measures and PEAQ, 251–54
Quantization, 20–34
Dolby AC-3, 382–85, 394–95
Error from, 34–38
Floating Point, 29–34
MPEG Layer I, 298–300
MPEG Layer II, 301–4
MPEG Layer III, 304–7
MPEG-2 AAC, 346–50
Nonuniform, 26–29
Uniform, 22–26
Quantization Error, 34–38

S

Sampling Theorem, 61–63, 68–69
Scale Factors
Basic Ideas, 29–34
Dolby AC-3, 382–85
MPEG Layer I, 298–99
MPEG Layer II, 301
MPEG Layer III, 304–5
MPEG-2 AAC, 346–50
Signal Summary Properties, 53–59
Signal to Mask Ratio (SMR), 162
Sound Pressure Level (SPL), 150
Spectral Flatness Measure, 210–12
Stereo and Multichannel Coding, 307–10, 318–21, 358–59, 385–89
Subjective Difference Grade (SDG), 431

T

Temporal Noise Shaping (TNS), 355–58

U

Up Sampling, 83–84

W

Windows, 105–10
  Hanning, 107–8

Kaiser-Bessel, 108–10
Kaiser-Bessel Derived, 117–19
Rectangular, 105–6
Sine, 106–7

Z

Z Transform, 77–84